Gene Expression and Carcinogenesis in Cultured Liver

ACADEMIC PRESS RAPID MANUSCRIPT REPRODUCTION

*Proceedings of an International Symposium
held at the University of California, Los Angeles,
Los Angeles, California, May 8–10, 1974*

Gene Expression and Carcinogenesis in Cultured Liver

edited by

L. E. Gerschenson

*Laboratory of Nuclear Medicine and Radiation Biology
and Department of Pathology, School of Medicine,
University of California, Los Angeles, California*

E. Brad Thompson

*Laboratory of Biochemistry, National Cancer Institute,
National Institutes of Health, Bethesda, Maryland*

Academic Press, Inc. New York San Francisco London 1975
A Subsidiary of Harcourt Brace Jovanovich, Publishers

COPYRIGHT © 1975, BY ACADEMIC PRESS, INC.
ALL RIGHTS RESERVED.
NO PART OF THIS PUBLICATION MAY BE REPRODUCED OR
TRANSMITTED IN ANY FORM OR BY ANY MEANS, ELECTRONIC
OR MECHANICAL, INCLUDING PHOTOCOPY, RECORDING, OR ANY
INFORMATION STORAGE AND RETRIEVAL SYSTEM, WITHOUT
PERMISSION IN WRITING FROM THE PUBLISHER.

ACADEMIC PRESS, INC.
111 Fifth Avenue, New York, New York 10003

United Kingdom Edition published by
ACADEMIC PRESS, INC. (LONDON) LTD.
24/28 Oval Road, London NW1

Library of Congress Cataloging in Publication Data

International Symposium on Gene Expression and Carcino-
 genesis in Cultured Liver, University of California
 at Los Angeles, 1974.
 Gene expression and carcinogenesis in cultured liver.
 Bibliography: p.
 Includes index.
 1. Oncology, Experimental—Congresses. 2. Carcino-
genesis—Congresses. 3. Gene expression—Congresses.
4. Liver—Congresses. I. Gerschenson, L. E.
II. Thompson, Edward Bradbridge, (date) III. Title.
[DNLM: 1. Cells, Cultured—Congresses. 2. Carcinogens
—Congresses. 3. Genes—Congresses. 4. Liver—
Cytology—Congresses. 5. Liver neoplasms—Etiology—
Congresses. 6. Neoplasms, Experimental—Etiology—
Congresses. WI735 G326 1974]
RC267.1536 1974   616.9'94'071   75-12744
ISBN 0-12-281150-X

PRINTED IN THE UNITED STATES OF AMERICA

CONTENTS

Contributors ix
Preface xv

Cellular Derivation of Continuously Cultured Epithelial Cells from
Normal Rat Liver
   Joe W. Grisham, Sara B. Thal, and Anneli Nagel   1

Some Characteristics and Functions of Adult Rat Liver
Parenchymal Cells in Primary Culture
   Robert J. Bonney and Frank Maley   24

Protein Synthesis and Excretion in Single Cell Suspensions from
Liver and Morris Hepatoma 5123 TC
   G. Schreiber and M. Schreiber   46

Studies on Normal and Neoplastic Liver Cells in Culture: Contact
Behavior, Cellular Communication and Transformation
   Carmia Borek   62

Retention and Loss of Certain Enzymes in Various Primary
Cultures and Cell Lines of Normal Rat Liver
   *Martine Chessebeuf, Aline Olsson, Paulette Bournot, Jean
   Desgres, Michel Guiguet, Gabrielle Maume, Bernard F. Maume,
   Bernard Perissel, and Prudent Padieu*   94

Microsomal Functions and Pheonotypic Change in Adult Rat
Hepatocytes in Primary Monolayer Culture
   D. M. Bissell and P. S. Guzelian   119

Biochemical, Autoradiographic and Electron Microscopic Studies
of Adult Rat Liver Parenchymal Cells in Primary Culture
   Michael W. Pariza, James D. Yager, Jr., Stanley Goldfarb,

James A. Gurr, Susumu Yanagi, Steven H. Grossman, Joyce E. Becker, Thomas A. Barber, and Van R. Potter ... 137

Differentiated Functions in Clonal Strains of Hepatoma Cells
Armen H. Tashjian, Jr., U. Ingrid Richardson, Robert Strunk, and Peter Ofner ... 168

Effect of Glucocorticoids on the Ultrastructure of Cultured Liver Cells
Judith A. Berliner ... 181

Hormonal Regulation of Amino Acid Transport in Rat Hepatoma Cells in Tissue Culture
Thomas D. Gelehrter, William L. Risser, and Samuel B. Reichberg ... 190

Regulation of Specific Protein Synthesis in Cultured Hepatoma Cells by Analogs of Cyclic AMP
Wesley D. Wicks, Kay Wagner, Michael D. Roper, Ben H. Leichtling, and Jayantha Wimalasena ... 205

Hormonal Effects on Two Rat Liver Cell Lines Cultured in Chemically Defined Medium
L. E. Gerschenson ... 220

On the Specificity of the Induction of Tyrosine Aminotransferase
H. Kröger ... 229

Isozyme Patterns of Branched Chain Amino Acid Transaminase in Cultured Rat Liver Cells
Akira Ichihara, Jiro Sato, and Masayoshi Kumegawa ... 232

The Phosphorylation Region of Lysine-Rich Histone in Dividing HTC Cells
Daryl Granner, David Sherod, Rod Balhorn, Vaughn Jackson, and Roger Chalkley ... 249

Protein Degradation in Liver Compensatory Growth
Oscar A. Scornik ... 264

CONTENTS

Factors Influencing Growth of Cells from Regenerating Liver
    D. M. Hays, Y. Sera, Y. Koga, H. B. Neustein, E. F. Hays, and
    M. O. Nicolson   282

Studies on the Control of Growth in Cultured Primary Fetal Rat
Liver Cells
    Dieter Paul   286

Metabolic and Growth-Promoting Properties of Serum Tripeptide
and its Synthetic Analog
    M. Michael Thaler and Loren R. Pickart   292

The Use of Selective Markers in the Study of Differentiated Gene
Function in Normal and Malignant Cells of Hepatic Origin
    David Rintoul and John Morrow   311

Regulation of the Corticosteriod Inducibility of Tyrosine
Aminotransferase in Somatic Cells Hybrids
    Carlo M. Croce, Gerald Litwack, and Hilary Koprowski   325

Expression of Hepatic Functions in Somatic Cell Hybrids
    Gretchen J. Darlington, Hans Peter Bernard, and
    Frank J. Ruddle   333

Extinction, Re-Expression and Induction of Liver Specific
Functions in Hepatoma Cell Hybrids
    Mary C. Weiss   346

Expression of Liver-Specific and Other Differentiated Functions
in Hybrids between Cultured Hepatoma Cells and L Cells
    E. Brad Thompson and Marc E. Lippman   358

Expression of Aryl Hydrocarbon Hydroxylase Induction in
Liver-and Hepatoma-Derived Cell Cultures
    Ida S. Owens, Akira Niwa, and Daniel W. Nebert   378

Susceptibility of Mammalian Cells *in Vitro* to Neoplastic
Transformation by Chemical Carcinogens
    J. A. DiPaolo   402

CONTENTS

The Enzymology of DNA Repair and Its Relation to
Carcinogenesis
  J. L. Van Lancker and T. Tomura — 412

Certain Aspects of Chemical Carcinogenesis *in Vitro* Using Adult
Rat Liver Cells
  P. T. Iype, T. D. Allen, and D. J. Pillinger — 425

Mechanisms of Chemical Carcinogenesis Analyzed in Rat Liver
and Hepatoma Cell Cultures
  *I. Bernard Weinstein, Nobuo Yamaguchi, Jan Marc Orenstein,
  Ronald Gebert, and M. Edward Kaighn* — 441

The Protective Effect of 7, 8-Benzoflavone and Steroid Hormones
against Aflatoxin B_1 and 7, 12-Dimethylbenz (a) Anthracene-
Induced Cytotoxicity in Cultured Rat Cells
  Arthur G. Schwartz — 460

The Study of Chemical Carcinogenesis Using Cultured Rat
Liver Cells
  Gary M. Williams — 480

International Symposium on Gene Expression and Carcinogenesis
in Cultured Liver: Summary and Some Perspectives
  Emmanuel Farber — 488

Subject Index — 493

CONTRIBUTORS

T. D. Allen, Paterson Laboratories, Christie Hospital and Holt Radium Institute, Manchester M20 QBX, England

Rod Balhorn, Departments of Medicine and Biochemistry, University of Iowa and VA Hospital, Iowa City, Iowa 52242

Thomas A. Barber, Department of Pathology, University of Wisconsin Medical School, Madison, Wisconsin 53706

Joyce E. Becker, McArdle Laboratory for Cancer Research, University of Wisconsin Medical School, Madison, Wisconsin 53706

Judith A. Berliner, Department of Pathology and Laboratory of Nuclear Medicine and Radiation Biology, School of Medicine, University of California, Los Angeles, Los Angeles, California 90024

Hans Peter Bernard, University of Basel, Basel, Switzerland

D. M. Bissell, Department of Medicine, University of California, San Francisco, San Francisco, California 94143

Robert J. Bonney, Division of Laboratories and Research, New York State Department of Health, New Scotland Ave., Albany, New York 12201

Carmia Borek, Radiological Research Laboratories, Department of Radiology, College of Physicians and Surgeons of Columbia University, New York, New York 10032

Paulette Bournot, Groupe de Biochimie de la Différenciation Cellulaire, Faculté de Médecine, Université de Dijon, 21033 Dijon, France

Roger Chalkley, Departments of Medicine and Biochemistry, University of Iowa and VA Hospital, Iowa City, Iowa 52242

Martine Chessebeuf, Groupe de Biochimie de la Différenciation Cellulaire, Faculté de Médecine, Université de Dijon, 21033 Dijon, France

Carlo M. Croce, The Wistar Institute, Philadelphia, Pennsylvania 19104

Gretchen J. Darlington, Yale University, New Haven, Connecticut 06520

Jean Desgres, Groupe de Biochimie de la Différenciation Cellulaire, Faculté de Médecine, Université de Dijon, 21033 Dijon, France

J. A. DiPaolo, Cytogenetics and Cytology Section, Biology Branch, Carcinogenesis Program, Division of Cancer Cause and Prevention, National Cancer Institute, National Institutes of Health, Bethesda, Maryland 20014

CONTRIBUTORS

Emmanuel Farber, Fels Research Institute and Departments of Pathology and Biochemistry, Temple University School of Medicine, Philadelphia, Pennsylvania 19140

Ronald Gebert, Institute of Cancer Research and Departments of Medicine and Pathology, College of Physicians and Surgeons of Columbia University, New York, New York 10032

Thomas D. Gelehrter, Departments of Human Genetics, Medicine and Pediatrics, Yale University School of Medicine, 333 Cedar Street, New Haven, Connecticut 06510

L. E. Gerschenson, Laboratory of Nuclear Medicine and Radiation Biology and Department of Pathology, School of Medicine, University of California, Los Angeles, Los Angeles, California 90024

Stanley Goldfarb, Department of Pathology, University of Wisconsin Medical School, Madison, Wisconsin 53706

Daryl Granner, Departments of Medicine and Biochemistry, University of Iowa and VA Hospital, Iowa City, Iowa 52242

Joe W. Grisham, Department of Pathology, University of North Carolina School of Medicine, Chapel Hill, North Carolina 27514

Steven H. Grossman, McArdle Laboratory for Cancer Research, University of Wisconsin Medical School, Madison, Wisconsin 53706

Michel Guiguet, Groupe de Biochimie de la Différenciation Cellulaire, Faculté de Médecine, Université de Dijon, 21033 Dijon, France

James A. Gurr, McArdle Laboratory for Cancer Research, University of Wisconsin Medical School, Madison, Wisconsin 53706

P. S. Guzelian, Department of Medicine, University of California, San Francisco, San Francisco, California 94143

D. M. Hays, Research Laboratories, Children's Hospital of Los Angeles and Department of Surgery, University of Southern California School of Medicine, Los Angeles, California

E. F. Hays, Laboratory of Nuclear Medicine and Radiation Biology, University of California, Los Angeles, Los Angeles, California 90024

Akira Ichihara, Institute for Enzyme Research, School of Medicine, Tokushima University, Tokushima, Japan

P. T. Iype, Paterson Laboratories, Christie Hospital and Holt Radium Institute, Manchester M20 QBX England

Vaughn Jackson, Departments of Medicine and Biochemistry, University of Iowa and VA Hospital, Iowa City, Iowa 52242

M. Edward Kaighn, W. Alton Jones Cell Science Center, Lake Placid, New York

CONTRIBUTORS

Y. Koga, Research Laboratories, Children's Hospital of Los Angeles and Department of Surgery, University of Southern California School of Medicine, Los Angeles, California

Hilary Koprowski, The Wistar Institute, Philadelphia, Pennsylvania 19104

H. Kröger, Robert Koch-Institut, 1 Berlin 65, Nordufer 20, Germany

Masayoshi Kumegawa, Department of Anatomy, Josei Dental College, Sakato, Saitama, Japan

Ben H. Leichtling, Department of Pharmacology, University of Colorado Medical Center, Denver, Colorado 80220

Marc E. Lippman, Laboratory of Biochemistry, National Cancer Institute, National Institutes of Health, Bethesda, Maryland 20014

Gerald Litwack, Fels Research Institute and Department of Biochemistry, Temple University School of Medicine, Philadelphia, Pennsylvania 19140

Frank Maley, Division of Laboratories and Research, New York State Department of Health, New Scotland Ave., Albany, New York 12201

Bernard F. Maume, Groupe de Biochimie de la Différenciation Cellulaire, Faculté de Médecine, Université de Dijon, 21033 Dijon, France

Gabrielle Maume, Groupe de Biochimie de la Différenciation Cellulaire, Faculté de Médecine, Université de Dijon, 21033 Dijon, France

John Morrow, Department of Biochemistry, Texas Tech University School of Medicine, Lubbock, Texas 79409

Anneli Nagel, Department of Pathology, University of North Carolina School of Medicine, Chapel Hill, North Carolina 27514

Daniel W. Nebert, Section on Developmental Pharmacology, Laboratory of Biomedical Sciences, National Institute of Child Health and Human Development, National Institutes of Health, Bethesda, Maryland 20014

H. B. Neustein, Research Laboratories, Children's Hospital of Los Angeles and Department of Pathology, University of Southern California School of Medicine, Los Angeles, California

M. O. Nicolson, Research Laboratories, Children's Hospital of Los Angeles and Department of Pediatrics, University of Southern California School of Medicine, Los Angeles, California

Akira Niwa, Section on Developmental Pharmacology, Laboratory of Biomedical Sciences, National Institute of Child Health and Human Development, National Institutes of Health, Bethesda, Maryland 20014

Peter Ofner, Children's Hospital Medical Center and Tufts University School of Medicine, Boston, Massachusetts 02111

CONTRIBUTORS

Aline Olsson, Groupe de Biochimie de la Différenciation Cellulaire, Faculté de Médecine, Université de Dijon, 21033 Dijon, France

Jan Marc Orenstein, Institute of Cancer Research and Departments of Medicine and Pathology, College of Physicians and Surgeons of Columbia University, New York, New York 10032

Ida S. Owens, Section on Developmental Pharmacology, Laboratory of Biomedical Sciences, National Institute of Child Health and Human Development, National Institutes of Health, Bethesda, Maryland 20014

Prudent Padieu, Groupe de Biochimie de la Différenciation Cellulaire, Faculté de Médecine, Université de Dijon, 21033 Dijon, France

Michael W. Pariza, McArdle Laboratory for Cancer Research, University of Wisconsin Medical School, Madison, Wisconsin 53706

Dieter Paul, Cell Biology Laboratory, The Salk Institute, Post Office Box 1809, San Diego, California 92112

Bernard Perissel, Groupe de Biochimie de la Différenciation Cellulaire, Faculté de Médecine, Université de Dijon, 21033 Dijon, France

Loren R. Pickart, Department of Pediatrics, University of California, San Francisco, San Francisco, California 94143

D. J. Pillinger, Paterson Laboratories, Christie Hospital and Holt Radium Institute, Manchester M20 QBX England

Van R. Potter, McArdle Laboratory for Cancer Research, University of Wisconsin in Medical School, Madison, Wisconsin 53706

Samuel B. Reichberg, Departments of Human Genetics, Medicine and Pediatrics, Yale University School of Medicine, 333 Cedar Street, New Haven, Connecticut 06510

U. Ingrid Richardson, Harvard School of Dental Medicine and Harvard Medical School, Boston, Massachusetts

David Rintoul, Department of Biological Sciences, Stanford University, Stanford, California 94305

William L. Risser, Departments of Human Genetics, Medicine and Pediatrics, Yale University School of Medicine, 333 Cedar Street, New Haven, Connecticut 06510

Michael D. Roper, Department of Pharmacology, University of Colorado Medical Center, Denver, Colorado 80220

Frank J. Ruddle, Yale University, New Haven, Connecticut

Jiro Sato, Cancer Institute, Okayama University Medical School, Okayama, Japan

G. Schreiber, The Russell Grimwade School of Biochemistry, University of Melbourne, Melbourne, Australia

CONTRIBUTORS

M. Schreiber, The Walter and Eliza Hall Institute of Medical Research, Melbourne, Australia

Arthur G. Schwartz, Fels Research Institute and Department of Microbiology, Temple University Medical School, Philadelphia, Pennsylvania 19140

Oscar A. Scornik, Department of Biochemistry, Dartmouth Medical School, Hanover, New Hampshire 03755

Y. Sera, Research Laboratories, Children's Hospital of Los Angeles and Department of Surgery, University of Southern California School of Medicine, Los Angeles, California

David Sherod, Departments of Medicine and Biochemistry, University of Iowa and VA Hospital, Iowa City, Iowa 52242

Robert Strunk, Children's Hospital Medical Center and Tufts University School of Medicine, Boston, Massachusetts 02111

Armen H. Tashjian, Jr., Harvard School of Dental Medicine and Harvard Medical School, Boston, Massachusetts

Sara B. Thal, Department of Pathology, University of North Carolina School of Medicine, Chapel Hill, North Carolina 27514

M. Michael Thaler, Department of Pediatrics, University of California at San Francisco, San Francisco, California 94143

E. Brad Thompson, Laboratory of Biochemistry, National Cancer Institute, National Institutes of Health, Bethesda, Maryland 20014

T. Tomura, Department of Pathology, School of Medicine, University of California, Los Angeles, Los Angeles, California 90024

J. L. Van Lancker, Department of Pathology, School of Medicine, University of California, Los Angeles, Los Angeles, California 90024

Kay Wagner, Department of Pharmacology, University of Colorado Medical Center, Denver, Colorado 80220

I. Bernard Weinstein, Institute of Cancer Research and Departments of Medicine and Pathology, College of Physicians and Surgeons of Columbia University, New York, New York 10032

Mary C. Weiss, Centre de Genétique Moléculaire du C.N.R.S., 91190 Gif-sur-Yvette, France

Wesley D. Wicks, Department of Pharmacology, University of Colorado Medical Center, Denver, Colorado 80220

Gary M. Williams, Fels Research Institute and Department of Pathology, Temple University School of Medicine, Philadelphia, Pennsylvania 19140

Jayantha Wimalasena, Department of Pharmacology, University of Colorado Medical Center, Denver, Colorado 80220

CONTRIBUTORS

James D. Yager, Jr., McArdle Laboratory for Cancer Research, University of Wisconsin Medical School, Madison, Wisconsin 53706

Nobuo Yamaguchi, Institute of Cancer Research and Departments of Medicine and Pathology, College of Physicians and Surgeons of Columbia University, New York, New York 10032

Susumu Yanagi, McArdle Laboratory for Cancer Research, University of Wisconsin Medical School, Madison, Wisconsin 53706

PREFACE

There is increasing interest in the use of cultured hepatic cells in studies on many aspects of carcinogenesis. To us, it seems that the general problem of gene control in cultured cells of hepatic origin and the problem of carcinogenesis in hepatic (and other epithelial) cells are in fact the same problem. Too often, however, experts in these two "fields" attend different meetings because of the arbitrary designations placed on their work. This symbposium, therefore, was organized to provide an opportunity for workers in fields fundamentally related, but diverse in detail, to talk to one another.

This conference was the offspring of the workshop on liver cell culture held at the McArdle Laboratory for Cancer Research, University of Wisconsin Medical Center, in 1972, and reported on by its organizer, Van R. Potter, in *Cancer Research* 32 (1972), 1998. That meeting was unusually useful in stimulating and focusing the research efforts of those who attended; thus it was decided that another similar conference was warranted. One can see in the papers in this volume the direct and fruitful results of investigation into many of the issues concerning the culture of hepatic cells raised two years earlier.

To the authors of these articles, we give our thanks. The high quality of their work speaks for itself. To them and to the readers of this book, we apologize for errors in form or substance which we, despite our best efforts, may have inadvertently allowed to be included.

We are indebted to Drs. J. DeVellis, H. Eagle, I. Harary, O. R. Lunt, K. Nagy, Van R. Potter and J. Ward for their invaluable assistance in organizing this symposium. We also wish to thank Mr. J. Johnson, Mrs. M. Allen, J. Yang and Ms. M. Kussel for their excellent work in coordinating the many details required in a meeting of this kind. The efficient help of Mrs. C. Negrych and C. Stram in editing the papers is also gratefully acknowledged. This symposium was cosponsored by the Atomic Energy Commission; the American Cancer Society; the Department of Pathology, University of California, Los Angeles; the Associated Western Universities; and the Mogul Corporation.

L.E. Gerschenson
E. Brad Thompson

CELLULAR DERIVATION OF CONTINUOUSLY CULTURED EPITHELIAL CELLS FROM NORMAL RAT LIVER

Joe W. Grisham, Sara B. Thal and Anneli Nagel

Introduction

Several investigators have independently established continuous cultures of diploid epithelial cells from normal livers of juvenile or adult mammals (1-8). The nature of the tissue cells *in vivo* from which these cultured cells derive is not established, although it is usually assumed that they arise from mature hepatic parenchymal cells (hepatocytes). Liver tissue contains several types of cells (10) from which cultures potentially may be established. Although most continuously culturable rat liver epithelial cells do not closely resemble hepatocytes morphologically being structurally more simple (8-9), they often possess some hepatocyte-like biochemical properties that appear to exclude their nonparenchymal origin (11).

Functionally indifferent tissue culture cells may theoretically originate from fully differentiated tissue cells which "dedifferentiate" in culture or from tissue cells that are not specifically differentiated *in vivo* (12). According to this hypothesis cytologically and functionally simplified liver epithelial cells might originate from fully differentiated hepatic parenchymal cells or from partially differentiated hepatocytic precursor cells (stem cells) in liver *in vivo*. Differentiated hepatocytes may undergo functional and structural simplification (modulation) in culture as a consequence of their new microenvironment *in vitro*. Alternatively, undifferentiated hepatocytic stem cells, which express characteristic hepatocytic structure and function only partially, may grow preferentially in culture. In this paper, we detail the results of experiments which indicate that isolated, differentiated normal hepatocytes proliferate imperfectly *in vitro* and rarely or never give rise to clones of liver epithelium. In culture, clones of liver epithelial cells developed only from single isolated cells which were morphologically unlike hepatocytes. Cultured liver epithelial cells appeared to arise from a cell that cycled *in vivo* in synchrony with hepatocytes and proliferated rapidly in culture, resulting in its selection *in vitro*. The *in vivo* precursors of these cultured cells may represent hepatocytic stem cells.

Materials and Methods

Livers from juvenile or adult rats of both sexes were used for all studies. Single-cell suspensions were prepared in either 0.025% trypsin (4) or in a dissociation solution containing 0.1% collagenase, 0.1% hyaluronidase, and 1% albumin (13). Dispersed cells were counted in a Neubauer chamber using phase optics. Mass cultures were established by plating from 5×10^4 to 5×10^5 cells suspended in 5 ml of tissue culture medium into 60 x 10 mm plastic dishes. Primary cell cloning was performed by diluting dispersed cells to a concentration of 10 to 30 cells /ml, and 0.1 ml was added to each well of several plastic dishes containing 48 10 x 10 mm wells. Three ml of tissue culture medium were then added to each well. After allowing time for cells to settle and attach, wells containing only one cell were identified and marked and the cell photographed with phase optics. Marked wells were examined biweekly for clonal growth.

All cell cultures were grown in Ham's F-10 medium buffered with 25 mM Hepes buffer (pH 7.4) and supplemented with 10% fetal calf serum or 10% horse serum, 3 mg/ml glucose, 6×10^{-7} M hydrocortisone, and antibiotics (25 units penicillin and 25 μg streptomycin each/ml). Cultures were grown at 37.5°C in humidified air and were fed two to three times a week. For transfer, cells were removed from dishes by briefly incubating them either with 0.125% trypsin in Ca^{++} and Mg^{++}-free salt solution or with the dissociation solution containing collagenase, hyaluronidase, and albumin. Transfers were made when cells in dishes or wells were confluent. Cells were prepared for electron microscopy by routine methods.

Explant cultures were prepared as previously described (14) from livers of 5-day old rats whose hepatocytes had been tagged by giving them three doses of ^{3}HTdR (0.5 Ci/gm/dose) intraperitoneally during the second 12 hr after partial hepatectomy. This procedure allows the preferential labeling of hepatocytes and of any other liver cells that proliferate synchronously with hepatocytes following partial hepatectomy (15). It has previously been shown that following partial hepatectomy, peak proliferation of all easily recognized nonparenchymal cells occurs about one day after peak proliferation of hepatocytes (15). Autoradiograms were made from sections of liver obtained at the time of culture and the proportion of different types of labeled liver cells was determined. Autoradiograms were also prepared from whole mounts of explant cultures and the percents of labeled hepatocytes and simple epithelial cells, as well as the numbers of silver grains overlying these labeled cells were determined after various culture intervals. Autoradiograms were produced by dipping tissue sections or culture whole mounts in Kodak NTB-2 liquid emulsion and preparing them subsequently by routine methods.

Results

A. Isolation of cells

Suspensions of single liver cells prepared with trypsin by the methods used here, were enriched in nonparenchymal cells, as compared to cellular populations in liver in vivo (Table 1). This may have resulted from a relatively more marked trypsin-induced lysis of hepatocytes than of other liver cells. Plasma membranes of hepatocytes frequently "bubbled" actively in the presence of low concentrations of trypsin, whereas membranes of other liver cells appeared to be more resistant to this enzyme. Supporting this opinion was the presence in trypsin digests of numerous extracellular mitochondria of large size (typical of those from hepatocytes), as well as many severely damaged, ghost-like hepatocytes. Also more than two-thirds of all "intact" hepatocytes in trypsin digests were more-or-less severely damaged ultrastructurally; this was not true of nonparenchymal cells. Although use of the collagenase-hyaluronidase-albumin dissociation solution resulted in less apparent diminution in the fraction of hepatocytes in cellular suspensions (Table 1), more than half were ultrastructurally damaged.

Isolated hepatocytes were readily distingusihed from nonparenchymal cells by phase microscopy (Fig. 1 and 2). Isolated hepatocytes maintained their usual polygonal shape, measured from 14 to 25 microns in smallest diameter, and contained a large amount of phase granular cytoplasm (Fig. 1). The ratio of cytoplasmic area to nuclear area was always three or more. That these cells were hepatocytes was confirmed by electron microscopy (Fig. 3), which showed their cellular ultrastructure to be typical of hepatocytes (16). Nonparenchymal cells as a group were smaller than hepatocytes (3 to 8 microns in diameter), they were typically ovoid or spindle-shaped and they contained a relatively scant amount of nongranular cytoplasm (Fig. 2). The ratio of cytoplasmic area to nuclear area was usually between one and two. Electron microsocopic examination of mixtures of nonparenchymal cells revealed cells with a variety of ultrastructural characteristics, only one of which is illustrated here. Occasionally occurring in groups joined by attachment plates (Fig. 4) or

	Hepatocytes	Other Cells
	(Percent)	
Liver in vivo*	60.2	39.8
Trypsin digest	32-39	61-68
Collagenase, hyaluronidase, albumin digest	53-58	42-47

*From Daost (10)

TABLE 1. PROPORTIONS OF HEPATOCYTES IN RAT LIVER IN VIVO AND IN CELL SUSPENSIONS PREPARED ENZYMATICALLY

	Hepatocytes	Other Cells
Number Observed	858	122
Clones Formed (Cloning Frequency)		
Total	0 (0)	5 (.041)
Epithelial	0 (0)	2 (.016)
Fibroblastic	0 (0)	3 (.025)

TABLE 2. CLONOGENIC CAPACITY OF HEPATOCYTES AND OTHER CELLS IN LIVER CELL SUSPENSIONS

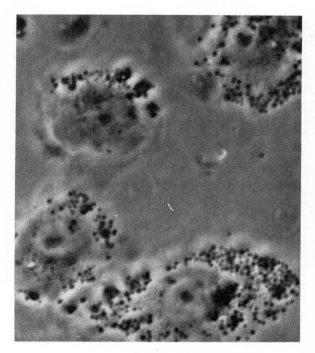

FIG. 1:

<u>Isolated hepatocytes dispersed by trypsinization</u>. The cytoplasm is densely granular. Phase micrograph. Mag. X 1950.

FIG. 2:

<u>Isolated nonparenchymal liver cells dispersed by trypsinization</u>. Phase micrograph. Mag. X 1950.

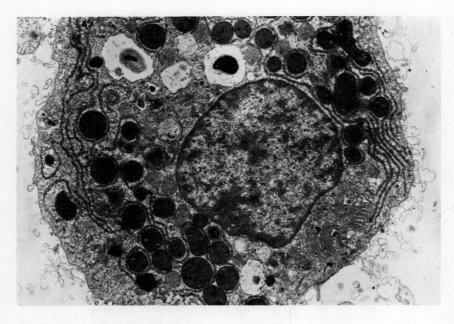

FIG. 3:

<u>Electron micrograph of an isolated, trypsin-dispersed hepatocyte</u>. Although this cell evidences some damage it is readily identifiable as a parenchymal liver cell because of its characteristic ultrastructure. Note the free mitochondrion and fragments of membrane outside this cell. Mag. X 8400.

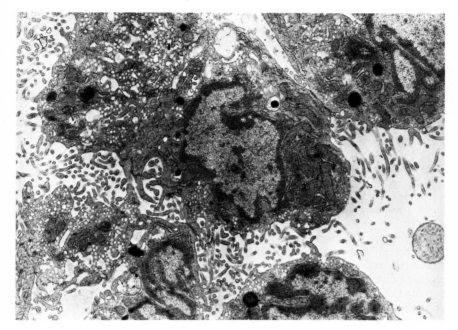

FIG. 4:

<u>Electron micrograph of one type of nonparenchymal liver cell in a trypsin digest.</u> These cells have prominent surface microvilli and cytoplasmic organelles are relatively sparse. In this photograph three adjacent cells are joined by attachment complexes. Mag. X 8700.

more frequently as single cells (Fig. 5), these cells had
prominent surface microvilli and scant cytoplasmic organelles.
Ultrastructurally they resembled the cells lining terminal
ductules in rat liver (17). This opinion was heightened by
the occasional chance observation of a similar small cell
joined by attachment complexes to a hepatocyte.

Other nonparenchymal cells, included phagocytes, various
blood leukocytes, and ultrastructurally nondescript, spindle-
shaped cells which probably were fibroblasts. Cells which
were phagocytic had engulfed hepatocyte mitochondria present
in the suspension of liver cells; they often contained large
residual bodies and probably represented Kupffer cells.

B. Mass Cultures

Primary mass cultures of cell suspensions prepared by
either of the enzymatic dissociation methods contained prominent foci of hepatocytes overlaid on a background of cells
growing in fibroblastic (Fig. 6) or epitheloid patterns.
Electron microscopic examination of these parenchymal foci
corroborated that they were composed of ultrastructurally
typical hepatocytes. Individual foci in primary cultures
contained from less than 10 to more than 100 hepatocytes
and some dishes contained more than fifty such cellular foci.
Cultures were dispersed enzymatically and replated at approximately weekly intervals and after each replating the numbers
of hepatocytic foci declined markedly. After three to four
enzymatic transfers, virtually no hepatocytes remained and
most cultures consisted only of complex monolayers of cells
growing in fibroblastic and epitheloid patterns.

C. Primary Cloning

No single cells identifiable as hepatocytes by phase
microscopy formed clones in this study (Table 2). Five of
122 nonparenchymal cells (small nongranular, oval to spindle-
shaped) formed clones (Table 2). Three of these had fibroblastic growth patterns and formed collagen fibers in confluent cultures. These clones are not considered further in
this report, but preliminary evidence suggests that they are
authentic fibroblasts. Two cells gave rise to clones which
grew in epitheloid pattern (Table 2). One clonogenic cell,
its first two progeny cells, and the primary clone of epithelial cells formed by it are shown in Fig. 7-9. The phase
microscopic characteristics of cells subcultured from this
clone are shown in Fig. 10. Electron microscopic examination of cells from this clone showed them to be ultrastructurally similar to liver epithelial cells established by
other investigators; they formed a continuous monolayer

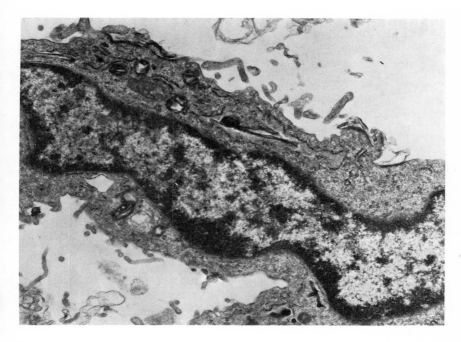

FIG. 5:

Electron micrograph of a single, trypsin-dispersed cell which is ultrastructurally similar to those shown in Fig. 4. Note surface microvilli and sparse cytoplasmic organelles. Mag. X 15,700.

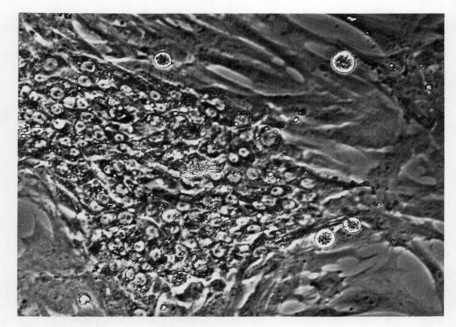

FIG. 6:

<u>A focus of hepatocytes in a primary mass culture of enzymatically isolated liver cells</u>. Surrounding the hepatocytes are other cells growing in a fibroblastic pattern. Phase micrograph. Mag. X 480.

sheet and were joined by attachment complexes. Cytoplasmic organelles were notably more sparse than in hepatocytes.

D. Explant Cultures

Hepatocytes composed more than 90 percent of all the liver cells tagged during the second 12 hr after partial hepatectomy of 5 day-old rats (Table 3). Previous studies have shown that proliferation of nonparenchymal cells is not augmented at this time after partial hepatectomy (15). Both hepatocytes and simple epithelial cells, the latter similar to the epithelial cells grown in continuous cultures derived from rat liver, grew out from explants (14). Nearly 100 percent of both types of cultured cells were tagged and, on the second culture day, the grain density over nuclei of both types of cells was similar (Table 4). With increasing time in culture the number of simple epithelial cells increased greatly (14) and the proportion of these cells that were labeled (and the number of grains overlying them) decreased precipitously (Table 4), indicating that they had passed through several division cycles. However, the number of grains overlying tagged hepatocytes decreased only modestly during this interval, denoting limited proliferation, which corroborated earlier observations (18,19). These results suggest that the simple epithelial cells were derived by rapid consecutive proliferation in culture of cells that were labeled in vivo synchronously with hepatocytes. Hepatocytes cycled only a few times after being placed in culture.

Discussion

Continuous culture of normal hepatocytes (that is, hepatic parenchymal cells competent to perform the myriads of hepatic functions in culture) has not been accomplished. However, structurally simple cells which often maintain some limited, but specific, functional or biochemical properties of hepatocytes can be cultured routinely by plating dispersed mammalian liver cells (1-8). These cultured liver cells typically grow in an epitheloid pattern, although hepatofunctional cells from human livers may grow in either fibroblastic or epitheloid patterns (6). Liver-derived epithelial cells usually form continuous monolayer sheets in which adjacent cells are joined by attachment complexes (8,9). Occasionally extracellular spaces, reminescent of bile canaliculi, have been observed between two adjacent groups of attachment complexes (8,11). Ultrastructurally the cytoplasm of continuously cultured epithelial cells is considerably less complex (8,9) than the cytoplasm of hepatocytes either in vivo (15) or in vitro in explant cultures

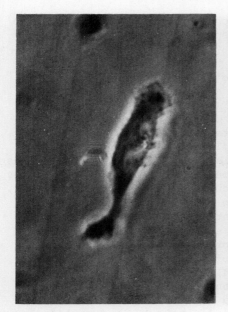 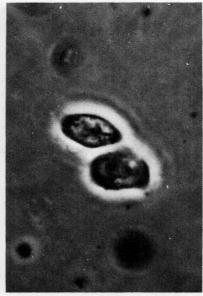

FIG. 7:

A single nonparenchymal liver cell which proliferated to form an epithelial clone. Phase micrograph. Mag. X 2000.

FIG. 8:

Daughter cells of the nonparenchymal liver cell illustrated in Fig. 7. This photograph was made after one week in culture. Phase micrograph. Mag. X 2000.

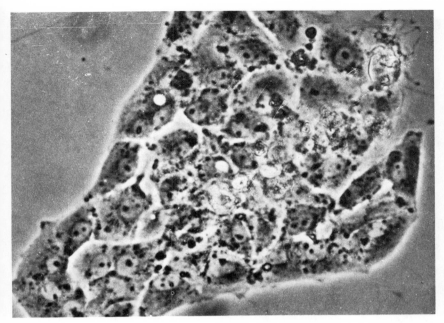

FIG. 9:

The young primary clone derived from the cells shown in Figs. 7 and 8. This photograph was made after three weeks in culture. Phase micrograph. Mag. X 1850.

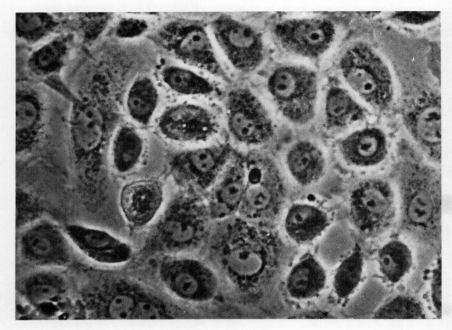

FIG. 10:

<u>Cells subcultured from the clone illustrated in Fig. 9.</u>
These cells were in their third passage when this photograph was made. Phase micrograph. X 2250.

	Hepato cytes	Kupffer cells	Duct cells	Other cells
Percent labeled in vivo	80±11	11±2	5± 1	4±0.5
Relative proportion of the total labeled population (percent)*	92.2	7.5	.02	.03

*Based on the relative populations of different types of cells in rat liver in vivo as determined by Daost (10).

TABLE 3. THE FRACTION OF HEPATOCYTES AND OTHER LIVER CELLS LABELED BY ^{3}HTdR IN LIVERS OF 5-DAY OLD PARTIALLY HEPATECTOMIZED RATS

	Days in Culture					
	4		6		12	
	Hepato*	SLEC*	Hepato	SLEC	Hepato	SLEC
Percent labeled	90±13	96±11	98±10	72±5	98±8	10±15
Nuclear grains/2	5.8±0.6	6.1±0.8	4.1±0.8	2.3±0.9	3.0±1.0	-

*Hepato and SLEC indicate hepatocytes and simple liver epithelial cells, respectively.

TABLE 4. PERCENT LABELED CELLS AND GRAIN COUNTS OVER HEPATOCYTES AND SIMPLE LIVER EPITHELIUM CULTURED FROM EXPLANTS OF REGENERATING LIVER TAGGED IN VIVO

(14) or in short-term nonproliferating monolayer cultures (20). Specifically all organelles tend to be more sparse than in hepatocytes and mitochondria are usually smaller also (2).

Liver tissue contains a variety of types of cells, several of which could serve as source of cultured cells (10). The precise *in vivo* cytologic origin of cultured liver epithelial cells has not been established. Nonparenchymal cells appear to be eliminated as a possible source since most cultured liver cell lines possess some liver-like properties (11). It is hypothetically possible that cultured liver epithelial cells derive from fully differentiated hepatocytes or from presumptive, undifferentiated hepatic parenchymal precursor cells (hepatocytic stem cells) (12). Hepatocytes may lose their differentiated structural and functional features as a consequence of isolation and culture. The present study indicates that enzymatic techniques of isolation, commonly used to form single cells suspensions from liver tissue, damage many hepatocytes. The microenvironment in culture differs greatly from that *in vivo* and this difference might allow or cause hepatocytes to undergo a modular shift in structure and function. Continuous cell proliferation, which is a requirement of propagable cultures, may prevent hepatocytes from maintaining their differentiated function and structure (21). However, studies on the structure and function of proliferating hepatocytes in liver regenerating after partial hepatectomy suggest that the structure and at least some aspects of the function of cycling hepatocytes are nearly normal (for review, see 22).

Culture liver epithelial cells may develop from a small *in vivo* population of culturally hardy hepatocytic stem cells, which proliferate selectively under the conditions of culture and which only incompletely express the structural and functional properties of differentiated hepatoctyes. Such presumed stem cells have been described at the portal tract periphery in livers of fetal and young adult rats (23) and recently a similarly located, nondescript cell has been suggested as the site of production of alpha fetoprotein (24). Structurally simple cells which form the smallest ramification of the intrahepatic biliary duct system (17) are said embryologically to derive from the same source as hepatocytes (25) and to retain the potential to differentiate into typical hepatocytes (26-28). These types of cells are not numerous in liver, probably accounting for less than 1 percent (24), and in order to populate cultures would have to plate and grow with high efficiency.

The many studies on the culture of dispersed liver cells have provided little data which specifically relates to the in vivo cell of origin. Although primary cloning of dispersed liver cells has been reported as a technique to efficiently develop continuously propagable lines of liver cells (2), identification of the cells from which clones developed was apparently not attempted. In a study in which differential adhesion to glass was used as a means of partially separating different types of dispersed liver cells, the cells pictured 24 hr after starting the culture and which were said to originate the cell line do not resemble hepatocytes structurally (4). Mass primary cultures of dispersed liver cells clearly contain foci of morphologically identifiable hepatocytes (13,29). Such cultures are capable of growing in an arginine-free medium presumably because the hepatocytes included in them synthesize this amino acid (29). Normal hepatocytes can survive and function for up to 4 days in monolayer cultures if proliferation is suppressed (21). Although proliferation was suspected to be incompatible with maintenance of specific hepatic function, no direct data bearing on this contention was presented (21).

Outgrowths from explant cultures of rat liver contain, in addition to hepatocytes, monolayers of epithelial cells that are morphologically identical and biochemically similar to liver epithelial cells in continuous cell cultures (14). Hepatocytes and simple liver epithelial cells (sometimes called clear epithelial cells in these studies because of the phase translucency of their cytoplasm) appear to represent distinctively different types of liver cells, both of which are present in vivo (14). Both hepatocytes and cultured liver epithelial cells emerge from explants soon after cultures are initiated and they form separate outgrowth areas which may be adjacent but do not intermingle (14). Ultrastructural study of such outgrowths and of the explants from which they were derived failed to produce any evidence to suggest a direct morphologic transition between hepatocytes and simple liver epithelium (14). Hepatocytes always appeared structurally normal or, in declining cultures, showed only evidence of degeneration and necrosis (14). Even in senescent cultures hepatocytes did not show any evidences of structural simplification (14). These observations, together with those of Bissell (21), suggest that hepatocytes are sufficiently robust to withstand microenvironmental changes associated with culture in vitro.

Hepatocytes and simple liver epithelium in explant outgrowths proliferated rapidly during their first day in vitro, but after the second day in culture the hepatocyte growth fraction declined precipitiously and by culture days 6 to 8, hepatocytes had virtually ceased proliferating (18, 19). Simple liver epithelial cells continued to proliferate vigorously and the culture population of these cells continued to increase until they often became more numerous than hepatocytes (18). These results suggest that hepatocytes proliferate poorly in culture and that, after an initial postexplantation spurt of growth, no further growth occurs. In contrast, simple liver epithelial cells, appear to thrive under the culture conditions.

The results of the present study directly augment the earlier observations and provide further evidence that hepatocytes have a limited potential for in vitro proliferation when grown as explants, when plated as mixtures of cells in mass cultures, or when plated as single cells (primary cloning). The enzymatic methods used clearly damaged many hepatocytes and may have reduced their capacity to grow in culture. However, hepatocytes grow poorly in explant cultures, even without enzymatic pretreatment. In sharp contrast, simple liver epithelial cells grow readily in any type of culture and independently of the use of enzymatic dispersion. The impaired capacity of hepatocytes to proliferate in culture, combined with the absence of any morphologic evidence of transitions between hepatocytes and simple liver epithelial cells (14) suggests that a culture induced modular shift in the form and function of hepatocytes cannot explain the origin in culture of simple liver epithelium. Our experiment on the label patterns in cultured hepatocytes and simple liver epithelium following the labeling of replicating DNA in vivo during the first day after partial hepatectomy strengthens this view. Both labeled simple liver epithelium and labeled hepatocytes appeared in outgrowths at approximately the same time and when they were first noted the number of silver grains per square micron of nuclear surface was nearly identical. This suggests that both types of cells were present in liver tissue in vivo and that both were proliferating and were labeled at the same time after partial hepatectomy.

At the time of their initial appearance in culture outgrowths, both hepatocytes and simple liver epithelial cells proliferated, as is shown by earlier studies in which proliferating cells were tagged with tritiated thymidine in vitro (18,19) and by the present studies which quantitated the decline in grains over nuclei. Hepatocytes cycled only once or twice in culture and soon stopped dividing (18, 19) whereas simple liver epithelial cells continued to cycle almost constantly for several days (18). Some simple liver epithelial cells cycled so many times that nuclear label was diluted until it was no longer autoradiographically detectable. In the same interval the nuclear label of the average hepatocyte declined only by about 50 percent, indicating that they each had cycled only once on average since being placed in culture. These results suggest that simple liver epithelial cells are derived from liver cells distinct from hepatocytes morphologically, but which proliferate synchronously with hepatocytes after partial hepatic resection. Furthermore, the data indicate that simple liver epithelial cells proliferate continuously in culture and that, by that means, their number is selectively augmented.

Single hepatocytes isolated in wells were never observed to be clonogenic in this study. Since the cloning efficiency of freshly isolated epithelial cells is low (2), this total failure may have resulted from the relatively small number of hepatocytes which we have analyzed for cloneforming capacity. Future studies may ultimately show that a small fraction of hepatocytes is able to establish clones in vitro. Of more interest is our observation that a relatively large fraction of cells not morphologically recognizable as hepatocytes formed clones that grew in an epithelial pattern and morphologically resembled rat liver epithelial cells established in culture by other investigators. Our preliminary observation that these cells produce alpha fetoprotein suggests their close functional relationship to hepatocytes. Unfortunately we can only speculate on the exact ultrastructural and functional characteristics of clone-forming cells in fresh isolates since the nonparenchymal cells were structurally heterogenous and since it is obviously impossible to analyze the structural and functional properties of a single cell and at the same time have it proliferate to form a clone. By phase microsocpy, clonogenic cells giving rise to epithelial offspring were ovoid, measured approximately 3 by 8 microns in shortest and longest diameters, and had scant nongranular cytoplasm. These characteristics clearly distinguish them from isolated hepatocytes, which were polygonal and the smallest of which measured at least 14 microns in shortest diameter. Additionally, hepatocytes had copious granular cytoplasm by phase microscopy.

The results of this study together with our analysis of the literature combine to provide the potential characteristics of the liver cell that grows in culture to originate propagable clones of liver epithelium. 1) It

proliferates in synchrony with hepatocytes after partial
hepatectomy but is structurally and functionally simpler,
though it may be capable of some liver-like functions. 2)
It is probably present in liver tissue in vivo in small
numbers relative to the population of hepatocytes, but it
has a very high primary cloning efficiency and proliferates
readily and continuously in culture, allowing formation of
large populations. 3) It performs a few hepatocyte-like
functions in vitro and different clones may perform different functions.

These characteristics are similar to those of specific
stem cells in other tissues and they suggest that hepatocytic stem cells are probably present in rat liver. Several
investigators have suggested the presence of cells in liver
that meet the criteria of stem cells, that is, specifically
undifferentiated cells that are capable of dividing and
differentiating to form hepatocytes when properly stimulated
(24-28). However, autoradiographic studies on rat liver
regenerating after partial hepatectomy indicate that stem
cells are not prominently involved in the increase in the
hepatocyte population that occurs after this operation (30).
Hepatic resection is apparently not the proper stimulus to
induce this differentiative change but such a transition has
been described in certain reactions of the liver to toxins
(27,28) and following bile duct ligation (30). Both of the
investigators cited identified the stem-like cells as those
forming the smallest radicals of the intrahepatic duct system (ductules), which connect directly with hepatocytes (30-
32). Embryologically, these ductular cells and hepatocytes
are both derived from the same precursor cells (25). Prominently disposed, imperfectly differentiated cells which
develop into hepatocytes and ductular cells have been detected at the portal periphery in embryonic animals (23). It
is possible that a few such cells remain in the liver in the
adult state. This situation has been suggested by a recent
study on the cytologic localization in liver of estrogen-
binding cells (which presumably synthesize alpha fetoprotein)
(24). Such cells were mainly located in livers of newborn
animals at the portal periphery and as single cells within
portal tracts (24). These estrogen binding cells may represent stem cells. It was suggested that similar cells were
present, but functionally inactive, in adult livers, where
they probably have the form of ductular cells (24). We have
infrequently observed in enzymatic digests of liver, hepatocytes joined by attachment plates to one or more small, structurally simple cells. We speculate that these ductular cells
may have stem cell properties and give rise in vitro to

cultured liver epithelium.

These results suggest one possible cellular source in vivo for the origination of continuously propagable liver epithelial cells. Production of clones from stem cells may explain the generally poor, incomplete and variable (from clone to clone) expressions of hepatic function by these cells. These results do not exclude the possibility that in other experimental situations it may be possible to establish euploid cultures directly from normal hepatocytes, which may retain more of their functional characteristics in vivo. This desirable goal has not been reached and our results suggest that it may not be obtainable using normal hepatocytes that have been separated with proteolytic enzymes. An alternative might be to define culture conditions that would allow differentiation of liver epithelial cells in vitro. Some studies have suggested that such differentiation may occur in confluent cultures of liver epithelial cells (2,4,11).

Summary and Conclusions

In this study single, enzymatically dispersed hepatocytes never developed into clones in vitro. Hepatocytes had a limited potential for growth in vitro when cultured as single dispersed cells, as heavy concentrations of dispersed cells (mass cultures), or as tissue explants not treated enzymatically. Epithelial clones developed from small oval to spindle-shaped, nongranular cells which could be distinguished from hepatocytes by phase microscopy. Liver epithelial clonogenic cells may be derived from presumtive hepatocytic stem cells. Since such cells represent only a small fraction (less than 1%) of the total cellular population of liver tissue they must have a very high cloning efficiency in vitro. The cells originating clones in vitro cycle in synchrony with hepatocytes in vivo and they proliferate regularly for the first few days in culture, giving rise to sizable derivative populations.

Supported by grant AM 17595 from the National Institutes of Health, grant BC-142 from the American Cancer Society and by a grant from the John A. Hartford Foundation, Inc.

References

1. Katsuta, K. and Takaoka, T. Japan J. Exp. Med. 33 (1963) 265.
2. Coon, H.G. Carnegie Inst. Washington Yearbook 67 (1969) 419.
3. Gerschenson, L.E., Anderson, M., Molson, J. and Okigaki, T. Science 179 (1970) 858.
4. Williams, G.M., Weissburger, E.L. and Weissburger, J.H. Exp. Cell Res. 69 (1971) 106.
5. Iype, P.T. J. Cell Physiol. 79 (1971) 281.
6. Kaighn, M.E. and Prince, A.M. Proc. Nat. Acad. Sci. U.S. 68 (1971) 2396.
7. Lambiotte, M. Susor, W.A. and Cahn, R.D. Biochimie 54 (1972) 1179.
8. Diamond, L., McFall, R., Tashiro, Y. and Sabatini, D. Cancer Res. 33 (1973) 2627.
9. Coon, H.G. and Manasek, F.J. Carnegie Inst. Washington Yearbook 68 (1970) 540.
10. Daost, R., In: Brauer, R.W. (ed) Liver Function, American Institute of Biological Sciences, Washington (1958) 3.
11. Potter, V.R. Cancer Res. 32 (1972) 1998.
12. Grobstein, C. In: Willmer, E.N. (ed) Cells and Tissues in Culture, Vol 1, Academic Press, London and New York, (1965) 463.
13. Gerschenson, L.E., Berliner, J. and Davidson, M.B. In: Colowick, S.P. and Kaplan, N.D. (eds), Methods in Enzymology, Vol. 32B, Academic Press, London and New York, In press, 1974.
14. Alexander, R.W. and Grisham, J.W. Lab. Invest. 22 (1970) 50.
15. Grisham, J.W. Cancer Res. 22 (1962) 842.
16. Bruni, C. and Porter, K.R. Am. J. Path. 46 (1965) 691.
17. Steiner, J.W. and Carruthers, J.S. Am. J. Path. 40 (1962) 253.
18. Grisham, J.W. J. Cell Biol. 47 (1970) 79a.
19. Rabes, H. and Jantsch, B. Z. Zellforsch 112 (1971) 414.
20. Chapman, G.S., Jones, A.L., Meyer, U.A. and Bissell, D.M. J. Cell Biol. 59 (1973) 735.
21. Bissell, D.M., Hammaker, L.E. and Meyer, U.A. J. Cell Biol. 59 (1973) 722.
22. Grisham, J.W., Tillman, R.L., Nagel, A. and Compagno, J. In: Bucher, N.L.R., Lesch, R. and Reuttner, W. (eds), Liver Regeneration after Experimental Injury. International Medical Book Corp., New York, In press 1974.

23. Wilson, J.W., Groat, C.S. and Leduc, E.H. Ann. N.Y. Acad. Sci. 111 (1963) 8.
24. Uriel, J., Aussell, C., Bouillon, D., deNechaud, B. and Loisillier, F. Nature New Biology 244 (1973) 190.
25. Bloom, W. Am. J. Anat. 36 (1926) 451.
26. Steiner, J.W., Carruthers, J.S. and Kalifat, S.R. Exp. Mol. Path. 1 (1962)1962.
27. Steiner, J.W. and Carruthers, J.S. Lab. Invest. 12 (1963) 471.
28. Wilson, J.W. and Leduc, E.H. J. Path. Bact. 76 (1958) 441.
29. Leffert, H.L. and Paul, D. J. Cell Biol. 52 (1972) 559.
30. Klinman, N.R. and Erslev, A.J. Proc. Soc. Exp. Biol. Med. 112 (1964) 338.

SOME CHARACTERISTICS AND FUNCTIONS OF ADULT RAT LIVER
PARENCHYMAL CELLS IN PRIMARY CULTURE

Robert J. Bonney and Frank Maley

Introduction

Several advances in the methodology of adult liver parenchymal cell isolation have been described recently (1-12). It is now possible to recover nearly half of the total liver parenchymal cells as a homogeneous population. These cells can be isolated under aseptic conditions, and several investigators have initiated primary tissue cultures with such cells (13-17). The structural integrity of the plasma membranes of the freshly isolated cells has been reported to be abnormal (18, 19). However, after several days in culture with daily medium changes, the cells resemble normal intact parenchymal cells (15, 16, 20). For example, the level of ATP per mg of protein was low in freshly isolated cells but returned to normal after one day in culture (15).

During the first week of culture there is little or no DNA synthesis or cell division among the parenchymal cells (14-16), and one can demonstrate a variety of biochemical activities that are characteristic of normal liver cells in vivo. These include the induction of tyrosine aminotransferase (TAT) (16, 20) and ornithine decarboxylase (17), the maintenance of adult liver aldolase activity (20), gluconeogenesis (15), the biosynthesis of glycogen and albumin (15), and the presence of two liver microsomal enzymes (15). However the inducibility of O-demethylase (15) and TAT (16, 20) appears to decrease, for unknown reasons, after several days in vitro. Other properties of liver cells, including vitamin metabolism and storage and the active transport of amino acids, have not yet been demonstrated in the primary cultures.

Our studies from this laboratory have focused on the regulation of thymidylate synthetase in intact regenerating liver (21) and in cultured parenchymal cells isolated from regenerating liver (22). The metabolism of folate and folate derivatives may be related to the control of this enzyme (21), and we wished to determine if cells in culture are capable of metabolizing folic acid into the conjugates which are considered to be normal storage forms for liver folates. In this investigation, aside from the potential metabolic alteration of cells maintained in primary culture when compared to their tissue of origin, changes in such structural components as the cell membrane had to be considered. These changes in the membranes may be reflected by the manner in which the cells actively transport specific amino acids.

This report deals with three metabolic events in cultured parenchymal cells: the regulation of TAT, the metablism of folic acid, and the active transport of an amino acid, α-aminoisobutyric acid (AIB).

Materials and Methods

Materials

Ham's F12 medium was purchased in dry powder form from Grand Island Biological Company and fetal calf serum from Microbiological Associates, Inc. Rat serum was obtained from adult Wistar rats at the Griffin Laboratory, New York State Department of Health, Albany. Collagenase, type I, cycloheximide, and dexamethasone were obtained from Sigma Chemical Co. Insulin was purchased from Eli Lilly Company and glucagon from Calbiochem. Radioactive [1-^{14}C] AIB was obtained from New England Nuclear and [4,5-^{3}H]L-leucine, [6-^{14}C]orotic acid, and [3',5'-^{3}H]folic acid were from Schwarz/Mann. Pteroylhepta-pteroylpenta-, and pteroyltriglutamic acid were obtained from Dr. J. Bertino, Yale University. [^{3}H]Pteroylmonoglutamic acid (folic acid) was purified by chromatography on a column of DEAE-cellulose (Whatman DE-52) (23). The purified material exhibited a single radioactive spot on cellulose thin layer chromatography (polygram cell 300 UV$_{354}$) in 0.1 M potassium phosphate, pH 6.2, which corresponded to folic acid.

Animals

Albino rats, Wistar strain, weighing 120-130 g, were partially hepatectomized according to the procedure of Higgins and Anderson (24) and were allowed to recover from the operation for at least 4 weeks.

Isolation of Liver Parenchymal Cells

Parenchymal cells were isolated from regenerated liver by a variation (12) of the Crisp and Pogson technique (7). Since that report (12), however, we have found it advantageous to perform all filtrations, centrifugations and washings at room temperature. In addition, the cells were washed with the incubation medium rather than Hanks' medium. The entire isolation procedure required approximately 1 h, and with a trichannel polystaltic pump (Buchler Instruments) two rat livers could be perfused simultaneously.

Culture Conditions

The culture conditions were essentially as described elsewhere (16) with the following modifications. The standard medium used in the present study was Ham's F12 with 10 mU of insulin per ml and with 15% instead of 25% fetal calf serum (16). The medium was changed 6 h after the initial plating of the cells and every 24 h thereafter. In some experiments, cells were cultured for 48 h in standard medium. The medium was then changed to Ham's F12 with insulin but without fetal calf serum for 24 h preceding the study.

Analytical Methods

Tyrosine aminotransferase was assayed as described previously (16). Protein synthesis was measured as follows: After a 6-h incubation of the cells in the presence of $[4,5-^3H]$L-leucine the medium was aspirated and the plates were rinsed twice with saline. Four ml of saline were added to the plates, and the cells were removed with a rubber policeman and transferred to test tubes. The cells were then frozen and thawed three times and finally placed in an ice bath. An equal volume of 20% trichloroacetic acid (TCA) was added to the tubes. The insoluble material was centrifuged at 2,000 x g for 10 min and washed with 10% TCA until the radioactivity in the supernatnat fluid was less than twice background. One ml of 1 N NaOH was added to the tubes which were incubated at $37^\circ C$ for 1 h. Aliquots were then taken for determination of protein by the Lowry method (25) and for estimation of radioactivity.

RNA synthesis was determined by measuring the incorporation of $[6-^{14}C]$ orotic acid into RNA. Total RNA was estimated by the method of Munro and Fleck (26).

The uptake of AIB was measured by determining the amount of $[^{14}C]$AIB inside the cells after specified periods of incubation. The medium was aspirated from the plates and its radioactivity determined. The cells were washed repeatedly with saline until the radioactivity in the rinse was less than twice background. One ml of 1 N NaOH was added to the plates, and the cellular material was removed with a rubber policeman and transferred to tubes. Aliquots were taken for the determination of radioactivity and protein. The data are also expressed as dpm/ml inside the cells divided by dpm/ml of the medium (distribution ratio). Cellular volumes were estimated in standard hematocrit tubes.

The formation of polyglutamates from pteroylmonoglutamate in normal liver was determined as follows: The rat was sacrificed 48 h after the intracardiac injection of 10µC of [^{3}H]pteroylmonoglutamate. The liver was blanched by perfusion with ice-cold phosphate-buffered saline through the inferior vena cava. Five grams of liver were cut into small pieces and added to 15 ml of 2% ascorbate in a 95°C H_2O bath to prevent autolysis of the conjugated folate forms. After 10 min the material was cooled in ice, homogenized with a Potter-Elvehjem tissue homogenizer, and centrifuged at 12,000 x g for 10 min. The supernatant was collected and an aliquot was removed for the determination of radioactivity; the remainder was lyophilized.

For the determination of polyglutamate formation in liver culture, 1.0 µC of [3_H]folic acid was added to each of four plates, each containing 10 ml of standard medium and approximately 2×10^6 cells per 100 mm plate. After a specified time (12 or 48 h), the cells were harvested as follows: The medium and two saline rinses were aspirated from the plates and discarded. The contents were removed from the four plates in 1% ascorbate with a rubber policeman, combined, heated to 95°C for 5 min, and centrifuged. An aliquot of the supernatant fluid was removed for the determination of radioactivity, and the remaining solution was lyophilized

Separation of the polyglutamates was performed by the method of Shin et al. (27). A Sephadex G-25 column (0.9 x 200 cm) was equilibrated with phosphate buffer (0.1 M phosphate, pH 7.0, containing 0.2 M 2-mercaptoethanol) and calibrated with the following standards: pteroylhepta-, pteroylpenta-, and pteroyltriglutamic acid, $5CH_3$-H_4-pteroylmonoglutamate, and folic acid. The lyophilized material from intact liver was dissolved in 5 ml of the equilibration buffer and that from the cultured cells was dissolved in about 1 ml of buffer. One ml of each solution was chromatographed in separate experiments. The flow rate rate was 9.5 ml/h and 1.75-ml fractions were collected. Radioactivity was determined in each fraction. The peaks eluted from the column were assayed for the presence of folate or folate derivatives by the standard microbiological assay with Lactobacillus casei (28).

Results

A sine qua non for the successful cultivation of liver parenchymal cells is that they be isolated viable, homogeneous and in high yields. Much progress has been made

Steps	This Report	Seglen (3)	Howard et al. (2)	Ingebretsen and Wagle (5)	Quistorff et al. (8)
Initial Perfusion Medium	Ca^{++} free Hanks' (37°C) with BSA (0.5%) insulin (10 mU/ml) and ($O_2 + CO_2$)	Buffer (8.3 g NaCl, 0.5 g KCl, 2.4 g Hepes, 5.5 ml NaOH in 1 liter H$_2$O)	Ca^{++} free Hanks' (cold)	Ca^{++} free Hanks' (37°C) with BSA (1.5%)	Locke's minus glucose
Flow Rate	10 ml/min in situ	50 ml/min in perfusion Apparatus	Liver in perfusion apparatus	100 ml/15 min in recirculating perfusion apparatus	5 ml/min in situ
Digesting Medium	Sigma I collagenase 0.05% (37°C)	Sigma I collagenase, 0.05% with CaCl$_2$ (37°C)	Worthington collagenase 0.05% and Sigma hyaluronidase, 0.1% (cold)	Sigma I collagenase 0.05%	1. Collagenase 0.08% and hyaluronidase, 0.05% in Hanks'. 2. Locke's with 2 mM EDTA

TABLE 1: Comparison of Liver Parenchymal Cell Isolation Procedures

TABLE 1: (continued)

Mechanical Treatment	Digested liver is dispersed with wide-bore pipette	1. Digested liver is dispersed with stainless steel comb and filtered 2. Cell suspension is incubated at 37°C for 30 min with shaking	Liver slices are incubated at 37°C for 55 min with shaking (1.26 mM $CaCl_2$ added for final 20 min)	Digested liver minced with scissors	1. Digested liver is homogenized in Potter-Elvehjem homogenizer 2. Cell suspension incubated at 37°C with shaking
Filtration	1. 253-µm nylon mesh 2. 61-µm nylon mesh	1. 250-µm nylon mesh 2. 100-µm nylon mesh	1. Stocking nylon 2. 61-µm nylon mesh	None	100-µm nylon mesh
Centrifugation	1. Twice at 50 x g for 2 min 2. Once at 100 x g for 5 min at room temperature	Four times at 200 rpm in a bench centrifuge	1. Once at 50 x g for 1 min 2. Twice at 20 x g for 1 min	600-800 rpm for 15-30 sec	1. Once at 80 x g for 1 min 2. Once at 50 x g for 1 min 3. Once at 80 x g for 1 min
Yield (Viability)	400-600 x 10^6 nuclei per 300 g rat (90-95%)	36.2% of liver wt recovered as cells (93.2%)	5.4 x 10^6 cells /g wet weight (85-98%)	230-300x10^6 cells per 100 g body weight (90-95%)	1.5-2.5 ml packed cells per 250 g rat

toward this goal since the initial studies of Howard et al. (29) and Berry and Friend (1) where enzymes rather than simple mechanical disruption were used to obtain liver cell suspensions. Many investigators have suggested variations in the original techniques (2-12). Some of the current procedures are listed in Table 1, which is not intended to be complete but which does illustrate the similarities and differences among a representative sample of published procedures.

The livers may be perfused in situ or removed from the carcass and perfused in a recirculating apparatus with various initial perfusion media (Table 1). However, in most of the studies, crude collagenase is used in the digesting medium, and several workers have claimed that the addition of hyalurondidase does not increase the yield of cells (3,5,12). The effect of the amount of mechanical treatment on the yields and condition of the isolated cells is difficult to assess since the data are reported in different units. However, procedures utilizing the least mechanical treatment appear to give the highest yields of viable cells. In four of the procedures listed in Table 1, the cells are filtered through nylon mesh. The exception is Ingebretsen and Wagle's procedure (5), and since their yield is among the highest reported, the necessity for this step may be questionable. One of the most important steps in isolating homogeneous populations of liver cells appears to be centrifugation. Most workers centrifuge the cell suspensions at 20-80 x g for short periods, allowing the contaminating nonparenchymal and broken parenchymal cells to remain in the supernatant. The final preparations of parenchymal cells by all methods listed are from 85-98% viable and are suitable for biochemical analysis.

Cell Attachment

The attachment of the hepatocytes to Falcon plastic dishes has been reported to be dependent on the presence of serum in the medium (16) unless the plates are first coated with collagen (17). The effect of adult rat serum or fetal calf serum can be seen in Fig. 1. From 6 to 14% of the cells were attached to the dishes after 6 h of incubation in the absence of serum. The addition of only 5% adult rat serum or fetal calf serum permitted the attachment of more than 50% of the cells. The maximum attachment occurred at concentrations of 10% for rat serum and 15% for fetal calf serum. After 20 h of incubation there appears to be a slight increase in the absence of serum, but there is a significant decrease in the presence of serum. Cells maintained their

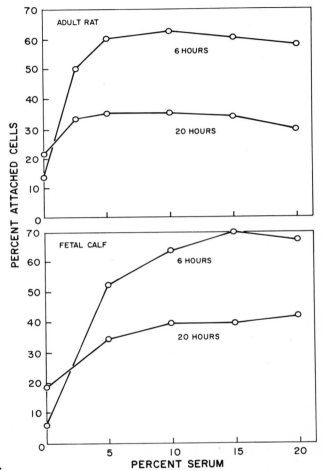

FIG. 1: Effect of serum on attachment of cells to Falcon dishes. Isolated parenchymal cells (2.5×10^6) were suspended in 4 ml of Ham's F12 medium with 10 mU/ml of insulin and varying amounts of adult rat serum or fetal calf serum. The suspensions were plated on 70-mm plates and harvested at 6 and 20 h. Each point is an average of duplicate plates.

characteristic liver activities as long in medium containing 15% fetal calf serum as in that containing 15% adult rat serum.

Tyrosine Aminotransferase Induction

It has been previously demonstrated that parenchymal cells in culture are capable of inducing TAT in response to the addition of dexamethasone to the culture (16, 20). This increase was similar during the second, third, and fourth day in culture but the response decreased thereafter (Fig. 2 and references 16 and 20). The possible explanations for this decreased cellular response include a decrease in the synthesis of protein or of RNA by the cells after a given time in culture.

The rate of incorporation of radioactive leucine into protein was used to measure protein synthesis. The amount of protein, which is proportional to the number of cells, decreased almost 35% between the first and second day, and more gradually between the second and seventh day (Fig. 3). Incorporation of labeled leucine into protein increased during the first 3 days and remained relatively constant for the next 4 days (Fig. 3)

The amount of RNA per plate and the rate of RNA synthesis are illustrated in Fig. 4. As expected, the amount of RNA per plate decreased in a manner similar to the amount of protein (Fig. 3). However, the rate of incorporation of [^{14}C]orotic acid reached a maximum during the second day, remained constant over the next day, and decreased slowly during the next 4 days (Fig. 4). This decrease in RNA synthesis may explain the decreased induction of TAT by dexamethasone after 4 days of culture, as synthesis of messenger RNA is reportedly required for cortisol induction of TAT (30,31).

Transport of α-Aminoisobutyric Acid

Uptake of AIB has been used as an index of cell viability for hepatoma cells (31,32) and hepatocytes maintained in filter-well culture (32). Uptake of this nonmetabolizable amino acid appears to involve energy-dependent active transport (33). We therefore tested the ability of the parenchymal cells in primary culture to concentrate AIB and the effects of various hormones on this process.

The uptake of AIB was found to be linear for at least 80 min (Fig. 5). Plotting the data as dpm of ^{14}C per mg of protein versus time eliminates the small difference in the number of cells from plate to plate. The distribution

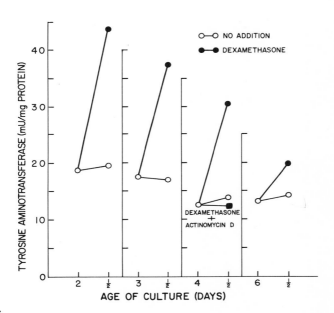

FIG. 2:

Induction of tyrosine aminotransferase as a function of culture age. Liver parenchymal cells were cultured in the standard medium. On the day indicated, the enzyme activity was determined at zero time and after 12 h in the absence or presence of dexamethasone (5×10^{-6} M). On day 4, dexamethasone and actinomycin D (6 µg/ml) were added to an additional set of duplicate plates, and the enzyme was assayed after 12 h. Each activity reported is an average of duplicate determinations. Some of these data appeared in tabular form in a previous report[20].

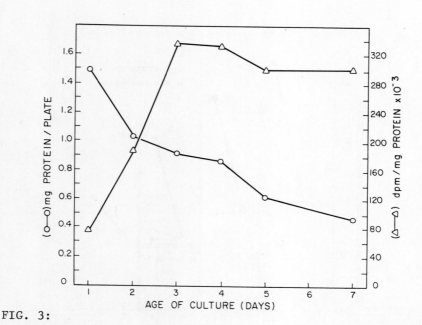

FIG. 3:

<u>Incorporation of [^{3}H]Leucine into protein as a function of culture age</u>. Parenchymal cells (6 x 10^6) were plated in 10 ml of standard medium on 100-mm-diameter dishes. Approximately 10 µC of [4-5^{3}H] L-leucine were added to each plate at zero time and after daily medium changes. The cells were harvested after 6 h, and the amount of protein and radioactivity was determined as described in Materials and Methods.

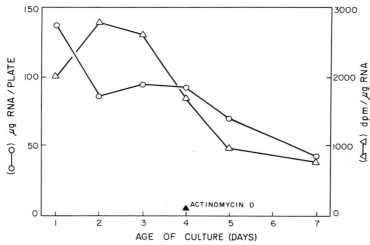

FIG. 4:

Incorporation of [^{14}C]Orotic acid into RNA as a function of culture age. Parenchymal cells (6 x 10^6) were plated in 10 ml of standard medium. Orotic acid, 1 x 10^{-6} M containing 5 μC of [6-^{14}C] orotic acid, was added to each plate at zero time and after daily medium changes. The cells were harvested after 6 h, and the amount of RNA and radioactivity was determined as described in Materials and Methods.

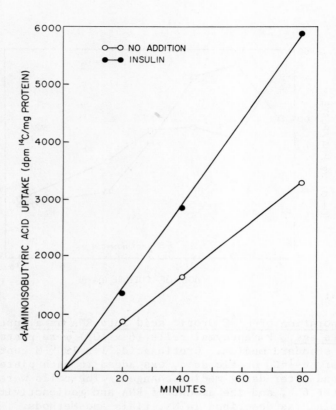

FIG. 5:

Uptake of α-aminoisobutyric acid by parenchymal cells in culture at various times after addition. Parenchymal cells were maintained for 48 h in 4 ml of standard medium with two medium changes. The medium was then changed to Ham's F12 without insulin or fetal calf serum. Half of the plates were then supplemented with 40 mU/ml of insulin and incubated for 18 h, after which 1×10^{-5} M [^{14}C]AIB containing 0.15 µC/ml, was added to the plates. The cells were harvested at the indicated times, and the radioactivity was determined. Control sets of plates were used to determine the number of cells per plate and the packed cell volume under both conditions.

ratio (see Methods) at 30 min was 11.7. When the cells were preincubated for 18 h with insulin (40 mU/ml) but in the absence of fetal calf serum, the uptake of AIB over the 80-min period was greater than that of the nontreated cells (Fig. 5). When this experiment was repeated in the continuous presence of fetal calf serum, there was no difference in the uptake by the cells in the presence or absence of insulin.

Addition of increasing amounts of unlabeled AIB caused a decrease in the distribution ratio of the radioactive AIB (Table 2). This indicates that the transport system is saturable and that the uptake of AIB is not due to simple diffusion. Preincubation of the cells for 1 h in the presence of 5×10^{-4}M NaCN inhibited the initial uptake of AIB by approximately 27%. In addition, the initial uptake by the cells at 4°C was decreased by 79%. This transport process thus appears to be energy-dependent.

Several hormones have been shown to influence the transport of AIB in liver (34) and in hepatoma cells in culture (33). Transport of AIB by liver cells in primary culture was also found to be modulated by various hormones (Table 3). Glucagon at 10^{-6}M was more effective than insulin in increasing the initial uptake of AIB, but the effect of both of these hormones was inhibited by cycloheximide at concentrations which inhibit the incorporation of [^{3}-H] leucine into protein by 95%. On the other hand, dexamethasone decreased the initial uptake, as Risser and Gelehrter reported for HTC cells (33).

Metabolism of Pteroylmonoglutamate

Polyglutamate forms of folic acid have been shown to be the major storage forms of the natural folates in several animal tissues. In liver 85-90% of the folates have been identified as reduced pteroylpentaglutamates (35). We, therefore, compared the capacity of the liver cells in primary culture and those in intact liver to convert folic acid to the various reduced monoglutamate and polyglutamate forms.

Sephadex G-25 chromatography of the folate derivatives found in the liver extract 48 h after the injection of [^{3}H]folic acid is shown in Fig. 6. There were seven discernible peaks although some may be composed of more than one compound. Peaks I and II comprised 80-85% of the total radioactivity and were found to support the growth of L. casei. The material from these peaks is thus composed of folate or folate derivatives, perhaps reduced forms of pteroylpentaglutamate similar to those found by Shin et al. (35). Peak VII corresponds to folic acid and probably

Additions	Distribution Ratio
None	12.5
1×10^{-6} M AIB	12.8
1×10^{-5} M AIB	12.3
1×10^{-4} M AIB	10.9
1×10^{-3} M AIB	7.5
5×10^{-3} M AIB	6.4
5×10^{-4} M NaCN	9.1

TABLE 2. Effect of α-Aminoisobutyric Acid Concentration and of NaCN on the Initial Uptake of ^{14}C AIB*

*^{14}C AIB (0.6 C) was added to each plate and the cells were harvested after 30 min. The distribution ratio was determined as described in Methods.

Hormone	Cycloheximide (3 µg/ml)	Distribution Ratio
None	−	13.1
	+	10.2
Insulin, 40 mU/ml	−	17.4
	+	9.9
Glucagon, 10^{-6} M	−	27.6
	+	9.4
Dexamethasone, 10^{-6} M	−	9.5
	+	4.7
Dexamethasone plus insulin	−	13.0

TABLE 3: Effects of Cycloheximide on Hormone-Modulated AIB Transport*

*
The hormones and cycloheximide were added to the culture medium and the cells were incubated for 4 h. At this time 0.6 µC of ^{14}C AIB was added to each plate. The cells were harvested after 30 min.

represents the unchanged injected precursor.

The elution profile of the folate derivatives in extracts from cultured parenchymal cells 12 and 48 h after the addition of [^{3}H]folic acid to the medium is shown in Fig. 7. Nearly half of the precursor appears to be unreacted after 48 h. However, six out of the seven peaks from the 48-h extract were found at the same elution position as those from intact liver. There also seemed to be a progressive synthesis of the higher polyglutamate forms with time of incubation in the presence of the labeled precursor.

Discussion

A system for maintaining liver cells in culture for several days with functions characteristic of intact liver is an important tool in liver biochemistry. Results from different laboratories can be compared directly since the cultured cells are from an identical origin, are of a single cell type, and have the same karyotype. However, potential differences due to the different media used must be kept in mind.

For any new finding on the regulation and control of liver metabolism to be accepted on the basis of studies performed with cultured cells, it is necessary to show that they are similar to intact liver cells in as many ways as possible. This report adds to the list of liver functions that can be studied by the in vitro system. For example, although many studies have dealt with the regulation of TAT, those studies have been carried out in the intact animal (36), in hepatoma cells (3)), or in permanent liver cell lines (37, 38). Since the primary liver cell system is componsed of a homogeneous population of normal parenchymal cells and is capable of inducing TAT, among many other liver functions, in a rather simplified medium, it offers a means of verifying and perhaps extending experiments performed with intact liver.

These studies have demonstrated that parenchymal cells in primary culture are capable of actively transporting amino acids. The transport process appears to require energy and is influenced by the addition of various hormones to the medium. This membrane-related activity could be used as a marker for normal membrane function following various manipulations of the cells.

Finally, the cells have been found to convert the vitamin folic acid to the folate derivatives considered to be its storage forms in the liver of intact animals.

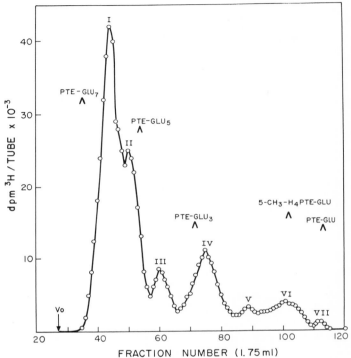

FIG. 6:

Folic acid derivatives in an extract of intact liver. The details of the Sephadex G-25 column chromatography are given in Materials and Methods. The positions of elution of the various standards are indicated. PTE-GLU, pteroylmonoglutamic acid; PTE-GLU$_3$, pteroyltriglutamic acid; PTE-GLU$_5$, pteroylpentaglutamic acid; PTE-GLU$_7$, pteroylheptaglutamic acid; 5-CH$_3$-H$_4$PTE-GLU, 5-methyltetrahydro-pteroylmonoglutamic acid.

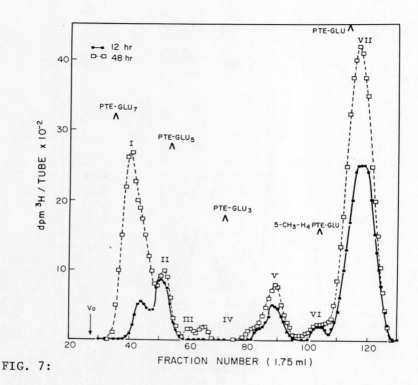

FIG. 7:

<u>Folic acid derivatives in cultured parenchymal cell extracts.</u>
The details of the Sephadex G-25 column chromatography are given in Materials and Methods. For abbreviations, see legend to Fig. 6.

This system could be used to extend the studies using whole animals and other culture systems, where folate pools have been shown to influence such folate-related enzymes as thymidylate synthetase (21) and serine transhydroxymethylase and dihydrofolate reductase (3)`. Of particular interest are the effects of folate derivatives on thymidylate synthetase, where a marked stabilization of this enzyme in cultured parenchymal cells has been obtained (22).

Summary

Adult liver parenchymal cells have been isolated by a collagenase-perfusion technique which involves a minimum of mechanical treatment and which provides viable cells in high yield (400-600 x 10^6/300 g rat). Approximately 60% of the cells are firmly attached to Falcon plastic dishes after 6 h in Ham's medium supplemented with serum from fetal calves or adult rats. A previous experiment showed that these cells, although viable for up to one week, do not divide in culture. Additional findings now indicate that the cultured liver cells behave in several ways similar to intact liver: (1) Tyrosine aminotransferase activity increased nearly 250% in 12 h when the cells were cultured in the presence of 5 x 10^{-6}M dexamethasone. This response to the hormone analog was similar during the second, third, and fourth days in culture but decreased during the sixth day. The decrease in induction of the aminotransferase in response to dexamethasone did not appear to be a defect in general protein synthesis but possibly reflects a decrease in RNA synthesis which began during the fourth day in culture. (2) α-Aminoisobutyric acid is actively transported by the parenchymal cells. This transport is increased by insulin and glucagon but decreased by dexamethasone. (3) The parenchymal cells converted labeled pteroylmonoglutamic acid into polyglutamates in a manner similar to that occurring in normal intact rat liver in vivo.

The authors wish to thank Dr. Joseph Bertino, Yale University, New Haven, Connecticut, for his generous gift of pteroylheptaglutamate, pteroylpentaglutamate, and pteroyltriglutamate.

References

1. Berry, M.N. and Friend, D.S. J. Cell Biol. 43 (1969) 506.
2. Howard, R.B., Lee, J.C. and Pesch, L.A. J. Cell. Biol. 57 (1973) 642.
3. Seglen, P.O. Exptl. Cell. Res. 82 (1973) 391.
4. Seglen, P.O. Exptl. Cell. Res. 76 (1973) 25.
5. Ingebretsen, W.R. and Wagle, S.R. Biochem. Biophys. Res. Commun. 47 (1972) 403.
6. Müller, M., Schreiber, M., Kartenbeck, J. and Schreiber, G. Cancer Res. 32 (1972) 2568.
7. Crisp, D.M. and Pogson, C.I. Biochem. J. 126 (1972) 1990.
8. Quistorff, B., Bondesen, S. and Grunnet, N. Biochim. Biophys. Acta 320 (1973) 503.
9. LaBrecque, D.R., Bachur, N.R., Peterson, J.A. and Howard, R.B. J. Cell. Physiol. 82 (1973) 397.
10. Hommes, F.A., Draisma, M.I. and Molenaar, I. Biochim. Biophys. Acta 222 (1970) 361.
11. Pretlow, T.G., II and Williams, E.E. Anal. Biochem. 55 (1973) 114.
12. Bonney, R.J., Walker, P.R. and Potter, V.R. Biochem. J. 136 (1973) 947.
13. Iype, P.T. J. Cell. Physiol. 78 (1971) 281.
14. Alwen, J. and Gallai-Hatchard, J.J. J. Cell Sci. 11 (1972) 249.
15. Bissell, D.M., Hammaker, L.E. and Meyer, V.A. J. Cell. Biol. 59 (1973) 722.
16. Bonney, R.J., Becker, J.E., Walker, P.R. and Potter, V.R. In Vitro in press 1974.
17. Pariza, M.W., Becker, J.E., Yager, J.D., Jr., Bonney, R.J. and Potter, V.R., Gann in press 1974.
18. Jungalwala, F.B. and Dawson, R.M.C. Biochem. J. 117 (1970) 481.
19. Solyom, A., Lauter, C.J. and Trams, E.G. Biochim. Biophys. Acta 274 (1972) 631.
20. Bonney, R.J. In Vitro in press 1974.
21. Labow, R., Maley, G.F. and Maley, F. Cancer Res. 29 (1969) 366.
22. Bonney, R.J. and Maley, F. Fed. Proc. in press 1974
23. Scrimgeour, K.G. and Vitrols, K.S. Biochemistry 5 (1966) 1438.
24. Higgins, G.M. and Anderson, R.M. Arch. Pathol. 12 (1931) 186.
25. Lowry, O.H., Rosebrough, N.J., Farr, A.L. and Randall, R.J. J. Biol. Chem. 193 (1951) 265.
26. Munro, H.N. and Fleck, A. Methods of Biochemical Analysis (ed. Glick, D.) 14 (1966) 113.

27. Shin, Y.S., Buehring, K.U. and Stokstad, E.L.R. J. Biol. Chem. 247 (1972) 7266.
28. Grossowicz, N., Mandelbaum-Shavit, F., Davidsoff, R. and Aronovitch, J. Blood 20 (1962) 609.
29. Howard, R.B., Christensen, A.K., Gibbs, F.A. and Pesch, L.A. J. Cell. Biol. 35 (1967) 675.
30. Tomkins, G.M., Gelehrter, T.D., Granner, D., Martin, D., Jr., Samuels, H.H. and Thompson, E.B. Science 166 (1969) 1474.
31. Bushnell, D.E., Becker, J.E. and Potter, V.R. Biochem. Biophys. Res. Commun. 56 (1974) 815.
32. Dickson, J.A. Exptl. Cell. Res. 61 (1970) 235.
33. Risser, W.L. and Gelehrter, T.D. J. Biol. Chem. 248 (1973) 1248.
34. Scott, D.F., Reynolds, R.D., Pitot, H.C. and Potter, V.R. Life Science 9 (1970) 1133.
35. Shin, Y.S., Williams, M.A. and Stokstad, E.L.R. Biochem. Biophys. Res. Commun. 47 (1972) 35.
36. Rosen, F.R., Harding, H.R., Milholland, R.J. and Nichol, C.A. J. Biol. Chem. 238 (1963) 3725.
37. Sellers, L.S. and Granner, D. J. Cell. Biol. 60 (1974) 337.
38. Gerschenson, L.E., Davidson, N.B. and Andersson, M. Eur. J. Biochem. 41 (1974) 139.
39. Chello, P.L., McQueen, C.A. and Bertino, J.R. Fed. Proc. 32 (1973) 928.

Note: We would like to thank the publishers of In Vitro for the permission to reproduce Table I.

PROTEIN SYNTHESIS AND EXCRETION IN SINGLE CELL SUSPENSIONS FROM LIVER AND MORRIS HEPATOMA 5123 TC

G. Schreiber and M. Schreiber

Introduction

Lysis of the intercellular matrix of liver or Hepatoma 5123 tc with 0.05% collagenase and 0.1% hyaluronidase is an appropriate method for the preparation of single cell suspensions. Average diameter and dry weight of obtained hepatoma and liver cells are 14μ and 28μ, and 17.2% and 10.5% of wet weight, respectively. The rough endoplasmic reticulum of the suspended cells is intact, even after incubation for one hour at 37°C. Hepatoma cells produce tumors upon reimplantation into rats of the same strain as that in which the hepatoma used for cell preparation has been grown.

Incorporation of L-leucine into protein of hepatoma cells is independent of the concentration of K^+, Mg^{2+}, and Ca^{2+} in the medium, whereas in liver cells distinct maxima of the incorporation rate are observed at 65 mM K^+, 3.5 mM Mg^{2+}, and 2.5 mM Ca^{2+}. The influence of changes in hydrogen ion concentration of the medium on protein synthesis in the cells is less pronounced in hepatoma cells than in liver cells.

The preparation procedure leads to the loss of the capacity for concentrative uptake of amino acids: both tumor and liver cell membranes become freely permeable to amino acids. Thus, after incubation with [^{14}C]-amino acids, net rates of protein synthesis can be calculated from the specific radioactivities of precursors and products. Hepatoma 5123 tc cells synthesize 1.4 μg protein/hr/mg cellular protein compared with 0.54 μg for liver cells. Protein synthesis in both types of cells is inhibited by very low doses of cycloheximide or puromycin. Mitochondrial protein synthesis contributes only a minor portion to total protein synthesis. Intact serum albumin synthesis can be demonstrated in suspended liver cells.

Radioactive protein is transferred into the medium of hepatoma or liver cell suspensions incubated with L-leucine-1-^{14}C. The kinetics of the transfer are similar to those observed for the secretion of serum protein by the liver into the blood stream in intact animals. This is in contrast to the lack of secretion of serum proteins found for hepatoma in vivo. The transfer of protein into the medium is not inhibited when protein synthesis is interrupted by addition of puromycin or cycloheximide or incubation under nitrogen.

Uridine, uracil and orotic acid are efficiently incorporated into RNA of hepatoma cells, whereas in liver cells only uridine is incorporated effectively.

Results and Discussion

1. *Methods for the Preparation of Single Cell Suspensions from Liver and Hepatoma.*

The methods for preparing single cells from liver have been critically discussed in a recent review (1). Continuous recycling perfusion with 0.1% hyaluronidase plus 0.05% collagenase (2) seems to be superior to other methods, in particular to those where the minced tissue is dissociated by shaking in solutions of potassium or calcium-binding compounds such as tetraphenylborate or ethylene diamine tetraacetate or by shaking in various enzyme solutions. Hyaluronidase does not seem to be absolutely necessary for the dissociation of the liver tissue (4,5). The action of collagenase may be reinforced by addition of 5 mM Ca^{2+} to the perfusion medium after previous non-recirculating perfusion with buffer free of Ca^{2+} (6).

In contrast to the results with liver cell suspensions, continuous recirculating perfusion is not mandatory to obtain suspended cells in good yield from hepatoma. It is sufficient to remove intravascular blood from tumor tissue by perfusing 60 to 80 ml of 0.09% NaCl of 25° through V. cava inferior. Then, tumors are excised, minced with scissors and rinsed with 0.9% NaCl at (2°C) in a plastic tea strainer. The minced tissue is shaken in a mixture of 0.05% collagenase and 0.1% hyaluronidase as described by Muller et al. (3). The results are comparable to those obtained for liver with the continuous recycling perfusion technique.

2. *Quality Control*

Exclusion of vital dyes such as trypan blue may be used as a simple and fast viability criterion. However, it is essential that conditions of staining are standardised. Results vary, e.g., with incubation time. Trypan blue solutions should be filtered before use and should be calibrated to avoid the influence of impurities which vary in concentration from one batch of dye to another (7). The calibration may be done with cells prepared by a procedure known to yield colored cells only, e.g. that of Jacob and Bhargava (8).

3. Morphology

The average diameter of suspended hepatoma cells is 14 µ, that of suspended liver cells 28 µ. The dry weight of hepatoma cells is 17.2%, that of liver cells 10.5% of the wet weight (9). In general, hepatoma and liver cells prepared by shaking in, or perfusing with collagenase and hyaluronidase, respectively, show a well preserved ultrastructure. Sometimes the rough endoplasmic reticulum appears to be slightly swollen, but all other cell organelles such as the mitochondria, microbodies, dictyosomes, myeloid bodies and cell membranes are intact (2,3,9,10). In suspended hepatoma cells even metaphase-to-anaphase stages of nuclear division can be observed. Immediately after preparation, mitochondria in hepatoma cells have a condensed conformation which changes to the orthodox appearance (for nomenclature see Hackenbrock (11)) upon incubation at 37° (9). In liver cells both types of mitochondrial conformation can be observed in the same cell in immediate vicinity to each other (3).

4. Reimplantation into Animals

Samples from cell suspensions from Morris hepatoma 5123 tc incubated for one hour or less produce tumors upon reimplantation into Buffalo rats in 100% of the re-implantations. The efficiency of tumor production by re-implantation decreases to 80% after incubation for 2 hours at 37° (9).

No tumors are obtained if 5123 tc cells are injected into strains of rats in which Hepatoma 5123 tc does not grow.

5. Reproducibility of Amino Acid Incorporation; Isolation of Radioactive Protein

One of the advantages of cell suspensions, if compared with slices or perfused organs, is the possibility of dividing the incubation mixture into various portions from which many homogeneous samples may be withdrawn. Proper shaking of incubation vessels is important to obtain good reproducibility since isolated liver cells have a tendency to re-aggregate. A fast and convenient procedure is necessary for the isolation of radioactive protein from the large number of samples. The method of Mans and Novelli (12) can be successfully applied to liver and hepatoma cell suspensions without homogenization of the samples prior to the application onto the filter paper discs. The standard deviation for samples withdrawn from different portions of cell suspensions is in the range of the pipetting error (1,13). A standard curve to obtain absolute counting efficiency of [^{14}C] -protein samples on filter paper discs can be easily prepared with [^{14}C] -labelled albumin (14).

6. Conditions of Incubation for Protein Synthesis

a. Buffers

The effect of varying the type and strength of buffers on the incorporation of L- [^{14}C] -leucine into protein has been studied in detail in liver cell suspensions (1). The most suitable buffer system seems to be a 3 mM to 12 mM Na-H_2PO_4/Na_2HPO_4 buffer of pH 7.17. Since this buffer is relatively weak, the pH of the incubation mixture has to be readjusted immediately before starting the reaction. Thereafter, the pH does not change more than 0.2 pH units during an incubation period of 120 min. Lower incorporation rates are obtained with bicarbonate-phosphate buffer, tris-HCl, tris-maleate and tris-phosphate buffer in the range of 10 mM to 20 mM. Incorporation is inhibited at buffer concentrations above 20 mM.

b. pH

A pronounced dependence of the incorporation of L- [^{14}C]-leucine into protein on the concentration of H^+ was found in liver cell suspensions, with a sharp maximum of the incorporation rate at pH 7.17. The width of the peak in the curve describing incorporation rate as a function of pH in the medium is 0.92 pH units (1,13). A broader pH optimum is obtained in hepatoma cell suspensions where the width of the curve describing rate of incorporation as function of pH extend from pH 6.40 to pH 8.35, i.e. 1.95 pH units (9).

c. Atmosphere During Incubation

Data are available for liver cells only. The same rate of incorporation of L-[^{14}C] -leucine into protein is obtained if the cells are incubated under air or under 95% O_2 + 5% CO_2. Withdrawal of samples from cell suspensions is, of course, technically much easier if the cell suspensions are incubated in air. No incorporation is observed if the cells are incubated under N_2 (1,13).

d. Addition of Energy-Providing Substrates

No effect on the rate of incorporation of L-[^{14}C]-leucine into protein of liver cell suspensions prepared from either fed or starved rats was detected following addition of 5 or 10 mM glucose.

e. Temperature

Data are available for liver cell suspensions only (1). The rate of incorporation of L-[$1-^{14}$C]-leucine and of L-[$1-^{14}$C]-methionine into protein increased threefold for a temperature increase of $10°C$ in the range $20°C$ to $37°C$.

f. Ion Concentration

The rate of incorporation of L-[$1-^{14}$C]-leucine into single cell suspensions from liver depends on the concentration of metal ions in the incubation medium. Maximal rates of incorporation are observed at a K^+ concentration of 65mM, a Mg^{2+} concentration of 3.5 mM (Fig. 1) and a Ca^{2+} concentration of 2.5 mM (1,13). Changing the concentration of K^+, or of Mg^{2+}, or of Ca^{2+} in the medium of hepatoma cell suspensions does not influence the incorporation of L-$1-^{14}$C - leucine into protein over a range of 0 to 116 mM for K^+ (Fig. 2), 0 to 20 mM for Mg^{2+} and 0 to 10 mM for Ca^{2+} (9). Ionic strength of the medium was kept constant by omission of an equivalent amount of NaCl.

g. Cell Density and Protein Synthesis

Increasing cell density in liver cell suspensions in the range from 0 to 4×10^6 cells per ml does not lead to a decrease in the rate of incorporation of L-[$1-^{14}$C]-leucine per mg cellular protein.

h. Amino Acid Concentration

Typical saturation curves are observed for the rate of incorporation of L-amino acids into the protein of liver or hepatoma cells in suspension (Fig. 3) (1,9,13). Straight lines are obtained if the reaction velocity is plotted against reaction velocity over substrate concentration (Fig. 4). This presentation, suggested by Hofstee (15) is

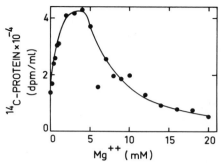

FIG. 1: Influence of the concentration of Mg^{2+} ions on the incorporation of L-[1-^{14}C]-leucine into protein of liver cells in suspension. Incubation conditions as described by Schreiber and Schreiber (1).

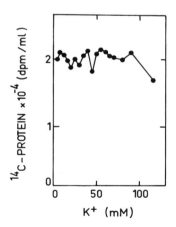

FIG. 2: Incorporation of L-[1-^{14}C]-leucine into hepatoma cell protein as function of K^+ concentration in the incubation medium. Incubation conditions as dexcribed by Schreiber et al. (9).

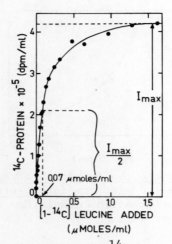

FIG. 3: Incorporation of L-[1-^{14}C]-leucine into protein of liver cells as a function of its concentration in the incubation mixture. After Schreiber and Schreiber (13).

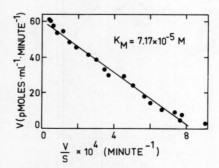

FIG. 4: Hofstee plot (15) for the incorporation of L-[1-^{14}C]-leucine into the protein of suspended liver cells as described in Fig. 3.

the most efficient for discriminating against non-linearity of the transformed hyperbola section. The values found for half-saturation of the amino acid incorporating system in liver cells are summarized in Table 1. Incorporation of L- $\lceil$ 1-^{14}C $\rfloor$ -leucine into the protein of hepatoma cells is half-saturated at a concentration of 107.5 µM (9).

7. Permeability of Cell Membranes to Amino Acids

The observation of saturation kinetics for the incorporation of L-amino acids into the protein of liver or hepatoma cell suspensions may be explained in various ways. Existence of a saturable transport system in the cell membrane, dilution of added radioactively labelled precursor in a non-labelled intracellular pool and saturation of the amino acid activating enzymes within the cells would all lead to the same type of saturation kinetics for incorporation of amino acids into protein. However, no saturation is observed if the uptake of L-amino acids into liver or hepatoma is measured directly (1,9,13). Furthermore, equilibration between extracellular and intracellular pools is reached almost instantaneously and independently of temperature. Both cell types seem to have lost the capacity for concentrative uptake of L-amino acids. It is likely, therefore, that the cell membranes have become freely permeable to amino acids and that the observed saturation kinetics reflect saturation of the amino acid activating enzyme system.

8. Rates of Net Protein Synthesis

Extracellular and intracellular 1-, $\lceil$ 1-^{14}C $\rfloor$ -leucine will have the same specific radioactivity when the cell membranes of the suspended cells are freely permeable to L-leucine. Net protein synthesis rates can then be calculated from the measured specific radioactivities of precursor and product. For such calculation, the average leucine content in hepatoma and liver protein may be assumed to be about 10% (14). The following values are then obtained for net rates of protein synthesis at half-saturation with amino acids: Suspended liver cells synthesize 0.54 µg protein per mg of cellular protein per hour compared with 1.4 µg in hepatoma 5123 tc cells. Thus, protein synthesis in hepatoma cells is 2.6 times faster than in liver cell suspensions.

Amino Acid	half-saturation at
	µM
L-alanine	900
L-arginine	517
L-asparagine	443
L-aspartate	670
L-cysteine	87
L-glutamate	450
L-glutamine	600
Glycine	610
L-histidine	449
L-isoleucine	28.1
L-leucine	71.7
L-lysine	150
L-methionine	51.4
L-phenylalanine	316
L-proline	287
L-serine	746
L-threonine	270
L-tryptophan	36.5
L-tyrosine	500
L-valine	190

TABLE 1. Half-saturation concentrations for the incorporation of ^{14}C-labelled L-amino acids into protein in single cell suspensions from rat liver (cf. (1)).

9. *Reutilization of Labelled Amino Acids Originating From Protein Breakdown*

Incorporation of L- $[1-^{14}C]$ -leucine into protein of liver or hepatoma cells is immediately interrupted by addition of an excess of non-labelled L-leucine. No release into the medium of radioactivity already incorporated in protein is observed when incubation is continued further (1,9). The result is independent of the addition to the medium of the other 19 L-amino acids which occur in protein.

10. *Inhibitors*

The incorporation of L- $[1-^{14}C]$ -leucine into protein of liver or hepatoma cells is inhibited by very low doses of antibiotics such as puromycin (Fig. 5). The concentration of various inhibitors at which protein synthesis is reduced by 25%, 50%, and 7% are summarized in Table 2.
Mitochondrial protein synthesis probably contributed only negligibly to total cellular protein synthesis since chloramphenicol inhibits amino acid incorporation only in very high concentrations.
Special caution is advisable in the interpretation of experiments on protein synthesis if no inhibition by low doses of puromycin or cycloheximide is observed (17,18).

11. *Effect of Alcohol on Cytoplasmic Protein Synthesis in Liver Cells*

Ethanol has been reported to influence protein synthesis in liver mitochondria (19,20). The data described in Table 3 demonstrate that addition of various alcohols also leads to a decrease of protein synthesis measured in liver cell suspension. From the high sensitivity towards cycloheximide and the relative resistance towards chloramphenicol one may assume that protein synthesis measured in suspended liver cells predominantly reflects cytoplasmic protein synthesis, and, hence, alcohols also influence cytoplasmic liver protein synthesis.

12. *Synthesis of Specific Proteins*

Demonstration of the synthesis of specific protein is desirable to show that liver cells in suspension retain the specific properties of the original organ. Synthesis of serum albumin (10) and fatty acid synthetase (21) by isolated liver cells has been reported.

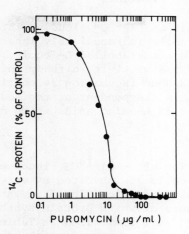

FIG. 5: <u>Effect of puromycin on the incorporation of L-$[1-^{14}C]$-leucine into protein in cell suspensions from hepatoma 5123 tc</u>. Incubation conditions as described by Schreiber et al. (9).

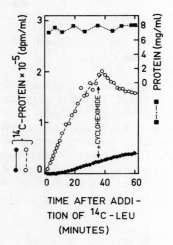

FIG. 6: <u>Synthesis and excretion of protein by rat liver cells in suspension</u>. After Schreiber and Schreiber (1).

		Dose (μM) producing an inhibition of		
Inhibitor		25%	50%	75%
Cycloheximide	Liver	--	0.21	0.68
	Hepatoma	--	0.32	1.2
Puromycin	Liver	0.83	2.6	8.3
	Hepatoma	4.6	13	19
Chloramphenicol	Liver	310	1300	--
	Hepatoma	340	680	960
Streptomycin	Liver	290	2600	--
	Hepatoma	--	--	--
Aurintri-carboxylate	Liver	75	170	650
	Hepatoma	1000	1800	3400
NaF	Liver	750	1400	2300
	Hepatoma	1500	2900	5800
Actinomycin D	Liver	25	110	--
	Hepatoma	--	--	--

TABLE 2. Inhibition of protein synthesis in single cell suspensions from liver and from Morris hepatoma 5123 tc)cf/ (9)).

Alcohol	Dose (vol %) producing an inhibition of		
	10%	50%	90%
Methanol	1.3	3.3	6.3
Ethanol	0.70	1.6	4.0
n-Propanol	0.19	0.33	0.78
n-Butanol	0.033	0.18	0.50
Ethylene Glycol	0.20	2.3	7.0
Glycerol		< 0.02	6.8

TABLE 3. Effects of alcohols on the incorporation of L-[1-^{14}C]-leucine into protein in single cell suspensions from rat liver (cf. (1)).

13. Albumin Content

The intracellular concentration of albumin in vivo is higher in hepatoma than in liver (22). Also, suspended hepatoma cells contain slightly more albumin than liver cells in suspension, the albumin content of hepatoma cell suspensions being 1.9 g of albumin per gram hepatoma cell protein and that of liver cell suspensions 1.1 g of albumin per gram of liver cell protein. This observation might be related to the previous finding of a lack of secretion of serum proteins in Morris hepatomas 5123 tc and 9121 (23).

14. Uptake of Macromolecules

The uptake of macromolecules into suspended liver or hepatoma cells may be studied conveniently by use of ^{14}C -albumin and ^{14}C -inulin. 70% of the cell water of suspended liver cells is found to be accessible to albumin after incubation for 15 to 30 minutes at 37°. Similar results are obtained for ^{14}C - inulin (1).

15. Release of Protein into the Medium

In vivo, serum proteins are secreted by the liver into the bloodstream. After intravenous injection of radioactively labelled amino acids, label appears in the serum proteins in the bloodstream after a lag period of 14 to 15 minutes (14, 23-31). This lag period is very constant. It does not change, for example, in nephrotic rats (14) in which the synthesis of serum proteins in the liver is greatly accelerated. It is shortened to about 10 minutes in regenerating liver (14), in which a large rearrangement of the rough endoplasmic reticulum takes place.

A 10 minute lag period can also be observed between addition of L- [1-^{14}C] -leucine to, and the appearance of radioactive protein in, the medium in cell suspensions from liver (Fig. 6) and hepatoma (9). After addition of 0.1 mg cycloheximide or 0.13 mg puromycin per ml of incubation medium protein synthesis is immediately interrupted, whereas the transfer of labelled protein from the cells into the medium continues. Concomitantly, the intracellular [^{14}C] -protein concentration decreases. Thus, the transfer does not depend on active protein synthesis; protein is further excreted even if its synthesis is blocked. The transfer is less effective at 2° than at 37°. It is not inhibited if the cells are incubated under nitrogen.

16. Incorporation of Radioactive Precursors into RNA

$[6-^{14}C]$-Orotate, $[2-^{14}C]$ - uracil and $[2-^{14}C]$ -uridine are efficiently incorporated into RNA of hepatoma cells, whereas in liver cells only $[2-^{14}C]$ - uridine is effectively incorporated. Half-saturation of orotate incorporation into hepatoma cell RNA is obtained at 55 µM. Incorporation of uridine into RNA of liver cells is half-saturated at 94 µM uridine. The incorporation of orotate into RNA in hepatoma cells is inhibited by 25%, 50%, and 75% by 0.052 M, 0.24 µM and 0.83 µM Actinomycin D, respectively.

References

1. Schreiber, G. and Schreiber, M. Sub-Cell Biochem. 2 (1973) 307.
2. Berry, M.N. and Friend, D.S. J. Cell Biol. 43 (1969) 506.
3. Muller, M., Schreiber, M., Kartenbeck, J. and Schreiber, G. Cancer Res. 32 (1972) 2568.
4. Ingebretsen, W.R. Jr. and Wagle, S.R. Biochem. Biophys. Res. Commun. 47 (1972) 403.
5. Ingebretsen, W.R. Jr., Moxley, M.A., Allen, D.D. and Wagle, S.R. Biochem. Biophys. Res. Commun. 49 (1972) 601; Seglen, O. FEBS Lett. 30 (1973) 25.
6. Seglen, P.O. Expt. Cell Res. 74 (1972) 450; 76 (1973) 25.
7. Forabosco, A., Zaffe, O. and Tosato, L. Boll. Soc. It. Biol. Splr. XLVIII (1971) 33.
8. Jacob, S.T. and Bhargava, P.M. Expt. Cell Res. 27 (1962) 453.
9. Schreiber, M. Schreiber, G. and Kartenbeck, J. Cancer Res. 34 in press (1974).
10. Weigand, K., Müller, M., Urban, J. and Schreiber, G. Expt. Cell Res. 67 (1971) 27.
11. Hackenbrock, C.R., Proc. Natl. Acad. Sci. U.S.A. 61 (1968) 598.
12. Mans, R.J. and Novelli, G.D. Biochem. Biophys. Res. Commun. 3 (1960) 540: Arch. Biochem. Biophys. 94 (1961) 48.
13. Schreiber, G. and Schreiber, M. J. Biol. Chem. 247 (1972) 6340.
14. Schreiber, G., Urban, J., Zahringer, J., Reutter, W. and Frosch, U. J. Biol. Chem. 246 (1971) 4531.
15. Hofstee, B.H.J. Nature 184 (1959) 1296.

16. Zydowo, M., Kaletha, K. and Dudek, A. Acta biochim. polonica 18 (1971) 367.
17. Haung, Y.L. and Ebner, K.E. Biochim. Biophys. Acta 191 (1969) 161.
18. Ranhotra, G.S. and Johnson, B.C. Proc. Soc. Exptl. Biol. Med. 132 (1969) 509.
19. Rubin, E., Beattie, D.S., and Lieber, C.S. Laboratory Investigation 23 (1970) 620.
20. Rubin, E., Beattie, D.S., Toth, A. and Lieber, C.S. Fedn. Proc. Fedn. Am. Socs. exp. Biol. 31 (1972) 131.
21. Burton, D.N., Collins, J.M. and Porter, J.W. J. Biol. Chem. 244 (1969) 1076.
22. Urban, J., Kartenbeck, J., Zimber, P., Timko, J., Lesch, R. and Schreiber, G. Cancer Res. 32 (1972) 1971.
23. Schreiber, G., Boutwell, R.K., Potter, V.R. and Morris, H.P. Cancer Res. 26 (1966) 2357.
24. Peters, T. Jr., J. Biol. Chem. 237 (1962) 1186.
25. Jungblut, P.W. Biochem. Z. 337 (1963) 267.
26. Majumdar, C., Tsukada, K. and Lieberman, I. J. Biol. Chem. 242 (1967) 700.
27. Le Bouton, A.V. Biochem. J. 106 (1968) 503.
28. Schreiber, G., Rotermund, H.M., Maeno, H., Weigand, K. and Lesch, R. Eur. J. Biochem. 10 (1969) 355.
29. Rotermund, H.M., Schreiber, G., Maino, H., Weinssen, U. and Weigand, K. Cancer Res. 30 (1970) 2139.
30. Morgan, E.H. and Peters, T. Jr. J. Biol. Chem. 246 (1971) 3500.
31. Peters, T. Jr. and Peters, J.C. J. Biol. Chem. 247 (1972) 3858.

STUDIES ON NORMAL AND NEOPLASTIC LIVER CELLS IN CULTURE: CONTACT BEHAVIOR, CELLULAR COMMUNICATION AND TRANSFORMATION

Carmia Borek

The most obvious phenotypic alteration seen in cultures of cells which have undergone neoplastic transformation is their ability to form multilayers and replicate to high densitites under conditions where their untransformed counterparts remain as monolayers and appear to be contact inhibited, Fig. 1 (1). The change which has occurred following transformation clearly involves a loss of responsiveness to growth control mechanisms mediated by cell to cell contact. The degree of escape from contact inhibition in cultures of transformed and tumor cells varies with the type of cell, the transforming agent and with the type of tumor (2-4).

While one cannot exclude genetic alterations as a primary factor in the loss of growth control there are a variety of structural and functional changes associated with the cell surface which strongly imply its role in the loss of contact inhibition (for review see ref. 5).

A number of mechanisms have been suggested in recent years to account for the restricted growth in cultures of normal cells, both normal primary and long term cell lines (for review see ref. 6).

One of the requirements for such control in culture is a system of intercellular communication between osculating cells (7,8). The movement of small ions or molecules from one cell to another through permeable intercellular junctions can be regarded as one mode of communication, and will be discussed in more detail later on.

Most studies on contact behavior between normal cells as opposed to transformed cells have been done with fibroblast-like cells. The only source of transformed epithelioid cells has been cultures derived from a carcinoma.

In the present chapter, I shall describe the development of a line of normal differentiated epithelial liver cells, the transformation of these cells into epithelioid hepatoma-like cells and the development of a hepatoma cell line derived from a differentiated hepatoma. I shall describe a variety of studies carried out on this system. The main questions asked throughout have been: What are some of the changes which occur in differentiated epithelial cells when they are transformed *in vitro*; and how do the changes compare to those in epithelial cells transformed *in vivo*, namely carcinoma cells? The subjects I shall deal with are karyotype, contact behavior

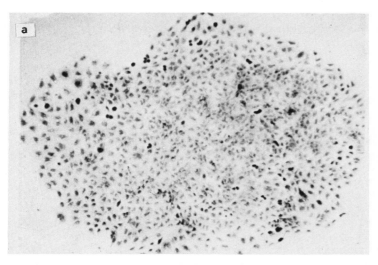

FIG. 1: <u>In a) and b) magnification x50. Enlargement on print x2.</u> a) <u>A 15 day old clone of contact inhibited liver cells (RLB)Giemsa stain x100.</u> From Borek et al. 1973 (4). b) <u>A 10 day old clone of rat liver cells transformed in vitro (RLT) Giemsa stain x 100.</u> Note the dense multilayering of the cells. From Borek et al. 1973 (4).

and cell communication, ultrastructural and chemical changes in the surface membrane, microstructural alterations using plant lectins as surface probes, and the effect of glucocorticoid hormones on these cells.

Establishment of liver cell lines

Normal hepatocytes (RLB)

A cloned line of differentiated epithelial-like cells that are cuboidal was established from normal liver of a 3 month old buffalo rat (8,9). The method used was a modified version of Coon's method (10). Fresh tissue was minced and dissociated progressively at $37^\circ C$ with 0.1% trypsin, 0.1% collagenase & 1% chick serum in Ca^{++} & Mg^{++} - free Hank's salt solution. The dissociated cells were removed at intervals of 40 min., centrifuged in a refrigerated centrifuge to remove the trypsin and resuspended in Ham's F_{12} medium supplemented with double the concentration of amino acid, 5% fetal calf serum and 1% antibiotic antimycotic solution. The cells of 4 consecutive dissociations were pooled, counted and their viability established by the Trypan blue exclusion method. The cells were then plated onto 100 mm petri dishes at seeding levels ranging from 10^4 to 10^6 cells per plate in 10 ml medium and were incubated at $36.5^\circ C$ in a humidified incubator with 5% CO_2 in air. Medium was changed 3 times a week. Within 20 days the cells grew into clones. In order to obtain a homogeneous cell population of epithelial cells free from spindle-shaped fibroblast-like cells, three consecutive clones were isolated using Puck's method (11). After the third isolation the clone of hepatocytes was grown into mass cultures (Fig. 2). The hepatocytes in the cell line were diploid with about 2% tetraploids. Liver normally contains tetraploid hepatocytes. The cells grew as monolayers with no cell overlap and with a generation time of 28 hours. As cell density increases cell size decreases until a supersaturated culture is attained (Fig. 3) where no more mitosis are observed nor is DNA synthesis detectable by autoradiography. The final saturation density obtained at controlled pH and regular feeding is 6×10^4 cells/cm^2 (4). Induction of DNA synthesis in these contact inhibited cultures can be obtained by streaking the monolayer with a micropipet and creating a "wound." (Fig. 4).

Cells of the cloned hepatocyte line are differentiated as indicated by their synthesis and secretion of serum proteins (9), glycogen synthesis and high glucose 6 phosphatase activity (4). They have very low requirement for serum and can undergo about 10 divisions in serum free medium. The liver cells have been now in culture for 48 months and still retain

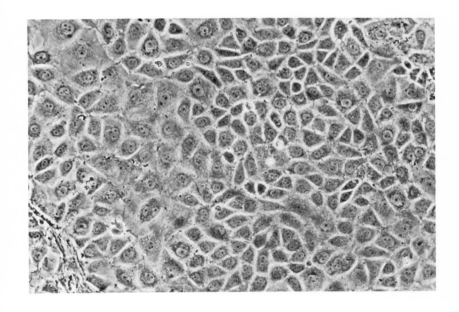

FIG. 2:

A confluent monolayer of the cloned hepatocyte cell line RLB.
Phase contrast x 125. From Borek et al. 1969 (8).

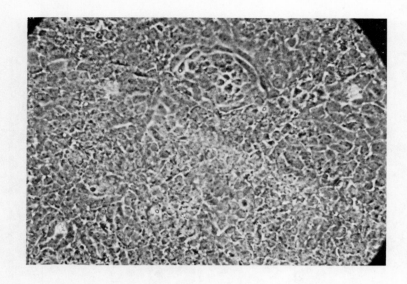

FIG. 3:

A dense culture of RLB at saturation density. Note the decrease in size of the cells and the lack of overlapping in the monolayer. Phase contrast, x 125.

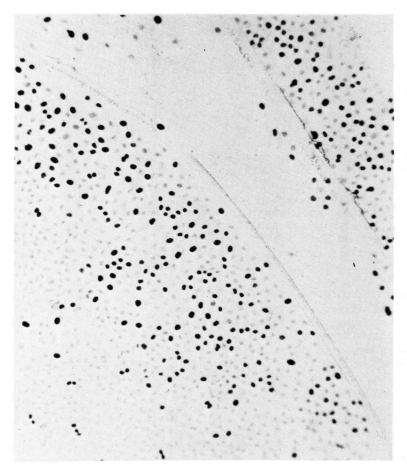

FIG. 4:

Autoradiography of a wound region made in the dense RLB culture of Fig. 3. Note the induction of mitosis (black spots) in areas close to the wound and the cells migrating into the wound region (white space). For autoradiography the cultures were incubated with 2 μCi of tritiated thymidine in 4 ml of culture medium overnight. The cells were then washed with medium, fixed with methanol, dried and then coated with Kodak emulsion N B2 and then incubated in the cold for 7 days and developed with Kodak D19. Magnification x125, enlargement on print x2.

their differentiated properties. It is, however, necessary to reclone them periodically and isolate healthy looking clones since some of the cells do undergo senescence.

Hepatoma cell line (H5123)

By the same method used for culturing hepatocytes a cloned hepatoma line was developed (8,9) from the Morris hepatoma No. 5123 (12) that is transplantable in Buffalo rats. The cells are epitheloid with large areas of contact in adjoining cells (Fig. 5). In contrast to the hepatocytes, they are not inhibited by contact with one another and are tumorigenic. With a generation time of 18 hrs, they multilayer to densities of $8 \times 10^5/cm^2$ at which point they begin to slough off and grow as floating suspensions. In spite of the fact that the tumor is considered a differentiated hepatoma, the cell line which was derived from a single clone synthesized only negligable amounts of serum proteins. Glucose-6-phosphatase activity was lower than that of the normal hepatocytes (4).

Transformed liver cell line (RLT)

The liver epithelial cells used in the transformation experiments had been continuously subcultured for 18 months after the establishment of the cloned line (9). The cells were always trypsinized and harvested when they reached semiconfluency so that they were never subject to crowded conditions. In 3 separate experiments transformation was obtained by suddenly subjecting these cultures to conditions of high saturation density and nutritional deprivation for a period 30-40 days (9). Under these conditions the cells were continuosly exposed to products of their own metabolism and a lowered pH which decreased to around 6.3. Transformed cells began to appear as isolated loci of refractile cells which rose above the monolayer and replicated to form "microtumors" (Fig. 6). These continued to grow as a second layer over the monlayer. No such loci were seen in control cultures which were fed regularly. The transformed cells remained epitheloid and grew in similar patterns to the H5123 cells (Fig. 7). Their generation time was reduced to 16 hrs, and their saturation density was increased to 4×10^5 cells/cm^2 at controlled pH. Similar to the hepatoma cells they begin to slough off at this point and grow in suspension. The RLC cells are aneuploid and are tumorigenic A subcutaneous injection of $10^6 - 10^7$ cells into syngeneic animals irradiated with 400 rads gives rise to carcinomas in 100% of the animals. As discussed later, the RLT cells are aneuploid. The transformed cells continued to synthesize serum proteins (9) but had a lower glucose-6-phosphatase activity than RLB (4).

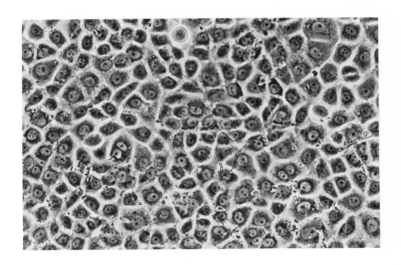

FIG. 5:

Rat hepatoma cell line cultured and cloned from Morris hepatoma H5123. The cells have an epithelioid morphology and can form multilayers in cultures similar to the RLT in Fig.7. Communication studies were carried out on cultures such as this one where cells were still as a monolayer before they began piling up. Phase contrast x 125.

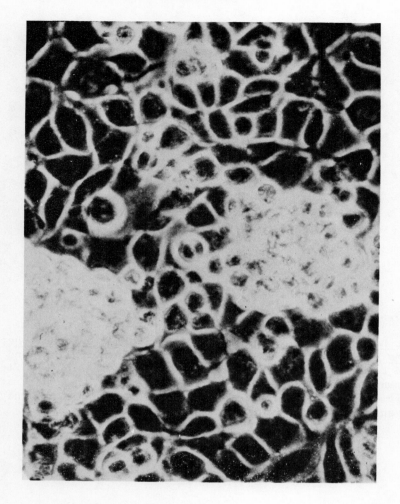

FIG. 6:

<u>Foci of transforming cells (microtumors) arising in dense cultures of the rat hepatocytes following exposure to conditions of nutritional deprivation (see text).</u> Phase contrast. Magnification x 250, enlargement on print x 2.

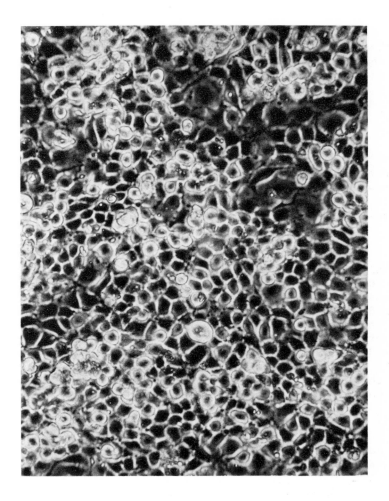

FIG. 7:

Rat hepatocytes transformed in vitro into neoplastic cells with epithelioid morphology. The cells are not contact inhibited, have a random pattern of growth, and can form multilayers in culture. Phase contrast, magnification x 125, enlargement on print x 2.

Karyotype analysis

The use of quinacrine fluorescence and Giemsa banding methods enables unequivocal identification of individual chromosomes. Karyotype analysis using these methods was carried out for the normal hepatocyte line RLB, the transformed RLT cell line (13) and H5123 (unpublished).

RLB

Metaphase chromosomes prepared from RLB cultures which had been 38 months in culture showed the same banding patterns as those from fresh embryo cultures (Fig. 8a). Chromosome number was essentially diploid (42 chromosomes) with 5% tetraploid.

RLT

While the banding patterns in the chromosomes of RLT were identical to those observed in RLB, transformation was accompanied by a marked increase in chromosome numbers to 60 or 80 per cell (Fig. 8b) with a new unidentifiable chromosome in some of the cells. Analysis of the chromosome complement of the transformed cells suggests that the first step was a doubling of the number of the chromosomes followed by a random loss. Experiments are in progress to establish whether this is the case or whether the tetraploid cells are the ones which are most susceptible to transformation.

H5123

Analysis of the hepatoma cell line (unpublished) indicates a normal banding pattern and a chromosome complement which is pseudo-diploid 44-52 chromosomes per cell. No specific markers are seen. Similar results have been found earlier with cells from the fresh tumor (14) and from the H5123 cell line (8) using the conventional prebanding staining methods.

Morphology, adhesiveness and growth in suspension

The normal hepatocytes RLB are flat, strongly adhering cells which grow as monolayers and will not replicate in suspension or in semi-solid medium such as 0.33% agar (9). The cells of the transformed epithelioid RLT, similar to the H5123 hepatoma cell line show a decreased adhesiveness to the surface on which they grow and one another, are removed

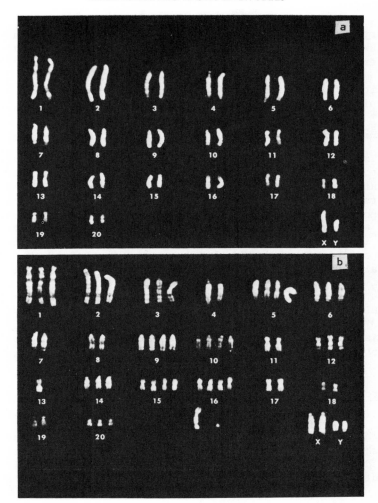

FIG. 8:

a) <u>Quinacrine fluorescent karyotype of a cell from a culture of the rat hepatocyte cell line RLB.</u> b) <u>Quinacrine fluorescent karyotype of a cell from the hepatocyte culture in (A) following neoplastic transformation in vitro into the RLT cell line.</u> From Miller et al 1972 (13).

by trypsin with great facility and can grow in suspension cultures and form colonies in agar (Fig. 9). The growth pattern of RLT and H5123 is multilayering of rounded cells, which is strikingly different from the crisscross pattern of fibroblasts transformed by chemicals, x-irradiation or oncogenic DNA viruses (2).

Contact behavior between RLB, RLT and H5123

In contrast to previous findings with fibroblasts transformed in vitro where the replication of cells transformed by various carcinogens was inhibited by normal cells and the reverse (2) both RLT (9) and H5123 (8) were not inhibited by the normal hepatocytes. RLT cells were able to replicate over RLB under conditions where the latter normal hepatocytes were in growing or in plateau phase. In addition, testing the possibility that "familiarity breeds contempt" two additional liver cell lines (from E.A. Kaighn) of different species were used with the same results. The transformed cells grew over the normal but did not affect their viability.
RLB cells prelabeled with tritium were seeded over RLT, H5123 and also on monolayers of RLB. From the autoradiography it was evident that the normal epithelial cells, similar to fibroblasts (2,6), were unable to replicate over the transformed cells, the tumor cells or over themselves.

Intercellular communication, contact behavior and transformation

The social behavior of normal cells in culture, namely their responsiveness to mechanisms of growth restriction at high densities, inherently suggests that some form of intercellular communication exists among the cells. On the other hand, the lack of responsiveness of tumor and transformed cells to their social environment where each cell freely replicates unrestricted by its neighbors, suggests that these cells have lost communication with their neighbors at intercellular junctions. One form of measurable intercellular communication is that of ionic communication (7,16).
The concept of ionic communication postulates the presence of low resistance pathways between cells in contact which facilitate intercellular transmission of ions. Thus junctional communication can be measured electrically by pulsing currents with a microelectrode between one cell interior and the exterior and by recording the resulting voltages with a second microelectrode inside a contiguous cell (8) (Fig. 10).
Intercellular communication between cells in contact can

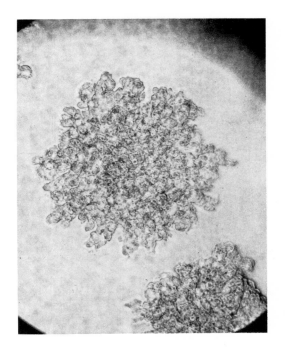

FIG. 9:

An 8 day old clone of transformed rat hepatocytes in 0.33% agar. Phase contrast x 125.

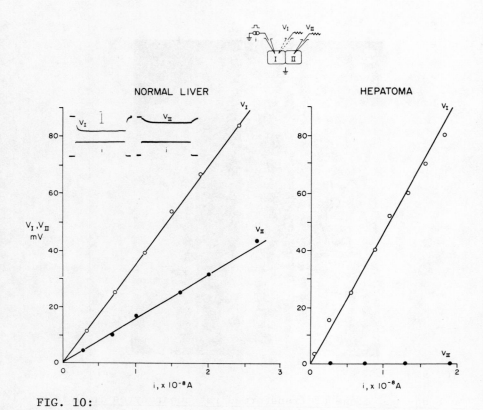

FIG. 10:

Membrane current-voltage in normal liver and Morris' H5123 hepatoma cells in culture. Current (i) is pulsed inward between a microelectrode inside one cell (I) and the outside (grounded). The resulting changes in membrane voltage (V) are recorded with a second microelectrode in a contiguous cell (II) and, in a subsequent measuring series, in cell I. Both cell cultures in fetal calf serum. Note the virtually zero V_{II}/i slope (transfer resistance) of the electrically uncoupled hepatoma cells against that of nearly 1.5×10^6 Ω of the coupled normal liver cells. Left inset: oscilloscope record samples of membrans currents (i=7×10^{-9} A, 100 msec pulse duration) and voltage in the normal liver cell cultures. Voltage calibration: 200 mV. From Borek et al 1969 (8).

also be demonstrated by injecting sodium fluoresceine into a cell and observing under dark field the movement of the fluorescent dye into contiguous cells (16).

Loewenstein reported (17) that in a study _in situ_ on normal liver and a hepatoma intercellular communication was present in the normal liver but absent from the tumor cells. The development of the differentiated RLB and the H5123 cloned cell lines were the closest one could get _in vitro_ to the _in vivo_ situation in order to test under defined conditions whether a correlation existed between regulation of cell replication (contact inhibition) and intercellular communication (8). Junctional communication was measured electrically as described above and in detail elsewhere (8).

The normal hepatocytes communicated well while the H5123 line and in addition the cloned hepatoma line H_4IIEC_3 (H_4) (18) did not communicate to any detectable degree. In the normal cells a considerable fraction of the probing ion current injected into one cell flowed into the contiguous cells (Fig. 10). This was both in sparse cultures as well as confluent ones. In the later ionic communication was detectable over a distance of several diameters from the current sources and any given cell was in communication with many, if not all, of its neighbors. In contrast the hepatoma cells in which communication was measured on cells while still as a monolayer (Figs 2 and 5) showed no junctional communication between cells and their osculating neighbors. The patterns of communication in homogeneous cell cultures were retained in a mixture of normal liver and the Morris hepatoma cells and also in normal liver and the H_4 cells. However, there was no communication between the normal cells to the cancerous cells (8). As discussed above there was no contact inhibition between them.

Since the tumor cell exhibited a lack of intercellular communication in contrast to the normal which were electrically coupled, it was of interest to see whether the transformation of the hepatocytes into the hepatoma like RLT involved a loss of junctional communication. The cells after transformation were found indeed to have no intercellular communication measured by ion current flows and in addition by movement of sodium fluoresceine molecules (9). The dye injected into one cell remained localized in the transformed cell in contrast to the non transformed hepatocyte where it spread into the neighboring cells.

The results thus indicated that in the case of our three lines of epithelioid tumor or epithelioid transformed cells, loss of communication was correlated with an escape from growth control and with malignancy. It should be mentioned here, though described in detail elsewhere (8), that such clear cut differences were not found between fibrosarcoma cells or fibroblasts transformed by a variety of carcinogens and their untransformed counterparts. The degree of cell communication in both normal and neoplastic fibroblasts varied with growth conditions and with the type of serum used. The only exception was a variant epithelioid cell line (8) derived from hamster embryo fibroblasts transformed by x-rays (19). These variant cells, which grow as multilayers, were karyotyped to ascertain their hamster origin and were found to be electrically uncoupled under all growth conditions tested, similar to the hepatoma cells.

From the above studies it seemed apparent that loss of contact inhibition of replication could be correlated with loss of ionic communication only in epithelioid tumor and transformed cells. As described earlier, transformed epithelioid cells have a completely different morphology and mode of growth than spindle shaped transformed fibroblasts, which constantly move over one another with ease and changing points of contact. Thus steric configuration and dissimilarities in adhesion between one cell to another and to the substrate on which they grow cannot be ruled out as relevant factors in the different patterns of communication between transformed fibroblasts and transformed epithelioid cells. To further increase our information about the intercellular contacts ultrastructural studies were made on the epithelial normal and hepatoma cells.

Electron Microscopy

Both types of cells were grown to saturation densities (see above) and were prepared for electron microscopy by Dr. E. Robbins. Using his methods (20), sectioning and ultrastructural studies were carried out by Dr. E.L. Benedetti using his reported methods (21). The findings, presented in Figs. 11 and 12, were similar to earlier ones obtained with preparations of freshly dissected normal liver and hepatoma tissue (21). In the normal cells tight junctions (Zanula occuludens) and intermediate junctions (Zanula adhaereus) and desmosomes (macula adhaereus) were present and found in fixed sequences along the junction (Fig. 11). In the hepatoma cells there is a marked decrease in mutual cell contact. No tight junctions were observed (Fig. 12). Desmosomes and intermediate junctions were sometimes present though less cemented together. Preliminary studies on the RLT cells indicate a similar lack of tight junctions.

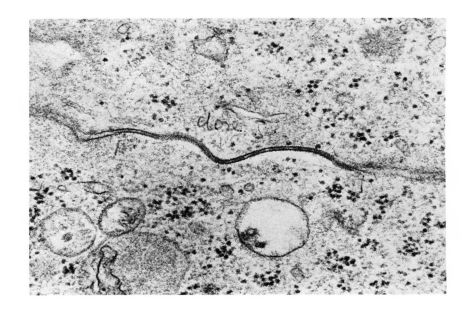

FIG. 11:

<u>Cell interaction between two normal hepatocytes from the RLB cell line</u>. Note the close-tight junction. x 43,000.

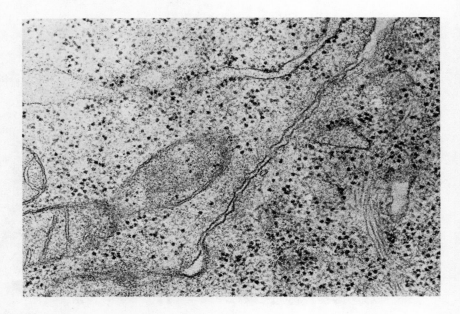

FIG. 12:

Cell interaction between two hepatoma cells of the H5123 cell line. No tight junctions are present and the membranes appear less cemented together. x 27,000.

Ganglioside composition

Gangliosides are sialic acid containing glycolipids which are largely found in cellular plasma membranes (22). Their composition in various fibroblast cell lines is altered following transformation by DNA viruses (22-24).

In the normal liver and hepatoma cells analysis of the gangliosides (25) revealed that their biosynthesis and composition were dramatically altered in the H5123 cells as compared to the normal liver. The changes were more striking than in transformed fibroblasts.

The gangliosides studied were N-acetylneuraminylgalactosyl-N-acetylgalactosaminyl (N-acetylneuraminyl) galactosylceramide (G_{D1a}), galactosyl-N-acetylgalactosaminyl (N-acetylneuraminyl) galactosylglucosylceramide (G_{M1}) and N-acetylneuraminylgalactosylglucosylceramide (G_{M3}). We have found that the mean molar ratios of gangliosides $G_{D1a}:G_{M1}:G_{M3}$ in the normal RLB cell line were 1:0.5:2.6, respectively while in the H5123 cell line the comparable ratios were 1:12:117. This change reflects a decrease in the amount of G_{D1a} coinciding with an increase in G_{M1} and G_{M3} in the hepatoma cell line. The total concentration of gangliosides in the H5123 cells was about 6-fold over that in the RLB in terms of equivalent amounts of cell proteins. Similar results have been obtained with the transformed RLT cells (work in progress). It was of interest that the epithelioid variant of the x ray transformed hamster cells, which like the hepatoma and RLT cells are electrically uncoupled, had a pattern of ganglioside composition similar to the H5123 cells.

Surface alterations detectable by plant agglutinins (lectins)

It is now well documented that fibroblasts which have been transformed by a variety of carcinogens and have escaped growth control are susceptible to agglutination by low concentrations of some plant lectins (3,5,26-29). The interaction between the cells and the agglutinin proceeds via specific carbohydrate containing receptor sites which become available on the cell surface membrane due to architectural rearrangements occurring concomitantly with the transformation. The receptors can be identified by finding the specific carbohydrates which by preincubation with the cells will prevent the cell interaction with the lectins, or will reverse the agglutination reaction after it has taken place.

The untransformed counterparts are not agglutinable at such low concentrations of the lectins unless they are mitotic cells or have been exposed to brief treatment with proteases (26-29),(for review see ref. 5).

No work had been done on transformed epithelioid cells and on their untransformed counterparts nor even on tumor cells of epithelial origin even though the preponderance of epithelial tumors (carcinomas) over fibroblastic tumors (sarcomas) is unequivocal. The system of the RLB (normal hepatocytes), RLT (transformed hepatocytes), H5123, and H_4 lines (hepatocarcinomas) offered a good model for this type of study (4). In view of their different modes of growth and lack in intercellular communication the question was would the epithelioid transformed and tumor cells differ from transformed fibroblasts in their pattern of agglutinability by a variety of lectins. Since the RLT, H5123 and the H_4 are malignant yet differentiated cells, we compared them on one hand to cells of the Novikoff hepatoma (30), which is an extremely undifferentiated and malignant tumor and on the other hand to the normal RLB cells. The results obtained are presented in Table 1.

Low concentrations of 5 agglutinins were able to agglutinate all the neoplastic liver cell lines. Growing and stationary phase cultures were equally agglutinable. Normal RLB were not agglutinable under those conditions.

The agglutinins used were wheat germ agglutinin (WGA) Phytohemagglutinin (PHA) Concanavalin A (ConA) Ricinus Communis (RC) Great Northern Bean (GNB) and Lens culinaris.

The following carbohydrates were inhibitory to the agglutination reactions and were able to reverse the agglutination by the specific lectins (in parentheses), dissociating already agglutinated clumps into single cell suspensions: d, N-acetyl chitobiose 2×10^{-3} to $4 \times 10^{-4}M$ (WGA); α methyl mannose 10^{-2} to $2 \times 10^{-3}M$ (conA); N acetyl galactosamine 10^{-2} to 2×10^{-3} (PHA); lactose 10^{-2} to $2 \times 10^{-3}M$ (ricinus communis); α methyl glucose and N-acetyl galactosamine 10^{-2} to $2 \times 10^{-3}M$ (lens culinaris). No inhibitors have yet been found for GNB agglutination.

In general the agglutination of the epithelioid transformed cells was comparable to that found with transformed fibroblasts (26). Some differences exist. H_4 are the most differentiated and the most "contact inhibited" of the neoplastic cells when grown in culture (cells do not slough off as easily at high densities as in the other lines). They were the least agglutianable by WGA, PHA, RC and GNB. In contrast, the Novikoff cells were most agglutinable. These cells are completely undifferentiated and have minimal inhibition by contact (most cells go into suspension having lost all anchorage dependence). ConA did not distinguish

NORMAL AND NEOPLASTIC LIVER CELLS

	WGA	ConA	PHA	Ricinus communis	GNB	Lens culinaris
H_4	6.5	11	13.2	7.5	3.1	0.10
H 5123	10.5	11	16	18.7	6.3	0.10
RLT	9.0	11	16	15	6.3	0.10
Novikoff solid or ascites	12.0	20	17.3	17.6	10.2	0.16
Tryp-RLB	17.9	16.5	16	23.5	10	1.6
Py-3T3	21.5	18.6	16.5	21.5	8.5	2.2

The concentration of a given agglutinin necessary to give rise to half maximal agglutination was extrapolated from 5 to 7 agglutinin concentrations. Data from 3 to 7 different experiments were pooled. The numbers given in the table, indicating how many times a given cell line agglutinates better than normal rat liver cells (RLB), were calculated by the following formula:

$$\frac{1}{\mu g/ml} \text{ for experimental cells} / \frac{1}{\mu g/ml} \text{ for normal rat liver cells (RLB)}.$$

Abbreviations for cell lines are found in Materials and Methods. Tryp-RLB are RLB cells released with EDTA and treated for 5 min with 30-50 µg/ml trypsin. Py-3T3 cells are 3T3 Swiss mouse fibroblasts transformed by polyoma virus and are included for comparison. Since Novikoff hepatoma cells derived from the solid tumor and from the ascites tumor behaved the same way, results were pooled. Agglutinin concentrations necessary to give rise to half maximal agglutination of normal rat liver cells (RLB) are for WGA 450 µg/ml, for ConA 550 µg/ml, for PHA 400 µg/ml, for Ricinus communis agglutinin 150 µg/ml, for GNB agglutinin 250 µg/ml, and for Lens culinaris agglutinin 50 µg/ml*

---

* *from Borek et al. 1973 (4)*

TABLE 1: AGGLUTINABILITIES OF EPITHELIOID TISSUE CULTURE CELLS

between H_4 and the other two differentiated lines but did interact with the Novikoff cells more than with the others.

The normal epithelioid RLB were not agglutinable by any of the 5 lectins but did interact with lens culinaris while the transformed cells did not. We suggested a possible swivel mechanism whereby certain receptors become available while others are concealed (4). To date normal fibroblasts have not been found to exhibit this interaction with lens culinaris.

Similar to untransformed fibroblasts (26-29) RLB can be rendered more susceptible to agglutination by mild treatment with protease. Incubation with 30-50 µg of crystaline trypsin results in agglutinability similar to that of Novikoff hepatoma cells (Table 1). When confluent RLB or mouse fibroblast cultures and to lesser extent growing cultures are incubated with 10 µg/ml Vitamin A alcohol for 2 hours or are treated roughly (scraped off petri dishes by rubber policemen) an increased protease activity in the medium is observed and the cells became agglutinable. The source of these proteases is probably a release from lysosomes whose membranes became more permeable under the influence of the above concentration of Vitamin A (29,31), or were disrupted by the rough treatment of the cells. Treatment of confluence and growing RLB or fibroblast cultures with cycloheximide (4×10^{-5}M) for 6 hours resulted in a similar effect only in the confluent cultures. Proteases were increased in the medium and the cells became agglutinable. By adding cycloheximide we actually disturbed the balance of membrane turnover. In confluent cultures membrane turnover is higher than in growing cultures (32,33); therefore inhibition of membrane synthesis resulted in an overall increased degradation. The cycloheximide effect could be inhibited by protease inhibitors (4). Six hours following removal of cyclohexamide its effect was completely reversed and the cells attained their original poor agglutinability.

Glucocorticoid induced supression of liver cell proliferation

In view of the known inhibitory effects of the glucocorticoid hormones on liver growth in vivo (34), it was of interest to see whether these same agents would result in an inhibition of cell proliferation in the more isolated system of the liver cell lines. Studies on the effect of glucocorticoid hormone were carried out on the three cloned lines RLB (normal), RLT (transformed) and H5123 (hepatoma) (35). While the majority of our studies have employed hepatoma cells, we found that the addition of glucocorticoid hormones is capable of producing profound supression of DNA synthesis in all three lines (Table 2). At low concentration of hormone (3 x 10^{-7}M hydrocortisone or 3 x 10^{-8} Dexamethasone) the incorporation of radioactive thymidine fell to 50% of control level by 36 hours, and at a higher concentration of hormone (e.g. 5 x 10^{-5}M hydrocortisone) inhibition can be noted as early as 12 hours and is complete by 24 hours (Fig. 13). This inhibition of radioactive thymidine incorporation reflects a true suppression of DNA synthesis and is accompanied by a corresponding inhibition of cell proliferation (Fig. 14). It is readily reversible upon removal of the hormone (35). No cell lysis or degradation of preformed DNA is observed, and even when ^{3}H thymidine incorporation into DNA is inhibited by 90% or more, incorporation of ^{14}C uridine into RNA proceeds with little change (35), which suggests that the DNA suppressive effect cannot be attributable to a generalized cytotoxic effect of the hormone. It was of interest to note that the capacity to respond to glucocorticoid with a decrease in DNA synthesis was not necessarily associated with inducibility for tyrosine aminotransferase synthesis (Table 2), another corticoid sensitive property demonstrated in many cell lines of liver origin (36-38).

Discussion

In the foregoing review, I have attempted to report some of the karyotypic and surface properties of a normal, differentiated, cloned line of epithelial liver cells (hepatocytes) and to discuss several changes which have occurred following neoplastic transformation of these cells _in vitro_ and _in vivo_ into malignant epithelioid cells. Table 3 provides a summary of these properties.

A cloned line of hepatocytes was derived from the liver of a 3 month old rat. The cells were epithelial-like diploid with 5% tetraploid, and were differentiated. The cells grew in monolayers characterized by a marked density inhibition of DNA synthesis (referred to here as contact inhibition). Transformation occurred in 3 separate experiments when hepatocytes that were cultured for 18 months at low cell densities and under optimal conditions were allowed to reach confluency and were maintained as supersaturated cultures without regular feeding for a period of 6 weeks. The cells therefore were exposed to prolonged high proximity, depleted medium, hypoxia, product of their own metabolism and a low pH. The transformed cells began rising above the surface as "microtumors" using the monolayer as a substrate and

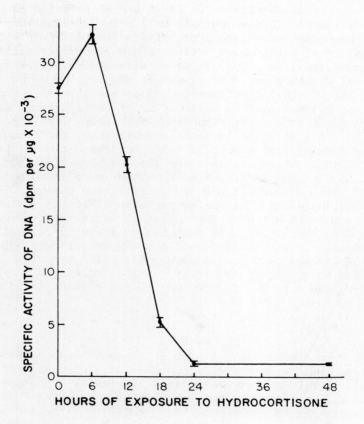

FIG. 13:

<u>Time course of inhibition of thymidine incorporation by H5123 hepatoma cells exposed to hydrocortisone</u>. Hydrocortisone was added to 1 day old cultures at various time intervals at a final concentration of 50 µM. 72 hours after the initial plating 10 µCi of (^{3}H) thymidine was added to each plate and the incorporation of the isotope determined 3 hours later. The abscissa shows the hours of exposure to hydrocortisone prior to the addition of isotope. Each point gives the mean ± SEM for five plates. From Loeb et al 1973 (35).

Cell Line	[^{3}H]Thymidine incorporation (dpm/ g of DNA)	TAT activity (units/mg of protein
Borek - Normal		
Control	12,600 = 1,100	2.96 = 0.24
Hydrocortisone	1,540 = 60	3.17 = 0.28
% of control	12%	107%
Borek - Transformed		
Control	14,600 = 800	1.60 = 0.11
Hydrocortisone	3,200 = 30	1.46 = 0.08
% of control	22%	93%
Morris Hepatoma #5123		
Control	43,100 = 700	2.31 = 0.18
Hydrocortisone	4,180 = 70	2.59 = 0.06
% of control	10%	112%

Five to seven plates of each of the above cell lines were treated with either hydrocortisone in ethanol (final concentration of hormone 50 µM) or with an equal volume of ethanol alone ("control"). After 36 hr 10 µCi of [^{3}H] thymidine was added to each plate, and the cells in each plate were harvested 3 hr later and assayed both for incorporation of isotope into DNA and for tyrosine aminotransferase activity. Although the addition of steroid results in a striking suppression of isotope incorporation, it is seen to have no effect upon the TAT activity exhibited by any of the cell lines.

From Loeb et al. 1973 (35).

TABLE 2. EFFECTS OF HYDROCORTISONE ON THYMIDINE INCORPORATION AND TYROSINE AMINOTRANSFERASE (TAT) ACTIVITY IN THREE DIFFERENT CELL LINES OF LIVER ORIGIN

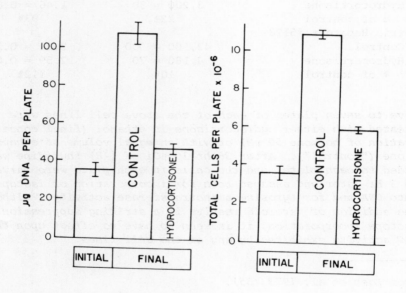

Fig. 14:

<u>Inhibition</u> <u>of</u> <u>cell</u> <u>proliferation</u> <u>and</u> <u>net</u> <u>DNA</u> <u>synthesis</u> <u>by</u> <u>hydrocortisone</u>. Three sets of petri dishes (6 to 7 plates per set) were seeded with 3.5×10^6 H5123 cells per plate; at zero time (24 hr later) the cells in one set were harvested and counted. Hydrocortisone in ethanol, final concentration, 50 μM was added to a second set and ethanol alone to a third set. Forty-eight hr later all cells were harvested and counted and the DNA assayed in all three sets of plates. Results show mean ± SEM. From Loeb et al 1973 (35).

forming a second and third layer of epithelioid cells
which remained differentiated but were aneuploid with their
chromosome complement varying from one cell to another.
A new unidentifiable chromosome appeared which was not
present in the original normal culture. The transformed
cells were no longer contact inhibited, and they grew in
suspension and in semi-solid medium and were malignant in
vivo. Their mode of growth in vitro was an unflattened
shape of cells which had very little adhesion to the substrate on which it grew and to its neighboring cells. The
transformed cells had acquired enhanced mobility and invasiveness and were able to replicate over normal cells. As seen
in Table 3, all characteristics of the transformed cells were
similar to those of cells cultured and cloned from a differentiated hepatoma. While the normal parental hepatocytes
exhibited ion communication at permeable intercellular junctions, no ion current flow could be detected between osculating transformed or hepatoma cells. The change in contact
behavior and loss of contact inhibition and malignancy thus
coincided with surface changes which led to a loss of
intercellular communication. There is insufficient evidence to state whether it is a change in membrane permeability of the cells or in the topographic relations of one cell
to another. The change is genetic since it is maintained in
the progeny of the cells. The decrease in intercellular
adhesion was confirmed by ultrastructural studies in the
hepatoma cells and in preliminary studies in the transformed
liver cells. No tight junctions were found. Correlations
between ultrastructural junctions and quality of ion communication has been reported (39). Dramatic changes were found
in ganglioside composition and biosynthesis in the hepatoma
cells, and in preliminary work in the transformed liver, as
compared to the normal cells. Since these sialic acid
containing glycolipids are mainly in plasma membranes (22)
their changed composition could modify cell permeability
at intercellular junctions and also contribute to decreased
cellular adhesion by altering cell surface charges (40).

Structural alterations in the transformed cell membrane
as manifested by their increased susceptibility to agglutination by lectins could also alter both permeability and
adhesion. Correlation between communication and agglutinability is seen also following treatment of normal hepatocytes and fibroblasts with proteases. Trypsin treatment
induces temporary loss of electrical communication which
can be recovered within 6 hours following the removal of
trypsin. In the same way exposure of the normal cells to
proteases or to conditions where proteolytic activity was

	Normal liver	Transformed cells	Hepatoma cells
Morphology	Epithelial	Epithelioid	Epithelioid
Cell generation time	28 hr	16 hr	18 hr
Karyotype	Diploid	Aneuploid	Aneuploid
Ability to manufacture serum protein	Yes	Yes	No
Contact Inhibition of replication	Yes	No	No
Pattern of growth in culture	Monolayers	Multilayers	Multilayers
Saturation density (cells/cm^2)	6×10^4	4×10^5	8×10^5
Ability to grow in suspension cultures	No	Yes	Yes
Colony formation in 0.33% agar	No	Yes	Yes
Agglutinability of the cells by concanavalin A and wheat-germ agglutinin	No	Yes	Yes
Communication at permeable intercellular junctions	Yes	No	No
Tight junctions	Yes	None	None
Average total gangliosides (moles/mg protein)	0.53	Work in progress	3.03
Ability to replicate over normal cells	No	Yes	Yes
Tumor production in vivo	No	Yes	Yes
Suppression of cell proliferation by glucocorticoids	Yes	Yes	Yes

TABLE 3. PROPERTIES OF CULTURED EPITHELIAL CELLS FROM RAT LIVER (HEPATOCYTES) AFTER TRANSFORMATION IN VITRO, COMPARED WITH A NONTRANSFORMED CULTURE AND WITH A CELL LINE CULTURED FROM RAT HEPATOMA H5123

increased (Vit A, rough handling, cycloheximide treatment, in confluent cultures) resulted in increased susceptibility to agglutination by lectins. This effect was reversible, and following removal of the protease environment, poor agglutinability was recovered within 6 hours. The coincidence is rather striking.

Although the molecular mechanisms of malignant cell transformation under various growth conditions in vivo and in vitro are still obscure, there are several events involving an increase in cellular activity of lysosomal enzymes (31,41-46) which are known to occur in cells exposed to prolonged abnormal contact, in stationary phase (41) starvation, (41,42) low pH (43) and hypoxia (44), and which could trigger the nonplastic conversion of the normal cells into fast dividing, aneuploid cells with new surface membrane properties characteristic of neoplastic cells (29,31,45). In particular, the occurrence of unequal division in mitosis that lead to formation of cells other than diploid have been reported in cells after a change of nutritional environment (42) and under conditions where activity of lysosomal hydrolases is increased (31). Vitamin A is known to increase the permeability of lysosomes and to facilitate the release of lysosomal enzymes towards the cell surface causing digestion of the cell coat (29,46). The increased agglutinability of the normal cells following treatment with Vit A (4,29) or after rough treatment of the cells which leads to cellular damage and lysosomal disruption (31) strongly implies the role of lysosomal proteases in modifying the cell surface architecture. This seems especially so since the stationary cells were more susceptible than the growing cells, and lysosomes are increased in stationary cells as compared to dividing cells (44). Glucocorticoids dramatically supressed cell proliferation in all three cell lines. The possibility that this inhibitory effect occurs concomitantly with surface alteration in the cells is currently under investigation. This is of interest especially in view of the fact that hydrocortisone stabilizes lysosomal membranes and prevents the release of their enzymes (47). The demonstration of liver cell transformation by means of nutritional stress into malignant epithelioid cells whose characteristics are nearly identical to those of a hepatoma induced in vivo would appear to have some obvious possible implications for some kinds of malignancies that develop in vivo and serve as a model for studying them.

References

1. Abercrombie, M. Cancer Inst. Monog 26 (1966) 249.
2. Borek, C. and Sachs, L. Proc. Nat. Acad. Sci. 56 (1966) 1705.
3. Borek, C. and Hall, E.J. Nature 243 (1973) 450.
4. Borek, C., Grob, M. and Burger, M.M. Exp. Cell Res. 77 (1973) 207.
5. Nicolson, G. in International Rev. of Cytology, (1974) (In press).
6. Stoker, M.G.P. Proc. Roy. Soc. Lond. 181 (1972) 1
7. Potter, D.D., Furshpan, E.J. and Lennox, E.S. Proc. Nat. Acad. Sci. 55 (1966) 328.
8. Borek, C., Higashino, S. and Loewenstein, W.R. J. Memb. Biol. 1 (1969) 274.
9. Borek, C. Proc. Nat. Acad. Sci. 69 (1972) 956.
10. Coon, H.G. and Weiss, M.C. Proc. Nat. Acad. Sci. 62 (1968) 254.
11. Ham, R.G. and Puck, T.T. Methods in Enzymol. 5 (1962) 90.
12. Morris, H.P., Sydransky, H., Wagner, B.P. and Dyer, H.H. Cancer Res. 20 (1960) 1252.
13. Miller, D.A., Dev, V.G., Borek, C. and Miller, O.J. Cancer Res. 32 (1972) 2375.
14. Nowell, P.C., Morris, H.P. and Potter, V.R. Cancer Res. 27 (1967) 1565.
15. Loewenstein, W.R. Ann. N.Y. Acad. Sci. 137 (1966) 441.
16. Azarnia, R. and Loewenstein, W.R. J. Memb. Biol. 6 (1971) 368.
17. Loewenstein, Nature 209 (1966) 1298.
18. Pitot, H.C., Periano, C. Morse, P.S. and Potter, V. R. Nat. Cancer Inst. Monog. 13 (1964) 229.
19. Borek, C. and Sachs, L. Nature 210 (1966) 276.
20. Robbins, E. and Gonatas, N.K. J. Cell Biol. 20 (1964) 356.
21. Benedetti, E.L. and Emelott, P.E. J. Cell Sci. 2 (1967) 499.
22. Sheinin, R., Onodera, K., Yogeeswaran, G., Murray, R.K. In the Biology of Oncogenic Viruses (2nd le Petit Symposium, Ed. Sievestri, L.G., Amsterdam, North Holland (1971) pp. 274.
23. Hakomori, S. and Murakami, W.T. Proc. Nat. Acad. Sci. 59 (1968) 254.
24. Mora, P.T., Brady, R.O., Bradley, R.M. and McFarland Proc. Nat. Acad. Sci. 63 (1969) 1290.
25. Brady, R.O., Borek, C. and Bradley, R.M. J. Biol. Chem. 244 (1969) 23.

26. Burger, M.M. Proc. Nat. Acad. Sci. 62 (1969) 994.
27. Inbar, M. and Sachs, L. Nature 223 (1969) 710.
28. Nicolson, G.L. Nature New Biol. 239 (1972) 193.
29. Poste, G. Exp. Cell Res. 73 (1972) 319.
30. Novikoff, A.B. Cancer Res. 17 (1957) 1010.
31. Allison, A.C. Lysosomes in Biology and Pathology (Eds Dingle, J.T. and Fall, H.B., North-Holland, Amsterdam) 2 (1969) 178.
32. Warren, L. and Glick, M.C. J. Cell Biol. 37 (1968) 729.
33. Baker, J.B. and Humphreys, T. Science 175 (1972) 905.
34. Henderson, I.C., Fischel, R.E. and Loeb, J.N. Endocrinology 88 (1971) 1471.
35. Loeb, J.N., Borek, C. and Yeung, L.L. Proc. Nat. Acad. Sci. 70 (1973) 3852.
36. Thompson, E.B., Tomkins, G.M. and Curran, J.F. Proc. Nat. Acad. Sci. 56 (1966) 296.
37. Gerschenson, L.E., Anderson, M., Molson, J. and Okigaki, T. Science 170 (1970) 859.
38. Butcher, F.R., Becker, J.E. and Potter, V.R. Exp. Cell Res. 66 (1971) 321.
39. Gilula, N.B., Reeves, O.R. and Steinbach, A. Nature 235 (1972) 262.
40. Abercrombie, M. and Ambrose, E.J. Cancer Res. (1962) 525.
41. Allison, A.C. and Young, M.R. in Lysosomes in Biology and Pathology (eds Dingle, J.T. and Fall, H.B., North-Holland, Amsterdam) p. 600.
42. Richards, B.M., Walker, P.M.B. and Deely, E.M. Ann. N.Y. Acad. Sci. 63 (1956) 831.
43. Sylven, B. Eur. J. Cancer 4 (1968) 559.
44. Ericsson, E.J. in Lysosomes in Biology and Pathology (ed Dingle, J.T. and Fell, H.B.) North-Holland Amsterdam 2 (1969) 345.
45. Weiss, L. The Cell Periphery, Metastasis and Other Contact Phenomena, North-Holland Amsterdam (1967).
46. Poste, G. Exp. Cell Res. 67 (1971) 11.
47. Weissman, G. and Dingle, J.T. Exp. Cell Res. 25 (1961) 207.

Note: We would like to thank the publishers of Exp. Cell Research, Proc. Nat. Acad. Sci. (USA), J. Membrane Biol., and Cancer Res. for their permission to reproduce Tables 1 and 2, and Figs. 1, 2, 3, 4, 7, & 8.

RETENTION AND LOSS OF CERTAIN ENZYMES IN VARIOUS PRIMARY CULTURES AND CELL LINES OF NORMAL RAT LIVER

Martine Chessebeuf, Aline Olsson, Paulette Bournot, Jean Desgres, Michel Guiguet, Gabrielle Maume, Bernard F. Maume, Bernard Perissel and Prudent Padieu

Introduction

   Analysis of cell differentiation can be carried out at the level of enzymatic differentiation of the cell. This functional differentiation is a late phenomenon which stamps the outcome of a differentiating program at the genetic level. Therefore, this type of study does not give any clue on the mechanisms of the prime cause by which starting with the first egg segmentations a cell will give rise to a definite tissue which will express and maintain a peculiar functional state. This development is linked to a differentiating program which the cells will follow during each mitotic cell cycle. Nevertheless, the study of the dynamic state of a differentiated cell is necessary: i) to take the census of specific cell markers; ii) to elucidate the regulatory process at the genome expression level and iii) to prepare the ground for eukaryote cytogenetics.
   Tissue culture is among the most attractive tools for studying the expression of differentiation through the maintenance of specialized functions in cells. But actually tissue culture is used with variable success. The difficulties in growing highly differentiated normal eukaryote cells and in recognizing the conservation of specific patterns are pitfalls from which one can escape by working with malignant or unspecific transformed cells. In this report, we want to show that "normal" rat liver cells can be obtained in primary culture and in cell lines in which they will retain some specific functions of normal liver, and that cell cultures allow to study primary expressions of a differentiated state free of the cellular interactions of the organism or under the action of effectors which cannot be easily mastered in investigations in vivo.
   Avian and mammalian liver are among the most frequently studied organs because of their key role in the metabolism of animal. It is not surprising that the earliest attempts in tissue culture have been focussed on liver as reported in 1911 by Margaret and Warren Lewis (1) and by Alexis Carrel (2). Among the most recent reports on liver cell cultures by Coon (3), Gerschenson et al. (4), Williams et al. (5),

Lambiotte et al. (6), Iype (7) and Bissel et al. (8), none gave information about the retention of specific functions which are lost during carcinogenesis like aldolase B isoenzyme activity. Aldolase B is the aldolase isoenzyme which is devoted in the liver to an efficient cleavage of fructose-1-phosphate (this carbohydrate being the second important hexose brought by food) and to synthesize fructose-1,6-diphosphate (Blostein and Rutter, (9)). Aldolase A, active mainly on fructose-1,6-diphosphate is present in human primary hepatoma (Schapira et al., 10,11) and in fast growing transplantable hepatoma (12). Walker et al. (13) reported the lost of hepatic isoenzyme of pyruvate kinase, hexokinase and aldolase in rat liver cells in culture. In the report of a recent Workshop on Liver Cell Culture (14) it was stated that the search for bona fide cultured hepatocyte was still to be done and maybe was the quest of the will o' the wisp.

Following our first reports on primary culture of postnatal rat liver cells (15,16), we want to report our detailed method to obtain primary cultures and liver cell lines which retain some specific properties of normal hepatocytes from the standpoint of morphology and biochemical markers. In both types of culture, these cells have shown an ultrastructure and tissue organization strikingly similar to hepatic tissue in vivo. Both clear and granulous dark epithelial cells were present in all cultures but with a high percentage for the granulous type which may be identical to the normal hepatocyte in vivo. To date specific biochemical markers investigated and found in our established cell lines were: an equal cleavage of fructose-1-phosphate and fructose-1,6-diphosphate by aldolase, and steroid metabolic pathways to be described (17,18).

Materials

A - Animals

Newborn animals from originally SPF Wistar rats were used for this study (inbred Wistar US/Commentry strain). Maximum care is taken to prevent contamination of the animal room strictly reserved to supply the tissue culture laboratory. Animals were from 18 days p.c. to 7 month-old and the sex was selected when necessary from pooled postnatal animals.

B - Tissue Dissociation Medium

Dissociation medium used to dissolve trypsin (EC 3.4.4.4.)

was made from a powder medium analogous to Ham F10 culture medium (HAM) (19) but without Ca^{++} and Mg^{++} ions. This H (hepatic) medium was prepared the day before or the morning of the tissue explantation and was filter sterilized together with the culture medium. A solution of 100 mg of trypsin (about 3000 NF unit/mg, Calbiochem A grade) and 44 mg of $MgSO_4$, 7 H_2O in 100 ml of H medium was made and sterilized by membrane filtration extemporaneously. Before use of any filtering membrane for medium sterilization the sterilized filtering unit (autoclaved 45 mm at 120°C and kept 15 min. under vacuum) was washed according the procedure described by Cahn et al. (20) to remove any remaining detergent. The trypsin solution was used immediately after checking the pH and adjusting it if necessary with N HCl, from 7.40 to 7.50.

C - Culture Medium

Only medium Ham F10 was used. This medium was especially prepared for us by Gibco without Ca^{++}, Mg^{++} and pyruvate. Therefore, it could be used to prepare the tissue dissociation medium by adding 1 mM pyruvate or the culture medium by adding 0.3 mM $CaCl_2$, 2 H_2O, 0.62 mM $MgSO_4$, 7 H_2O and the required amount of pyruvate (0.5 to 1 mM) and supplementing with 10% human and 10% fetal calf sera. The human serum was obtained from the local Blood Bank and ultracentrifuged before use or storage at -80°C. Penicillin and streptomycin were added to give final concentrations of 200 000 units and 0.2 g/l respectively, and the culture medium was sterilized by filtration through a Sartorius filter consisting of three membranes of decreasing pore size (1.2 µ, 0.6 µ and 0.2 µ) with a prefilter in between to prevent any clogging. The filter is pretreated as mentioned above after autoclaving. All the filtrations were performed in the sterile room.

D - Sterile Room

A 20 cubic meter room was equiped with a sterilizing air conditioner, recycling 75% of the room atmosphere and adding 25% of air taken on the roof of the four-story building establishing a positive pressure of 2 mm of mercury. The building is located far from any air polluting industries.

E - Tissue Desintegration Apparatus

A mechanically operated magnetic stirrer for six flasks has been built in a 100 l tissue culture incubator, the driving electric motor being outside to prevent heat accumulation. A rpm meter allows to adjust to the desired stirring speed. The inside temperature of this incubator is kept constant at 35°C to 36°C.

F - Tissue Culture Incubator

To keep the cells in a controlled atmosphere we used a Lequeux IGR incubator with six LWOFF (or IGR) air-tight boxes (Lequeux, 74 rue Gay-Lussac, Paris 5eme). Humidity was maintained by the presence of trays filled with distilled water. The IGR boxes were connected to a gas tank containing compressed air with 5% CO_2. The gas mixture was blown through a gas membrane filter for ten minutes when closing the door after each opening but thereafter no permanent gassing was done.

G - Photomicroscopic Investigation

A Wild microscope M40 with accessories for time lapse 16 mm movie camera-cinemicrography and video-magnetic tape cinemicrography were used.

Methods

A - Isolation of Liver Cells: Fractional Trypsinization

All the operations up to the first trypsinization were done as fast as possible by two persons. Six to ten day-old animals from one litter (8-10 rats) were decapitated outside the sterile room and each extirpated liver was transferred in a sterile dish containing the H solution and immediately minced with the scissors. This operation is repeated for all subsequent liver specimens. All the further operations were then done in the sterile room. The mince, rinsed four times with the H solution and then once with the trypsin solution was transferred into a stirrer flask with a suspended magnetic impeller and having an appropriate volume to immerse the impeller in twice the mince volume (Wheaton "Celstir", 25 or 50 ml, Gibco). An amount of trypsin solution equal to the volume of the mince was added by the side arm of the flask, generally 5 ml. The mixture was stirred at 36°C in the trypsinization incubator for 10 min. at 200 rpm. The flask was removed and the tissue was allowed to settle for two minutes. The supernatant was gently decanted by pouring off sideways through the special spout into a plastic culture tube (Falcon no. 3026) containing a volume of tissue culture medium twice the volume of trypsin solution for the purpose of stoping the trypsin action. The tube was centrifuged at no more than 600x g in a clinical conical centrifuge for 8 min. At the same moment the same amount of trypsin solution was added to the stirring flask for the second tissue desintegration. The same operation was repeated twelve times.

B - Cell Culture

Each separated pellet was resuspended in 4.3 ml of culture medium by gently tipping the tube, and the suspended cells were transferred in full into the culture dish (no. 3009 Falcon). The medium was removed 7 hours after the initial plating, and was then changed each 48 hours.

C - Subculture

The first subculture was performed when the culture was nearly reaching confluency which generally occured six to eight days after plating. Four ml of trypsin solution were added to each dish kept in the incubator at 37°C. After 10 min. the medium was gently pipetted several times to achieve the desintegration of the cell monolayer. The subsequent operations were carried out exactly like the primary cell explantations. The first subculture and sometimes the second were done with or without splitting the dish into halves depending on the overall aspect of the culture and ability to reach confluence. The subsequent subcultures were split progressively into more dishes to give 20 dishes out of one after the 8th culture.

Results and Discussion

A - Primary Cell Growth

A few hours after plating, three types of attached cells could be distinguished: few fibroblasts, dark flat thin cells with distinguishable nucleus, and spherical-rounded or spherical-triangular cells, very refringent, with a very thin regular membrane and a dark central nuclear mass. Mitosis could be easily detected. The amount of viable cells, the seeded cells which in less than 2-3 hours strongly attached to the dish, and the distribution among the different types were dependent on the trypsinized fraction. Up to the fourth there were not many cells which would give rise to epithelial monolayers. The sixth to the ninth pellets had a high percentage of flat and rounded attached cells while the tenth and subsequent pellets had fewer cells but with a higher percentage of fibroblasts. Constantly the seventh or the eighth trypsinization gave always the highest absolute amount and the best percentage of cells which would evolve to polygonal epithelial cells. Starting with 10^6 to 2.10^6 cells seeded in each 6th, 7th and 8th dishes 5.10^5 to 8.10^5 cells were attached 7 hours after plating and over 70 to 90% of the cells were epithelial polygonal cells with a granular and dense cytoplasma, and one or two clear nuclei with two very dark nucleoli (F. 1a, 1b, 1c, 1d).

The fibroblasts showed a slow development and were rather pushed outside of the area of polygonal cells making trabeculations between these areas. A third type of large rounded cell appeared with a less sharp delineation of the outer membrane and of the nucleus membrane and a clear cytoplasma. These cells which did not seem to be macrophages or Kupffer cells were surrounded by the polygonal cells developing rosette organizations (Fig. 1).

Primary cultures from liver of fetuses or from animals older than 20 days showed a high percentage (50% and over) of Kupffer cells which could be easily recognized by the morphology and also by their phagocytosis of iron granules (1 μ diameter). This population developed slowly and was overgrown at the time of confluency of the primary culture or during the first subculture. We have observed that the medium of primary culture with Kupffer cells was very clean, cleared of cellular debris and that consequently the growth of the epithelial cell culture was much helped (Figure 2a). When the monolayer was confluent, cells divided pushing the other cells to reattach but never piled up. Some old cells were forced to detach and die, floating away in the medium. At confluency, in primary culture as well as in subculture, new cell divisions started in the center of the cell areas where the cells are the oldest to spread out centrifugally. Healthy confluent primary or secondary culture have been kept for 45 days to sixty days in good condition changing the medium every two days (Figure 2a). During the logarithmic phase of growth the generation time was calculated to be about 16-18 hours. At confluency, time-lapse film showed that 6 to 8% of the cell population divided in 24 hours.

B - *Subculture (Figure 2b)*

It was found that the first subculture has to be done just before complete confluency, that is to say the sixth to the eighth day after the primary explantation, and that for

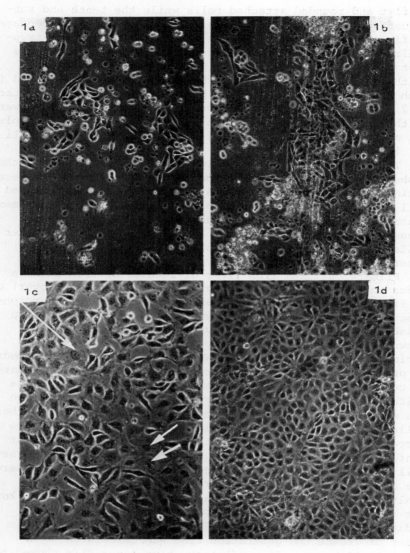

FIG. 1:

Development of primary culture of liver cells from 8 day-old rats; seeded cells represent the 8th trypinization (see text). Age of culture after seeding were, 1a: 18 hrs (x187), 1b: 70 hrs (x187), 1c: 4 days (x187), 1d: 8 days (x187).

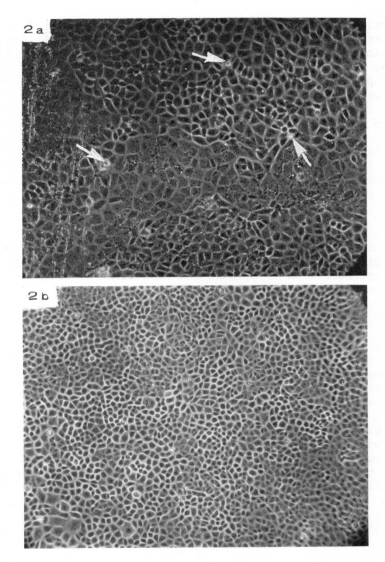

FIG. 2a:

Same type of primary culture 50 day-old (x187). Conspicuous are as of older large cells (in the middle) and younger darker cells (top and bottom of picture). Arrows show dividing cells. FIG. 2b: Typical subculture 7 day-old at the 5th passage with a splitting ratio of 6. Cells originate from the 7th trypsinization of 8 day-old rats' livers (x112).

starting the line safely one dish had to be seeded without splitting. We observed that not only did fibroblasts not develop but that they decreased in amount and disappeared after the fourth subculture. The second or third subculture was split in two or four dishes. After the tenth subculture 100 cells seeded in one 50 mm. diameter dish reached confluency in about 10 days giving rise to about $3.5 \ 10^6$ cells in 12 to 16 generations. After the third subculture cells were grown either in plastic culture dishes or plastic culture flasks. Presently, we have established lines from 18 day-old p.c. fetus to 7 month-old adult. Some lines have been propagated during 30 subcultures with a 1:20 splitting ratio (18 FR line).*

C - Age of Rats

Livers from 8 to 10 day-old rats were the easiest to put in culture. But once the culture method was well established we have succeeded to culture and subculture liver cells from rats of any age.

D - Electron Microscopy and Histochemistry Findings

Electron microscopy pictures were taken by Perissel et al. (21) from 8 FR line at 9th passage and 18 FR line at 22nd passage. Dark granular cells showed a striking rosette organization (Figure 3a). The nuclei of the cells were at the outer zone while the numerous organelles: rich dictyosomal and vesicular Golgi (Figure 3e), mitochondria, rough endoplasmic reticulum around flat elongated cisterns (Figure 3d) were distributed from the outer nucleus to the central zone of the rosette appearing as a clear space similar in some ways to a biliary duct (Figure 3a). Several forms of cell junctions have been found (Figures 3a, 3b, 3c, 3d): i) zonulea occludentes often as macula occludentes and sometimes as tight junctions, the surrounding cytoplasma showing many microfibrils, ii) zonulae adherent and desmosomes were seen more rarely, iii) relatively numerous nexus with some of them forming membranous hernia or bulb junctions. Other types of junctions were missing as were interdigitations and true mature desmosomes. Those types of junctions found were relevant to the noticeable cell motility as revealed by time-lapse microcinematography on confluent cultures (15).

*Our lines are numbered the following way: 2 digits or more for age, FR for "foie rat" and digits for number of subculture.

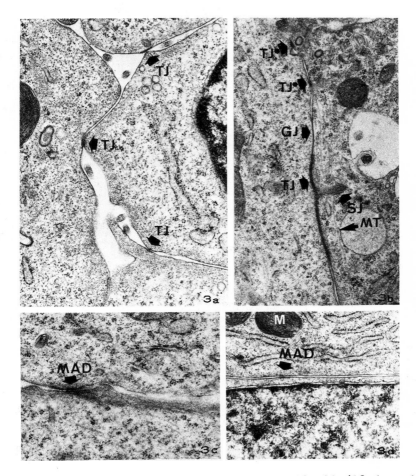

FIG. 3: Electron microscopy of cell line 18FR22 (18 day-old rat, 22nd passage, 200 cell generations). FIG 3a: Punctiform tight junctions (maculae occludentes((TJ) often located at the extremities of cytoplasmic expansions (x30,000). FIG. 3b: Junctiform tight-junction (*TJ) continuous tight-junction surrounded by dense fibrillar material (TJ), gap junction (GJ), studlike junction (SJ) and perijunctional microtubules. FIG. 3c: Macula adherens diminuta (MAD) with more differentiation due to the presence of microfibrils tied on both sides of the dense membrane material (x30,000). FIG. 3d: Macula adherens diminuta (MAD) or embryonic desmosome constituted of thickenings of plasmic membranes and a locally denser intercellular space, mitochondria (M). (x24,000).

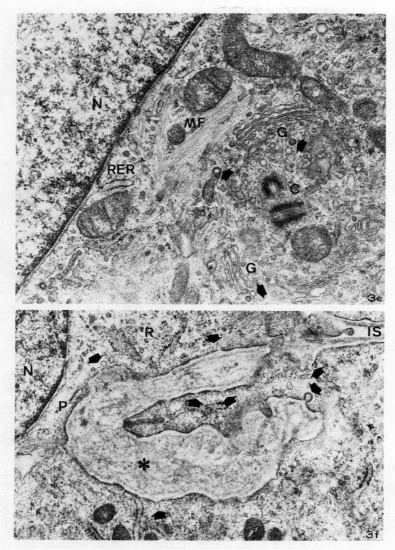

FIG. 3e: Association of Golgi (G) and centrosome frequently observed, dictyosomal Golgi (G) and vesicular Golgi (arrow), mitochondria, rough endoplasmic reticulum (RER) associated to external nuclear membrane and microfibrills. FIG. 3f: Intracellular cavity (*) prolongating the intercellular space (IS) lines by open rough or smooth vesicules (arrows), nucleus (N), polysomes (P) and ribosomes (R).

ATPase and thiamine pyrophosphatase activity have been detected by electron microscopy, the first on the cell membranes and the numerous villi which protruded from it and also on the membrane surrounding the clear space with a canalicular value.

A high succino-dehydrogenase activity and abundant glycogen by the McManus staining were observed using histochemical techniques and light microscopy.

Chromosome analysis displayed a majority of normal karyotypes: 40 + YY or XY for diploid and 84 for tetraploid cells, less than 5% of the karyotypes were abnormal (25).

F - Biochemical Findings

1° - <u>Aldolase</u>: fructose-1,6-diphosphate D-glyceraldehyde-3-phosphatelyase (EC 4.1.2.13).

Aldolase activities on fructose-1,6-diphosphate (F-1,6-diP) and fructose-1-P (F-1-P) were assayed (17,18) using the method of Blostein and Rutter (9). The cells obtained from three to four Cooper dishes were sonicated in 0.45 ml of Hank's or H solution. The extract was then centrifuged 1 h at 100,000 xg. On one set of dishes three assays of $NADH_2$ oxidation were performed: blank without substrate, with F-1,6-diP and with F-1-P. Liver organ and sometimes brain extracts were tested with each culture run with the same procedure.

Table 1 shows some of our results from different primary cultures and cell lines. The primary cultures can be divided in two groups, one exhibiting equal aldolase activities with both substrates relevant to the presence of aldolase B and the other showing aldolase cleavage restricted to F-1,6-diP. The same phenomenon was found on cell lines, the loss or the maintenance of Aldolase B activity.

We have expressed the activities of aldolase on F-1,6-diP or on F-1-P (Figure 4) in mIU per dish against protein content per dish. They both show that primary cultures or cell lines are distributed in two families, aldolase B type culture having the lowest range of protein contents, and aldolase A type culture in the highest range. The two activities on F-1,6-diP and F-1-P in either type of culture in mIU/mg protein were in the same order than <u>in vivo</u> and even sometimes higher for cultures with an aldolase ratio F-1,6-diP/F-1-P around 1. This observation can be explained by the fact that these cultures were consisted nearly completely of epithelial cells which is obviously not the case in liver specimens.

In the liver organ, we have found the following activi-

Aldolase A activity / Aldolase B activity

Primary culture

	8FR18	10FR	10FR	10FR	8FR		6-8FR	6-8FR	8FR	8FR	8FR	8FR	19FR	8FR
Ratio $\frac{F-1,6-diP}{F-1-P}$	>5	>50	>50	13	6.4	50	1	1.4	0.8	2.3	0.9	1.2	2.5	1.4	4
Activity on F-1,6-diP	32	49	58	45	30	41	52	73	16	34	23	50	58	90	36
Activity on F-1-P	0	0	0	3.5	4.7	0	49	52	21	15	27	40	23	64	9

Cell line

	8FR18*	14FR2*	2FR16*	8FR3*	18FR22*	8FR4*	2FR16*
Ratio $\frac{F-1,6-diP}{F-1-P}$	0.8	5.3	2.2	1.2	1.1	1.0	0.7
Activity on F-1,6-diP	2.7	40	27	21	38	19	19
Activity on F-1-P	3.3	7.5	12	17	19	12	27

	Liver Organ		Liver fibroblasts culture	
	miU/ml	miU/mg	mIU/ml	mIU/mg
Ratio $\frac{F-1,6-diP}{F-1-P}$	0.96 ± 0.15	0.95 ± 0.1	50	50
Activity on F-1,6-diP	39 ± 17	36 ± 5	30	33
Activity on F-1-P	39 ± 16	35 ± 7	0	0

TABLE 1. Activities of aldolase cleavage of fructose-1,6-diphosphate (F-1,6-diP) and fructose-1-phosphate and ratio of these activities F-1, 6-diP/F-1-P found in primary culture in mIU/mg protein (upper panel) and in cell lines in mIU/ml cell extract (lower panel) Ratio: 50 (50 = ratio in muscle) means undetectable amount of F-1-P aldolase activity.

In addition activities in liver organ expressed in both units (mean of 9 assays) and in liver "fibroblast" primary culture (see text for mean and standard deviation of primary cultures).

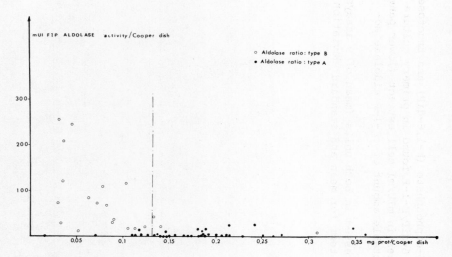

FIG. 4:

<u>Fructose-1-phosphate (F-1-P) aldolase activity of cultures versus protein content both expressed per culture dish.</u> Open circle (O) are primary cultures exhibiting equally high enzymatic activities on F-1-P and fructose-1, 6-diphosphate (F-1, 6diP) with an aldolase cleavage ratio (F-1, 6-diP/F-1-P) of 1.7±0.9. (●) are cultures with low or no aldolase activity on F-1-P while the aldolase activity on F-1, 6-diP was high and the cleavage ratio higher than 10 or more rrequently than 50.

ties: 36 ± 5 mIU/mg protein on F-1,6-diP, 35 ± 7 mIU/mg protein and an aldolase ratio 0.95 ± 0.1 typical of aldolase B activity. In our primary liver cell culture with the same type of ratio 1.7 ± 0.9, this activity on F-1,6-diP was 48 ± 22 mIU/mg protein and on F-1-P 34 ± 17 mIU/mg protein. In primary cultures with aldolase A activity, high aldolase ratio from 10 to infinite, the aldolase activity on F-1,6-diP was 42 ± 9 mIU/mg protein while on F-1-P it was near zero.

In the cell line RLC established by Gerschenson et al. (4), Walker et al. (13) found an aldolase activity of 50 mIU/mg protein on F-1,6-diP which is close to our value but no aldolase activity on F-1-P evidencing that this cell line has only aldolase A.

We have no explanation from our findings which are supported by almost twenty different cultures. A random selection of cells all of which express one of each gene from one culture to another is highly improbable. Rather we may assume that growth conditions in the culture may play a role on the sequential expression of A and B gene at a crucial moment of the cell cycle. We have not yet any fact which can support this hypothesis, but it can be supported by the observation that slow growing hepatomas show aldolase B while fast growing tumors exhibit an aldolase A activity (12, 26, 27, 28). We have cultured cells from primary explants or during subculture in medium with both D-glucose and D-fructose (1 g/l each). We have also grown cell lines on D-fructose only (1 g/l) from the very beginning of the culture, the tissue trypsinization. We could not find any correlation between aldolase B activity and the presence of D-fructose. Consequently if cells with aldolase A were not utilizing D-fructose at a high rate they should have thrived on serum lipids.

2° - Other Enzymes

A broad spectrum of enzymes has been assayed in cell line 8FR23 mainly for the glycolytic pathways common to all tissues. Enzymes were at normal level with the exception of lactate dehydrogenase and phosphoglycerate kinase which were decreased four times and phospho-hexose mutase which was diminished by a factor of 60, when the enzyme activity is expressed in mIU/mg protein. Glucose-6-phosphatase has been assayed in primary cultures of fetal liver (19 day-old p.c.) and of liver of 8 day-old rat (8FR1). The activities were around 3.7 mIU/mg of protein in 8 FR1 and 1 mIU/mg protein in 19pc FR1 cultures. In the line developed from the 8 FR1 primary culture the activity has been assayed from the 2nd to the 9th subculture, and was equal to 1 ± 0.54 mIU/mg protein. The activity of the enzyme has been assayed on liver and liver mince remaining in the flask after trypsinization of 8 day-old rat; the values were respectively 70 ± 4 mIU/mg protein and 69.4 ± 1.7 mIU/mg protein. This enzyme therefore is sharply decreased at the stage of primary culture and is quite completely lost at the first subculture. This loss of activity appears to be associated with the first mitosis undergone by the cultured cells.

3˚ - Steroid Metabolism

Liver is the site of important and fundamental metabolism of steroid hormones which is increasingly thought to be more than a catabolic process to help in the excretion of inactive steroid metabolites through bile urine. In rat steroid hormone metabolism is different between adult male and female and is sex-dependent as shown by the early observation of Forchielli et al. (29). Hormone metabolites arising from corticosterone, testosterone, androst-4-ene-3,20-dione and pregnenolone are very different in both sexes and are the result of a sexual differentiation of liver enzymes leading to male or female patterns. Much work has been done on the phenomenon of testosterone imprinting at birth which will reveal its effects only at adulthood (30, 31, 32). Recently, Gustafsson and Stenberg (33, 34) have demonstrated the existence of relations between gonads, hypothalamo-pituitary axis and liver in the induction of steroid hydroxylases specific either of male or female liver.

In an attempt to use cultured liver as a biological system to study the mechanism of the sexual differentiation, we have carried out studies of metabolic pathways of steroid hormones which should be present in cultured liver cells while in the mean time we were investigating the endogeneous steroid hormone metabolites in the organ in male and female before and after puberty (18, 35). For these studies, we have developed techniques of gas chromatography-mass fragmentography (GC-MF) (36) which allowed us to work on few centigrams of fresh tissue fractionating and quantitating methyloxime and trimethylsilyl derivatives of steroids at the sub-nanogram level.

A - Metabolism of Corticosterone

Corticosterone or compound B has been studied under its free form or conjugated form as 21-sulfate. Cell lines, 8 FR12 and 18 FR12 were incubated in serumless medium during 72 hours with 0.14 nM of $4-^{14}C$-corticosterone (56,7 mC/nM) per Cooper dish. In serumless medium the cells can survive for at least 5 days. Steroids were extracted from the medium after incubation and separated by TLC. Radioactive spots are located by radioscanning and derivatized steroids are fractionated and quantitated by CG-MF on specific ions.

No conjugated steroids have been found but the following free 11-dehydrocorticosterone appeared as the major metabolite and small amount of 5 α-dihydrocorticosterone (5α-DHB) (18).

Corticosterone $\xrightarrow{5\alpha\text{-reductase}}$ 5α-dihydrocorticosterone (5α-DHB)

$\xrightarrow{\boxed{11\beta\text{-dehydrogenase}}}$ 11-dehydrocorticosterone (compound A)

Incubation of 4-^{14}C-corticosterone-21-sulfate has been done the same way with 15 FR7 and 6 FR8 line in serumless medium. More complex pathways were followed by this substrate and they are still under investigation. Preliminary results (37) show that, depending on the cell line and the time of incubation compounds, we identified in addition to 5α-DHB and A all as 21 sulfate conjugates, three isomers of tetrahydrocorticosterone: 3α, 11β, 21-trihydroxy-5α-pregnan-20-one (3α, 5α-THB), 3α, 5α-THB and 3α, 5β-THB. Other compounds are still under identification like those in disulfate metabolites, and hydroxylated derivatives of corticosterone.

In all these experiments only 10% of initial corticosterone remained unmetabolized in the culture medium. These experiments show that liver cells in culture: i) cannot conjugate free corticosterone as 21-sulfate, ii) can metabolize 21-sulfate corticosterone in more varied metabolites than free corticosterone, iii) that different cell lines have different patterns, iv) that in the case _in vivo_, missing compounds in the metabolic pathways are due to complete utilization of intermediate substrate, v) that our liver cells maintain in culture most of the pathways of corticosterone metabolism.

B - *Metabolism of Testosterone*

This male hormone was actively metabolized by various cell lines in serumless medium: 8 FR and 18 FR lines at 7th, 11th and 17th passages and the 6 FR18 line. Quantitation of the amount of 4-^{14}C-testosterone metabolized and specific activity calculation have been done by CG-MF. By isotope ratio measurements (38), the ^{14}C label was monitored as heavy isotope compound with a mass equal to M + 2 and was compared to the cold compound of mass equal to M. For quantitative assay the metabolite formed during incubation was derivatized as a perdeuterated trimethylsilyl compound, i.e.: 5α-dihydrotestosterone-17-OSi $(CD_3)_3$ with a mass equal to M+9 while a known amount of internal standard i.e. 5α-dihydrotestosterone- 17-O-Si $(CH_3)_3$ with a mass equal to M was added to the sample. The comparison of the two peaks M and M+9 allowed to quantitate the amount of biological dihydrotestosterone.

-Incubation in the 8 FR and 18 FR line allowed to elucidate the following pathways (18):

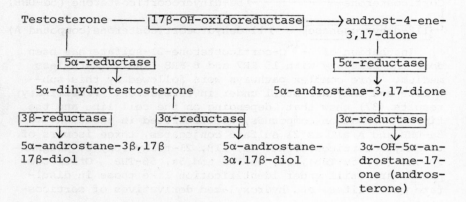

Measurement of specific activity (S.A.) by GC-MS have shown that compound had the same S.A. than 4-^{14}C-testerone. From 4.5 nM of labeled testosterone added per Cooper dish 60% remained unmetabolized, 20% was transformed in 5α-dihydrotestosterone, 10% in androsterone and 10% in the other metabolites.

-Incubation of 4-^{14}C-testosterone with another line, 6FR18, gave a richer metabolite pattern. Twenty % of testosterone has been metabolized in 5α- androstane-triol by reduction of the 3-oxo group (3α and 3β-reductase) and hydroxylation in three different positions 2β, 7α and 16α. Other compounds as androstanediolone and androstadione are under structural elucidation by gas chromatography - mass spectrometry.

These findings show that our cells behave according to a female pattern because compounds in majority are 3α and 5α hydrogenated isomers (31,35). In male liver the major route for reduction of testosterone is the 5β-reduction of the double bond of the glucuro-conjugate by microsomial enzymes (39). Nevertheless, in these cultures some male pattern was expressed through 2β and 16α hydroxylation,, 7α-hydroxylation being not sex specific.

It may be concluded from these preliminary results from corticosterone and testosterone incubation in cell line from rats between 6 to 18 day-old that a specific pattern cannot be expected. Cells at the time of incubation were at the 70th to 250th generation depending on the cell line from the first inoculum and have severed any dependence upon other tissues which may have regulatory effects: adrenals, gonads, hypothalamo-pituitary axis but are subjected to action of steroid hormones and other hormones brought by human and fetal calf sera. Nevertheless, these exogenous hormones are too low to have any detectable effect in diluting endogenous metabolites. Experiments performed by incubation of cells without added hormones did not show the presence of any appreciable amount of steroid hormone metabolites.

The metabolite production shows that at least 12 different enzymes are active in these post-natal and prepubertal rat liver cell culture and that many others will be elucidated. Hydroxylation reactions are significant metabolic process in addition to the classical reducing metabolism devoted to emunctory excretion. It is our conviction that, although more systematic investigations are needed the multiple steroid enzymatic metabolisms displayed augur profitable use of these cells to study the mechanism of hormone action as determinants of sex enzymatic differentiation.

C - Metabolism of Progesterone (25,40)

Progesterone is a key role steroid, i) as an intermediary metabolite in all steroid hormone secreting tissues where it is not excreted as an hormone, ii) as one of the most important hormones during gestation. Therefore, liver during its development is dedicated in various ways to the metabolism of this steroid. Especially during gestation fetal and maternal livers have different metabolisms towards progesterone. For this reason, we have developed cell lines from liver 18- day-old p.c. to the very moment of parturition day by day. Incubation of 100 ng/dish of $4\text{-}^{14}C$-progesterone (29, 3 mCi/mM) were done for 60 hours once in complete medium and thereafter in Ham F10 synthetic medium. Steroid metabolites excreted in the medium were extracted the usual way and submitted to TLC and GC-MF analysis. No conjugates were found.

-Incubation of 6 FR9 (25) line in complete medium: 40% of the initial radioactivity was recovered as a single metabolite: formed by reduction of the 4-5 double bond, the isomeric form α or β being not yet known. The remaining radioactivity was unmetabolized progesterone:

Progesterone ——— |5α-reductase| ———→ 5α-pregnane-3,20-dione

= Incubation with fetal liver cell lines. 19pc FR6 (serumless). All subsequent studies were done in synthetic culture medium. The metabolic route were:

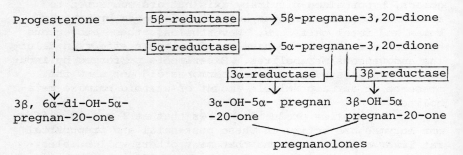

pregnanolones

Most interesting was the production of the 6α-hydroxylated compound which may be synthesized according to the following route (41) because 6α-hydroxylase does not exist.

Progesterone →6β-OH-progesterone→ pregn-4-ene-3,6,20-trione
———→6α-OH-progesteron ———→3β, 6α-diOH-5α-pregnane-20-one.

Ninety % of progesterone was metabolized leading to the production of 2/3 of 6α-hydroxylated metabolite and 1/3 or pregnanediones and pregnanolones for these last compounds 3β,5α form was the most important while the 3α,5α was at trace amount.

= Incubation with postnatal cell line 6 FR9 (serumless) (25). Same pathways were found with 90% progesterone metabolized in 50% of 6α-hydroxylated compound and 50% of pregnanediones and pregnanolones for these last compounds the ratio of 3β,5α/3α,5α isomers was equal to 2/1.

= Incubation with adult male rat cell line 7mnFR6 (serumless) (40). Again the same pathways were found except that 30% of the radioactivity found was representing the 6α-hydroxylated compound and that another metabolic peak not detected before was present which may represent another type of reduction. Quantitative and structural analysis are still in progress.

In conclusion, cultured liver keeps the functions of fetal liver: producing a 6α-OH derivation of progesterone.

Albeit, this metabolic pathway is normally absent in adult liver in vivo it was still function in adult liver cell culture. We can notice a fall of production from 70% in fetal cell culture to 30% in adult cell culture. The 5α-reduction is more prominent in female than in male rats but the 3β reduction is on the contrary a male characteristic (35). These 2 reductions are both expressed in the cultured cells. These findings may support the hypothesis that highly differentiated liver cells once they have severed connections with extracellular interactions which regulate the expression of their genome are able to express functions which normally are active at certain developmental periods of the organ.

Conclusion

In the field of methodology, these investigations have reached the stage that: 1) Liver cell cultures are now realized in a routine fashion yielding rapidly cell lines free of "fibroblast" which can be frozen in liquid nitrogen, thawed and plated again with nearly 100% success. 2) Metabolite analysis with the potent resources of gas chromatography - mass fragmentography given by the mass spectrometer allow to reach the range of separation, detection and quantitation required for work on a few culture dishes to elucidate the biochemical mechanisms linked to differentiation.

In the field of cell physiology, studies carried out in parallel in vivo and in vitro will be expected: 1) To reveal what cell and tissue interactions may remain concealed in vivo and how the presence or absence of organ-specific function in cultured cells is regulated. 2) To make progress in elucidating the complex cellular mechanisms of metabolic regulations owing to the reciprocal help arising from the study of both biological systems.

In the field of tissue culture, liver cells represent a system of choice to study direct action of effectors which can work at a high level of regulation, i.e. genome replication and/or transcription and translation which determine either normal enzymatic differentiation or abnormal disdifferentiation (28) in carcinogenesis.

This work was supported by an equipment grant from the Direction de l'Enseignement Superieur, through Vth and VIth National Plans and of research contracts from the Institut National de la Sante et de la Recherche Medicale 71.1.433.2 and 73.1.519.AU from the Delegation Generale a la Recherche Scientifique et Technique 72.7.0473, from the Centre National de la Recherche Scientifique, Equipe de Recherche Associee 267, from the Fondation pour la Recherche Medicale Francaise and from the Ligue Nationale Francaise contre le Cancer.

We gratefully knowledge the technical skill of Miss Odile Bazerolle, Mrs. Nicole Pitoizet and Mr. Rene Gautheron.

Our thanks are expressed to Dr. W.F.H.M.M. Mommaerts for helping us during the preparation of the manuscript and for his exemplary contributions to molecular interactions in eukaryotic cell physiology.

References

1. Lewis, M.R. and Lewis, W.H. Anat. Rec., 5 (1911) 277.
2. Carrel, A. Berl. Klin. Wschr 48 (1911) 1364.
3. Coon, H.G. Carnegie Institution of Washington Year Book 67 (1969) 419.
4. Gerschenson, L.E., Anderson, M., Molson, J. and Okigaki, T. Science 170 (1970) 859.
5. Williams, G.M., Weisburger, E.K., Weisburger, J.H. Exptl. Cell Res. 69 (1971) 106.
6. Lambiotte, M., Susor, W.A. and Cahn, R.D. Biochimie 54 (1972) 1179.
7. Iype, P.T. J. Cell Physiol. 78 (1971) 281.
8. Bissel, D.M., Hammacker, L.E. and Meyer, U.A. J. Cell. Biol. 59 (1973) 722.
9. Blostein, R. and Rutter, W.J. J. Biol. Chem. 238 (1963) 3280.
10. Shapira, F., Schapira, G. and Dreyfus, J.C. C.R. Acad. Sci. Paris 254 (1962) 3143.
11. Schapira, F., Dreyfus, J.C. and Schapira, G. Nature 200 (1963) 995.
12. Schapira, F., Hatzfeld, A. and Reuber, M.D. Cancer Res. 31 (1971) 1224.
13. Walker, P.R., Bonney, R.J., Becker, J.E. and Potter, V.R. In Vitro 8 (1972) 107.
14. Potter, V.R. Cancer Res. 32 (1972) 1998.
15. Padieu, P., Barbier, F., Chessebeuf, M., Cordier, D., Gerique, M., Lallemant, C. and Olsson, A. 1ere Conference Internationale sur la Differenciation Cellulaire, Nice 13-16 Septembre 1971.

16. Padieu, P., Lallemant, C., Barbier, F. and Chessebeuf, M. Proceedings 7th Meeting Fed. Europ. Biochem. Soc. 20-25 Sept., Varna (Bulgary) 863 p. 298A. (1971).
17. Chessebeuf, M. Olsson, A., Bazerolle, O. and Padieu, P. C.R. du Seminaire sur les aspects moleculaires de la differenciation des cellules eucaryotes en culture. Publication de l'INSERM, Paris (1974).
18. Bournot, P., Maume, G., Maume, B.F., Chessebeuf, M., Olsson, A. and Padieu, P. Proc. Int. Symp. on Mass Spectrometry in Biochemistry and Medicine, Frigerio, A. Ed. Raven Press, New York (1974) p. 151.
19. Ham, R.G. Proc. Nat. Acad. Sci. US, 53 (1965) 288.
20. Cahn, R.D., Coon, H.G. and Cahn, M.B. in Wilt, F.M. and Wessels, N.K. (eds) Methods in Developmental Biology, Thomas Y. Crowell, New York (1967) p. 493.
21. Perissel, B., Chessebeuf, M., Malet, P. and Padieu, P. C.R. Acad. Paris, 277 (1973) 2429.
22. Perissel, B., Malet, P., Charbonne, F., Padieu, P. and Chessebeuf, M. Colloque Soc. Franc. Microscopie Electronique. Rennes (France) 27-30 Mai 1974.
23. Perissel, B., Malet, P., Chessebeuf, M., Padieu, P. and Geneix, A. Monolayer Cell Cultures of Post-Natal Rat Liver: an Ultrastructural Study (in preparation).
24. Perissel, B., Charbonne, F., Joffroy, J.Y., Chessebeuf, M. et Turchini, J.P. Colloque Soc. Franc. Microscopie Electronique, Rennes (France) 27-30, Mai 1974.
25. Guiguet, M., Desgres, J., Turc, C. and Padieu, P. Table Ronde sur la culture de cellules, Strasbourg (France), 18-19 Janvier 1974.
26. Nordman, Y., Schapira, F., Eur. J. Cancer 30 (1967) 247.
27. Schapira, F., Reuber, M.D. and Hatzfeld, A. Biochem. Biophys. Res. Commun., 40 (1970) 321.
28. Sugimura, T., Matsushima, T., Kawachi, T., Kogure, K., Tanaka, N., Miyake, S., Hozumi, M., Sato, S. and Sato, H. in Weinhouse, S. and Ono, T. (eds), Isoenzymes and Cancer, University Park Press, Baltimore, (1972) p. 31.
29. Forchielli, E., Brown-Grant, K. and Dorfman, R.I. Proc. Soc. Exp. Biol. Med. 99 (1958) 594.
30. De Moor, P. and Denef, C. Endocrinology 82 (1968) 480.
31. Schrieffers, H., Ghraf, R. and Lax, E.R. Hoppe Seyler's Z. Physiol. Chem. 353 (1972) 371.
32. Begue, R.J., Gustafsson, J.A. et Gustafsson, S.A. Eur. J. Biochem. 40 (1973) 361.
33. Gustafsson, J.A. and Stenberg, A. J. Biol. Chem. 249 (1974) 711.

34. Gustafsson, J.A. and Stenberg, A. J. Biol. Chem. $\underline{249}$ (1974) 719.
35. Bournot, P., Maume, B.F. and Padieu, P. Biomed. Mass Spectrometry $\underline{1}$ (1974) 29.
36. Maume, B.F., Barbier, F., Bournot, P., Lhuguenot, J.C., Maume, G., Prost, P., Padieu, P. and Baron, C. Anal. Chem. $\underline{45}$ (1973) 1073.

MICROSOMAL FUNCTIONS AND PHENOTYPIC CHANGE IN ADULT RAT HEPATOCYTES IN PRIMARY MONOLAYER CULTURE

D.M. Bissell and P.S. Guzelian

Introduction

    The technique of tissue and cell culture has proven useful for serial cultivation of cells from most mammalian organs, including liver. Its value, however, frequently is limited by the fact that functional change commonly occurs in culture (1,2). This propensity for functional alteration is particularly evident in highly specialized cells such as those of the hepatic parenchyma, which <u>in vivo</u> carry out numerous synthetic and catabolic functions. Though seldom reported specifically, functional loss in hepatocyte culture can be inferred from the small number of processes that have been detected in cell lines either from hepatomas or from normal liver. Synthesis of serum proteins is exhibited by a number of liver cell lines (3-6) and appears to be a relatively stable function in culture. Similarily, steroid-inducible tyrosine aminotransferase activity has been found both in hepatoma cell lines (3,7) and in a line derived from normal liver (8). One hepatoma cell line with an unusually long generation time, also conjugates testosterone (3) or bilirubin (9). Findings such as these in a cell line, however, appear to be exceptional. As a rule, liver-specific function declines broadly and is frequently absent in cultured hepatocytes.

    For the goal of culturing widely differentiated hepatocytes <u>in vitro</u>, it would be important to characterize precisely the functional changes that occur. However, an established cell line may not be the best approach to the problem, since development of a liver cell line usually requires many mitotic cycles before the initial inoculum has grown to a size sufficient for biochemical analysis. Therefore, at the earliest possible point of study, the cell material is already days or weeks and several cell generations removed from the <u>in vivo</u> state. Adaptive events that occur early in culture may be missed; moreover, genetic change may supervene, further complicating the analysis of functional change.

    For these reasons—also, in order to study hepatic functions not usually expressed by liver cells in continuous culture—we have developed methods for primary culture of parenchymal cells from adult rat liver (10). By these techniques, fully confluent monolayer cultures are rapidly achieved, which contain ample cell material for frequent assay of specific functions from the time of plating of the cells.

In this paper, we review the method for preparation of adult rat hepatocytes in primary monolayer culture and certain characteristics of the cultured cells. Virtually every hepatic function investigated to date is measurable in this system, although the level of activity of individual functions may vary with reference to the liver in vivo. Some remain near the in vivo level and are stable for several days, while others change within hours of the time the cells are placed in culture. Examples of the latter phenomenon will be discussed as reflecting phenotypic change in liver cell culture. In this regard we have studied in some detail changes in the activity of the so-called microsomal drug metabolizing system and in the concentration of its central component cytochrome P-450. The findings appear to have implications both for liver cell culture in general, and for normal hepatic physiology.

Methods

Preparation of monolayer cultures of parenchymal cells from adult rat liver. The details of the procedures for monolayer preparation have been published elsewhere (10) and are summarized in Fig. 1. In brief, a male Sprague-Dawley rat, weighing 225-275 g, is subjected to a standard 2/3 partial hepatectomy. Four days later, after regeneration of the liver and return of the rate of DNA synthesis essentially to normal, the animal is anesthetized with ether and the regenerated liver is perfused with 0.05% collagenase. The perfusion is carried out essentially as described by Berry and Friend (11), except that hyaluronidase is omitted from the perfusion solution and phosphate buffer is used (10). Perfusion with collagenase reduces the liver to a suspension of mostly single cells, from which parenchymal cells are readily separated from debris and from other cell types by low-speed centrifugation. Aliquots of the cell isolate, containing about 5×10^6 cells, are plated without delay in 60 mm plastic tissue culture dishes, in a total volume of 3 ml. The plates are swirled carefully to ensure even distribution of the cells over the surface of the plate and are then incubated under air at 35° C.

Nutrient medium for monolayer formation and incubation. The monolayer system was designed for precise study of metabolic regulation in liver cells; thus, a medium containing

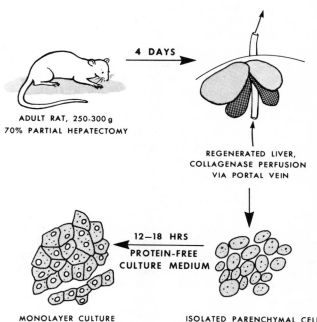

FIG. 1:

Procedure for preparation of primary monolayer cultures of adult rat hepatocytes. See text for experimental details.

serum and other poorly defined biological fluids was considered undesirable. Yet, serum usually is required for the attachment of primary cells to the surface of the culture vessel. This problem was solved by preparing the donor animal with a partial hepatectomy: cells from regenerated liver reliably formed monolayers in serum-free media. Cells from resting rat liver, prepared as shown in Fig. 1, failed to attach or to form a monolayer and rapidly lost viability, unless the medium contained 10% fetal calf serum. Even with added fetal calf serum, however, cells from resting rat liver showed reduced plating efficiency and less reliable monolayer formation compared to cells from regenerated liver. The optimal time for exploiting the effect of prior partial hepatectomy appears to be 4 or 5 days after the operation. Not only is monolayer formation most satisfactory at this time, but the functional perturbations that may accompany partial hepatectomy have largely subsided (10).

The medium employed initially, as the monolayer system was being developed, was Leibovitz' L-15, which is commercially available (Microbiological Associates, Inc., Bethesda, Md.). More recently, we have used several modifications of L-15, which we prepare in our own laboratory, using glass-distilled water and the best available grade of amino acids (Sigma Chemical Co., St. Louis, Mo., or Calbiochem, Los Angeles, Calif.). The carbohydrates (galactose and pyruvate) of standard L-15 may be replaced by 5 mM glucose; in addition, all of the nonessential amino acids, including glutamine, may be omitted, without affecting monolayer formation. The details of this work will be reported elsewhere.[1] In general, the maintenance of adult hepatocytes in primary culture may require nutritional conditions that differ qualitatively and quantitatively from those required for the continuous proliferation of typical liver cell lines.

Analytical procedures. Published methods were used to measure NADPH-cytochrome C reductase (12), cytochrome P-450 (13), aminopyrine N-demethylase (14), benzo(a)pyrene hydroxylase (15), aniline hydroxylase (16), and heme oxygenase (17). Other experimental details are included in the legends to the figures.

Results

Hepatocytes from adult rat liver were isolated and placed

---

[1] Guzelian, P.S., and D.M. Bissell, in preparation.

in culture, as described in Fig. 1. Microscopic inspection of the cell plates revealed that, with optimal plating density, the round cells lay as a single layer on the bottom of the plate tangentially in contact with adjacent cells (Fig. 2a). After 4 to 6 hours of incubation, the cells attached to the plate and, concomitantly, exhibited a morphologic change from spherical to cuboidal, coming into extensive contact with neighboring cells after 12 to 18 hours of incubation (Fig. 2b). A variable number of cells, usually less than 20% of the total, failed to attach but remained spherical and were found to be non-viable on testing with trypan blue dye (10). Such cells were largely removed with changes of the culture medium.

Synthesis of DNA and protein by cultured hepatocytes. When monolayers were examined for mitotic figures at intervals during the first 48 hours of incubation, only one mitosis in approximately 10^5 cells was found. Synthesis of DNA was assessed during this period by the incorporation of ^{3}H-thymidine into acid-insoluble material. The apparently high rate of DNA synthesis at the time of plating (Fig.3) may be an artifact, at least in part. The freshly isolated cells exhibit increased permeability, relative to cells established in monolayer culture, to a variety of small molecules (unpublished data). Thus, the intracellular concentration of labeled DNA precursor may be higher in the fresh cells than in the same cells after incubation for a few hours in monolayer. In fact, the radioactivity in the TCA-soluble ("pool") fraction of the first point shown in Fig. 3 was approximately twice that of the other points, which were all similar in magnitude. On the other hand, the actual rate of DNA synthesis may be increased, since the cells are derived from regenerated liver and, at 4 days after partial hepatectomy, may still synthesize DNA at a somewhat faster rate than does resting rat liver. In any case, the rate of DNA synthesis during the initial period of monolayer formation clearly shows no stimulation; on the contrary, it may fall significantly. However, the rate of incorporation of ^{14}C-leucine into protein by monolayers prepared from the same batch of cells remains constant or increases (Fig. 3). Thus, lack of proliferation is characteristic of hepatocytes cultured by this method, as, indeed, it is characteristic of hepatic parenchymal cells in vivo. The fact that the rate of protein synthesis appears to be independent of DNA synthesis is in accord with recent findings in other mammalian cell systems (18).

Functional adaptation of hepatic parenchymal cells to monolayer culture. Hepatocyte monolayers are viable from 4

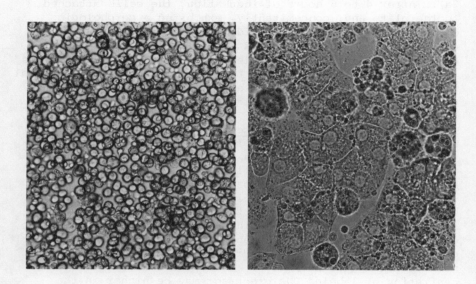

FIG. 2a:

<u>Photomicrograph of freshly isolated hepatic parenchymal cells, which demonstrates the plating density usually employed for monolayer formation</u> (unfixed, unstained preparation, 250X).

FIG. 2b:

<u>The same preparation after 12 hours of incubation in culture</u> (440X). The round, coarsely granular cells that overlie the monolayer (Fig. 2b) are non-viable and are largely removed with changes of the culture medium.

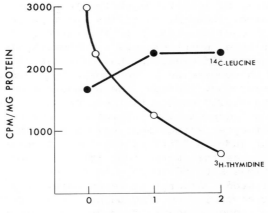

FIG. 3:

Synthesis of protein and DNA by hepatocytes in primary monolayer culture. Freshly isolated cells (Day "0") and monolayer cultures at various time points after cell plating were incubated with ^{3}H-thymidine or with ^{14}C-leucine for 30 minutes. At the end of the incubation, the labeled medium was removed; the cells were washed twice with cold, unlabeled medium and exposed to 10% trichloroacetic acid (TCA), at 4°C. The acid-precipitated material was washed twice with 5% TCA and, in the case of labeled protein, extracted once with ethanol:ether (1:1) (28). The washed precipitates were dissolved in 2% sodium carbonate in 0.1 M NaOH. An aliquot of this solution was assayed for radioactivity and for protein.

days to 2 weeks or longer, depending in part on the density of cells plated. Regardless of the total period of viability, functional adaptation is initiated as the cells are placed in culture and progresses with remarkable reproducibility among different batches of cells. These functional changes are unaccompanied by any morphologic change, as seen by light microscopy. Indeed, the appearance of the cultured cells is essentially constant, there being no growth of fibroblasts or morphologic conversion of the cells from a typical hepatocyte, a fact that serves to emphasize the necessity for judging "differentiation" in culture by criteria other than morphology (5). In Table I, a number of the functions studied to date in the hepatocyte culture system are listed. Several of these remain near in vivo activity for several days, including albumin synthesis and secretion, glucogeneogenesis from 3-carbon precursors, glycogen synthesis, responsiveness of the cells to insulin and glucagon and inducibility of p-nitroanisole O-demethylase, a microsomal mixed-function oxygenase (10). We have studied also bile salt conjugation, uptake and storage of sulfobromophthalein and heme catabolism to bilirubin, all of which appear to be maintained near the expected in vivo level.

By contrast with these functions which are relatively stable, other processes change almost from the time of plating and after a few hours of incubation exhibit marked deviation from initial levels, as determined by serial assays in a single batch of cells. A striking example of this phenomenon involves cytochrome P-450 and oxygenase reactions mediated by this microsomal hemoprotein (19). As shown in Table II, over the first 24 hours of incubation, the cellular concentration of cytochrome P-450 decreases to approximately 20% of its initial level. The ultrastructural correlate of this process is seen in electron micrographs of cells cultured for 3 days, which exhibit a decrease in endoplasmic reticulum, although other organelles appear normal both in number and in structural detail (20). Similarly, the rate of metabolism of aminopyrine and aniline, whcih are drug substrates for the osygenase system mediated by cytochrome P-450, decrease in parallel to the concentration of cytochrome P-450.[1]

While the changes in cytochrome P-450 and related processes are striking, they are not indicative of generalized metabolic deterioration of the cultured cells. The preservation of numerous functions, listed above, and of overall protein synthesis (Fig. 3) attest to the viability of the cells. Furthermore, on survey of other microsomal functions, it appears that loss of cytochrome P-450 and of the drug-

Albumin synthesis and secretion	Bile salt synthesis, conjugation, secretion
Glycogen synthesis (insulin responsive)	Sulfobromophthalein uptake and storage
Glycogenolysis (glucagon responsive)	Heme synthesis
Gluconeogenesis from lactate or pyruvate	Bilirubin production, conjugation
Cholesterol synthesis	Drug Metabolism
Fatty acid and triglyceride synthesis	

TABLE 1. SELECTED FUNCTIONS OF ADULT HEPATOCYTES IN PRIMARY MONOLAYER CULTURE

	Cytochrome P-450, nmoles/mg protein	Glucose-6-phosphatase, g P/15 min/mg protein	Heme oxygenase, nmoles bilirubin formed/min/10 mg protein
Whole liver _or_ Freshly isolated hepatocytes	0.14 ± 0.05	7.2 ± 1.3	0.12 ± 0.04
Monolayer culture, 24 h incubation in L-15 medium	0.03 ± 0.01 (21%)	6.0 ± 2.2 (83%)	0.52 ± 0.10 (430%)

All measurements were carried out in 10,000 x g supernatant fraction of extracts prepared as described previously (10). The figures in parentheses indicate percent change of concentration or activity in the incubated cells, relative to fresh cells.

TABLE II. MICROSOMAL ENZYME ACTIVITY DURING MONOLAYER FORMATION

metabolizing activities mentioned above is a selective process, not shared by several other microsomal systems. As shown in Table II, glucose-6-phosphatase changes relatively little in culture. Another enzyme, p-nitroanisole O-demethylase, which is believed to involve cytochrome P-450, exhibits variable activity, often showing little change in culture and is inducible by polycyclic hydrocarbons (Fig. 4), while cytochrome P-450 is not.[1] Heme oxygenase, another microsomal activity which appears to catalyze the conversion of heme to bilirubin (17), was also measured. In distinct contrast to the fall in cytochrome P-450, this activity increases approximately 4-fold during the first 24 hours of culture (Table II). The stimulation of p-nitroanisole metabolism by polycyclic hydrocarbons and the spontaneous rise in heme oxygenase during culture are both inhibitable by cycloheximide, which implies that these processes require protein synthesis.

Discussion

It is virtually axiomatic that liver cells in culture lose many of the characteristic features of the liver in vivo. Thus, the technique still lacks general acceptance as an in vitro method suitable for analysis of normal hepatic physiology or pathophysiology. Moreover, a systematic study of functional change in liver cell culture has not been possible, because of the problem of separating, in rapidly proliferating cells, genetic from phenotypic changes. The importance of this distinction is that phenotypic change is potentially reversible or preventable with appropriate modification of the conditions of culture. Primary culture of adult rat hepatocytes provides a way to study this problem, since the functional changes that occur in this cell system almost certainly are purely phenotypic. As shown by the studies of cytochrome P-450, they occur within a few hours after plating of the cells in vitro, which is outside the time frame of genetic change. In addition, the primary cultures are non-proliferating, so that it is highly unlikely that genetically altered cells in the plates would be numerous enough to dominate the overall functional or metabolic expression of the culture. Also, by virtue of the technique used for preparation of cultures, the system is well suited for study of early phenotypic change. A mass culture is achieved rapidly, with sufficient cell material being generated by a single rat liver to permit multiple or serial observations. Finally,

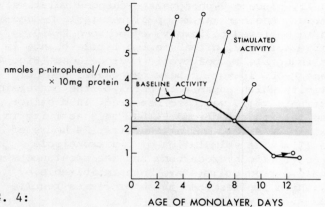

FIG. 4:

Stimulation of p-nitroanisole O-demethylase activity in cultures hepatocytes by benzo(a)pyrene. Between 2 and 11 days of incubation, the culture medium was supplemented at various times with benzo(a)pyrine, 1 µg/ml, and activity was assayed 36 hours later. The shaded area indicated the mean ± 1 S.D. of enzyme activity in liver similar to that from which the cultures were prepared. From Bissell, et al. (10).

the use of serum-free media makes possible precise definition of conditions that modify phenotypic change.

The data presented have indicated the nature and magnitude of the phenotypic change that involves microsomal function in primary hepatocyte culture. Mitochondrial and cytosolic functions also can be identified, which change comparably and with a time course similar to that of cytochrome P-450 (unpublished data). The occurrence of phenotypic change in cultured hepatocytes clearly has implications for conventional (proliferating) liver cell culture; in addition, it may have value as a novel tool for studying normal hepatic physiology.

Proliferating liver cell culture clearly differs from this primary hepatocyte system in many particulars. It should be noted, however, that a liver cell line is initiated as a primary culture. In fact, the crucial transition from in vivo to in vitro may be very similar for the two kinds of culture, since phenotypic change occurs to a similar degree under most standard conditions of culture (see below). Furthermore, without specific modification of the culture medium to prevent such changes, the altered function presumably would persist as the cells divided, unless genetic change supervened to modify it still further. Hence, the possibility exists that several of the specific deficiencies of liver cell lines might stem from phenotypic change in primary culture.

We have analyzed the microsomal fraction both from primary hepatocyte culture after 24 hours of incubation and from a permanent cell line, HTC cells, derived originally from a transplantable hepatoma (7). The levels of cytochrome P-450 and cytochrome b_5 and the activity of heme oxygenase and of NADPH-cytochrome C reductase in microsomes from both sources deviate markedly from those found in vivo; moreover, the relative magnitude of the change for each microsomal component is remarkably similar in the two preparations.[2] While not excluding a genetic mechanism, these data strongly suggest that phenotypic change underlies these altered microsomal functions of HTC cells. Thus, similar phenotypic change may occur in any primary liver cell culture, regardless of whether the initial cells are destined for primary culture only, or for proliferation and passage in vitro. Interestingly, a rough correlation can be drawn between those functions that are relatively stable in primary culture and those that often persist in established

---

[2] Bissell, D.M., and L.E. Hammaker, in preparation.

liver cell lines, e.g., albumin synthesis. Conversely, a function or cell constituent, such as cytochrome P-450, which decreases rapidly in primary culture, appears to be either very low or undetectable in liver cell lines.

We have investigated the effect of numerous modified conditions of incubation on the concentration of cytochrome P-450 in primary cultures of adult rat hepatocytes. Standard media other than L-15 were tested; calf, fetal calf, or adult rat serum was added; anti-oxidants and various hormones were included in the medium. The actual penetration of a given hormone into the cells in vitro was verified by its effect on a metabolic process not related to cytochrome P-450, e.g., stimulation of glycogenolysis by glucagon (10). None of these various manipulations had any effect on the rapid decrease in cytochrome P-450 observed over the first 24 hours of incubation.

The rationale for further studies was based on data from intact animals, although the precise way in which cytochrome P-450 is regulated in the liver in vivo remains unclear. It has been suggested that the abundance of this hemoprotein represents the need of the organism for a detoxification mechanism in dealing with a chemically hostile environment. Thus, the normal level of cytochrome P-450 may reflect steady-state induction of the hemoprotein by foreign compounds. Accordingly, the loss of cytochrome P-450 in culture might be due to the absence of these inducer substances from the incubation medium. However, exposure of the hepatocytes in culture to a number of different drugs and other inducer compounds for the microsomal oxygenase system-- with or without serum and other supplements mentioned above-- was without effect on the level of this hemoprotein. Penetration of drug into the cultured cells was confirmed by the appearance of drug metabolites during incubation. Therefore, the data suggest that, though hormones or drugs may modulate the regulation of cytochrome P-450 in vivo, the regulatory mechanism itself is primarily subject to the internal metabolic structure of the hepatocyte--perhaps determined by the relative activity of different routes of carbon flow or by key ratios such as $DPNH/DPN^+$ in various cell compartments (21). Our recent experiments, directed at this more fundamental area, have been fruitful in providing clues as to the intracellular regulatory mechanism(s) for cytochrome P-450. The addition of ascorbate to the medium significantly increases the level of cytochrome P-450. Also, simplification of the amino acid component of L-15 medium has a similar effect. The latter findings suggest that nutrient imbalance, rather than deficiency, may underlie some of the functional changes that occur in cultured hepatocytes, a problem we are actively pursuing at the present time.

With data from studies of cytochrome P-450 and from analysis of other processes, it should be possible to formulate a medium more supportive of liver cell function in culture than are those presently available; in addition, insight may be gained into the regulation of hepatic function. The appearance of an altered function, such as decreased cytochrome P-450, may be "abnormal" in culture in the sense of being a significant deviation from the in vivo situation. On the other hand, it may reflect a "normal" response of the cell to the artificial (and imperfect) conditions of culture. If physiologic regulation in fact governs phenotypic change, then primary hepatocyte culture would acquire major value as a method for studying the regulation of a variety of important functions, including cytochrome P-450.

Adult hepatocytes in primary monolayer culture are as closely related to the intact animal as is physically possible in a cell culture system. While it might appear reasonable to assume that metabolic responses are similar in this cell system and in the liver in vivo, correlation of in vitro and in vivo findings is of obvious importance. The present studies of cytochrome P-450 produced an unexpected finding but one readily approached in the intact animal: the association of decreasing concentrations of cytochrome P-450 with increasing activity of heme oxygenase (Table II). Since stimulation of heme oxygenase in vivo is readily accomplished by the administration of either hemeprotein (methemalbumin or hemoglobin) or of non-heme compounds such as endotoxin (22), its association with rapid turnover of cytochrome P-450 could be tested in the intact animal. The study was carried out by monitoring the rate of degradation of cytochrome P-450 in rat liver in vivo, as described elsewhere (23), with and without administration of a stimulator of heme oxygenase. The results confirmed the in vitro findings, in that accelerated turnover of cytochrome P-450 preceded increased activity of heme oxygenase, regardless of the mode of stimulation used. These data provide at least indirect evidence that the microsomal changes occurring in primary hepatocyte culture may be regulated by physiologic mechanisms.

While the discussion thus far has been limited largely to consideration of liver cell culture, the studies should and do have ramifications for hepatic physiology in vivo. For example, the apparently reciprocal relationship between the concentration of cytochrome P-450 and the activity of heme oxygenase in the liver had not been appreciated from previous studies in vivo; on the contrary, it was held that heme oxygenase, while being atypical in some respects (24), required cytochrome P-450 in the fashion of other mixed-function oxygenases (25). The data from the primary hepatocyte system, confirmed in the intact animal, makes this hypothesis largely untenable. On the other hand, alternative models of the mechanism of heme breakdown may now be envisioned. For example, a microsomal cytochrome other than P-450 may serve as electron acceptor; or the substrate, heme, after complexing with a membrane site ("heme oxygenase") may mediate electron flow for its own destruction, in the fashion of a coupled oxidation in vitro (26). Indeed, in retrospect such models are more consistent with all the existing data on heme oxygenase that the previously held hypothesis involving cytochrome P-450. The findings also suggest that exogenous heme (e.g., hemoglobin), administered in vivo, stimulates heme oxygenase only indirectly, through its effect on the turnover of cytochrome P-450, and not by "induction" as was formerly supposed (27). While such data do not rule out a role for heme oxygenase in the breakdown of hemoblobin heme, they at least raise the possibility that other mechanisms may be responsible for this process. Finally, the data suggest that heme oxygenase more accurately reflects turnover of hepatic cytochrome P-450 than flux of exogenous heme through the liver.

Conclusion

The study of hepatic physiology in vitro has been frustrated by the fact that specific functions are radically altered in most, if not all, systems of hepatocyte culture. However, in view of the extensive participation of the liver in vivo with nutrient flow, with humoral regulation, with functions of various organs and other processes, it should not be surprising, teleologically, that transfer of these complex cells to an artificial nutrient medium causes a rather profound modification of their usual functions. For liver cell culture, the more important problem is not that function changes but that so little information exists as to the nature and mechanisms of altered function. A culture system consisting of non-proliferating adult rat hepatocytes in primary culture has been developed to examine this problem. While numerous characteristic hepatic functions are present in this culture system and remain relatively stable for several days, other functions change rapidly, in a manner consistent with phenotypic alteration of the cultured cells. One such change involves cytochrome P-450, which diminishes to 20% of its normal in vivo level over the first 24 hours of incubation in standard culture media. Detailed study of this process has demonstrated that phenotypic change is preventable by appropriate specific modification of the culture medium. Moreover, the studies appear to provide important insight into hemoprotein function and regulation in vivo. Indeed, while the many normal characteristics of adult rat hepatocytes in primary culture are worthy of attention, phenotypic change may ultimately prove more useful than normal function for probing the still vast terra incognita of hepatic physiology.

Supported in part by USPHS Grants AM-11275 and GM-21042 and by the Walter C. Pew Fund for Gastrointestinal Research.

References

1. Davidson, E.H. Adv. Genet. 12 (1964) 143.
2. Waymouth, C. Nat. Cancer Inst. Monogr. 26 (1967) 1.
3. Tashjian, A.H., Jr., Bancroft, F.C., Richardson, U.I., Goldlust, M.B. Rommel, F.A. and Ofner, P. In Vitro 6 (1970) 32.
4. Borek, C. Proc. Natl. Acad. Sci. USA 69 (1972) 956.
5. Kaighn, M.E. and Prince A.M. Proc. Natl. Acad. Sci. USA 68 (1971) 2396.
6. Ohanian, S.H., Taubman, S.B. and Thorbecke, G.J. J. Nat. Cancer Inst. 43 (1969) 397.
7. Thompson, E.B., Tomkins, G.M. and Curran, J.F. Proc. Natl. Acad. Sci. USA 56 (1966) 296.
8. Gerschenson, L.E., Andersson, M. Molson, J. and Okigaki, T. Science 170 (1970) 859.
9. Rugstad, H.E., Robinson, S.H., Yannoni, C. and Tashjian, A.H., Jr. J. Cell Biol. 47 (1970) 703.
10. Bissell, D.M., Hammaker, L.E. and Meyer, U.A. J. Cell Biol. 59 (1973) 722.
11. Berry, M.N., and Friend, D.S. J. Cell Biol. 43 (1969) 506.
12. Masters, B.S.S., Williams, C.H., Jr. and Kamin, H. In: Vol X. Academic Press, New York (1967) 565.
13. Omura, T. and Sato, R. J. Biol. Chem. 239 (1964) 2370.
14. Poland, A. and Nebert, D.W. J. Pharm. Exp. Therap. 184 (1973) 269.

15. Nebert, D.W. and Gelboin, H.V. J. Biol. Chem. 243 (1968) 6242.
16. Gang, H., Lieber, C.S. and Rubin E. J. Pharm. Exp. Therap. 183 (1972) 218.
17. Tenhunen, R., Marver, H.S. and Schmid, R. J. Biol. Chem. 244 (1969) 6388.
18. Baenziger, N.L., Jacobi, C.H. and Thach, R.E. J. Biol. Chem. 249 (1974) 3483.
19. Gillette, J.R., Davis, D.C. and Sasame, H.A. Ann. Rev. Pharmacol. 12 (1972) 57.
20. Chapman, G.S. Jones, A.L., Meyer, U.A. and Bissell, D.M. J. Cell Biol. 59 (1973) 735.
21. Williamson, D.H., Lund, P. and Krebs, H.A. Biochem. J. 103 (1967) 514.
22. Gemsa, D., Woo, C.H., Fudenberg, H.H. and Schmid, R. J. Clin. Invest. 53 (1974) 647.
23. Bissell, D.M., Guzelian, P.S., Hammaker, L.E. and Schmid, R. Fed. Proc. 33 (1974) 1246.
24. Tenhunen, R., Marver, H.S. and Schmid, R. J. Lab Clin. Med. 75 (1970) 410.
25. Tenhunen, R., Marver, H., Pimstone, N.R., Trager, W.F. Cooper, D.Y. and Schmid, R. Biochemistry 11 (1972) 1716.
26. O'Carra, P. and Colleran, E. FEBS Letters 5 (1969) 295.
27. Pimstone, N.R., Engel, P., Tenhunen, R., Seitz, P.T., Marver, H.S. and Schmid, R. J. Clin. Invest. 50 (1971) 2042.
28. Odessey, R. and Goldberg, A.L. Amer. J. Physiol. 223 (1972) 1376.

Note: We would like to thank the publishers of the J. Cell Biology for the permission to reproduce Figure 4.

BIOCHEMICAL, AUTORADIOGRAPHIC, AND ELECTRON MICROSCOPIC STUDIES OF ADULT RAT LIVER PARENCHYMAL CELLS IN PRIMARY CULTURE.

Michael W. Pariza, James D. Yager, Jr., Stanley Goldfarb, James A. Gurr, Susumu Yanagi, Steven H. Grossman, Joyce E. Becker, Thomas A. Barber and Van R. Potter

Perfusion of adult rat liver in situ with collagenase, first introduced by Berry and Friend (1), appears to be the best available method for isolating adult liver parenchymal cells (2). This technique gives high yields of biochemically active single liver parenchymal cell suspension, virtually free of stromal cells (1,3-10). Such preparations have been used immediately in biochemical studies (1,2,7,9,10) or to prepare cultures of growing cells (12). However, both of these protocols are open to criticism and we wish to present evidence in favor of a third approach.

The use of freshly isolated hepatocytes for biochemical studies that require only 1 or 2 hour incubation periods may seem attractive because manipulation is held to a minimum and microbial contamination is not an issue. However, such cells have shown evidence of membrane damage (1,2,7) and some loss of response to hormonal stimulation (3,8,9). Furthermore, ATP levels in freshly isolated hepatocytes were depressed (3), amino acid transport may have been abolished (2), and some investigators (9) have reported metabolic data consistent with a loss of cell energy. These observations are not surprising since one might expect cells to incur some injury as a result of the isolation procedure, and it is reasonable to conclude that such cell injury could be reflected in metabolic behavior not characteristic of normal liver parenchymal cells in the adult animal.

Selecting for cells in the suspension capable of cell division (12) may also be an unsatisfactory approach for certain types of studies. Dividing cells derived from liver lack many differentiated functions (13-17), and cannot be expected to faithfully represent metabolic regulation of liver parenchymal cells in the adult. Moreover, maintaining serial cultures of dividing cells isolated from adult liver has been observed to result in malignant transformation (18-21). Nevertheless, prior to transformation, serial cultures of diploid liver cells (20,21), may yet prove capable of maturation to the fully adult stage and deserve further study.

Recently, methods for studying adult rat liver parenchymal cells in short-term non-replicating primary monolayer cultures have been developed in this laboratory by Bonney et al. (4-6) and Pariza et al. (8), and independently in the laboratory of Dr. D. Montgomery Bissell (3). The rationale behind this approach is:
  1. The cells are given time to repair functional injuries caused by the isolation procedure, and to re-associate with each other in monolayers resembling their former in situ state.
  2. Cells unable to form monolayers (and possibly dying or dead) can be discarded or studied separately.
  3. Metabolic studies lasting several days, instead of just a few hours as with perfused liver or liver parenchymal cell suspensions, can be conducted. With large numbers of replicate cultures, protocols requiring several hours can be repeated on successive days under standard or experimentally altered culture conditions.
  4. The primary monolayer culture method appears to offer a realistic approach to the problem of obtaining normal liver cells which can be made to modulate between the proliferating (fetal) and non-proliferating (adult) states, thus providing for the in vitro study of liver ontogeny and liver oncogenesis.

This report details our latest findings and is intended in part to address basic questions concerning the relationships between various biochemical measures of viability, the conditions of culture, and cell morphology. We will emphasize the desirability of monitoring protein and RNA labeling as a necessary objective test of improvements in cell isolation and culture conditions which may not yet be optimal. Experiments are analyzed from two points of view: Incorporation of radioactive label on a whole culture and per µg DNA basis, and as percent of individual cells labeled as determined by autoradiographic (ARG) techniques. Additionally, preliminary results on the retention of adult properties by cultured hepatocytes, and studies involving the culturing of cells from regenerating liver, will be presented.

Cell Isolation and Culture Conditions
  Using the technique of Berry and Friend (1) with some modification (4-6,8), we routinely obtain 300 to 450 million single liver parenchymal cells from the liver of one 200-250 gram rat. Generally, 70% or more of the cells exclude the vital dye, trypan blue (discussion of this property is included in a subsequent section). The cells are virtually free of non-parenchymal elements (4-6,8), and are inoculated at a

density of about 500,000 nuclei per ml of culture medium
The volume of medium plus cells added to different size
dishes is varied to maintain a density of about 1000 nuclei/
mm^2 of culture surface.

Our studies to date have predominantly dealt with cells
which attach to a suitable culture surface and form monolayers. This is because adult liver cells do not survive for
more than a few hours in suspension (10), and we have found
that cells which do not attach under favorable conditions
degenerate rapidly (4-6,8).

Chart 1 shows the percent of the cell inoculum attaching
in 24 hours under several different conditions. Two different culture media were employed: Ham's F-12 (22) plus various levels of fetal calf serum, and serum-free modified Waymouth's MB 752/1 (23) plus oleate and bovine serum albumin
(24) and supplemented with additional amino acids (8). We
have previously (8) referred to this medium as H1-WO/BA, the
commercial tradename for the basal medium, but in this communication we shall designate this medium $WO/BA-M_1$ to indicate modification.

In Chart 1, both Ham's F-12 and $WO/BA-M_1$ contained 13
milliunits (0.5 µg) insulin per ml. Additionally, tissue
culture plastic dishes alone and coated with a thin layer of
rat-tail collagen were tested for suitability as substrates
for cell attachment.

In Ham's F-12 medium plus insulin, less than 20% of the
cells attached to plastic culture dishes, and less than 30%
of the cells attached when collagen-coated plastic was used
(Chart 1). However, when fetal-calf serum (1%, 5%, 10% or
25%) was added to F-12 plus insulin, attachment was between
46% and 54% and appeared independent of collagen coating.
When serum-free $WO/BA-M_1$ plus insulin was used, attachment
was 47% on collagen-coated plastic, but less than 20% when
uncoated plastic was used (Chart 1).

As stated previously, attachment was assessed at 24
hours in Chart 1. However, when $WO/BA-M_1$ plus insulin is
used in conjunction with collagen-coated plates attachment is
very rapid and appears complete within two hours of cell inoculation (Pariza, M. W., and Yager, J. D. Jr., unpublished).
We have also found that cell attachment as well as biochemical parameters which we study are not dependent upon adding
additional glutamine to the basal medium ($WO/BA-M_1$ contains
6.8 mM glutamine), and we now use the original glutamine
concentration of Waymouth (23), 4 mM, and refer to the medium
as $WO/BA-M_2$ (Waymouth's MB752/1 (23) plus bovine serum
albumin and oleate (24) and supplemented with 0.5 mM serine
and 0.4 mM alanine (8)). The experiments to be described

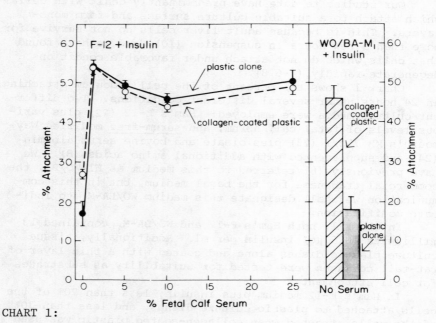

CHART 1:

Percent cell attachment at 24 hours in Ham's F-12 medium + 13 mU insulin/ml + various levels of fetal calf serum or in serum-free WO/BS-M, X X + 13 mU insulin/ml. Tissue culture plastic alone or coated with a thin layer of rat-tail collagen were the attachment surfaces. Redrawn from (8).

were performed with cells cultured in serum-free WO/BA-M$_2$ plus insulin on collagen-coated dishes, except where indicated to the contrary.

Tests of Viability

The cultured liver parenchymal cells from normal rats do not synthesize appreciable amounts of DNA and do not divide, thus behaving like normal liver parenchymal cells in vivo. Because of this property, biochemical tests of viability other than replication must be employed. We have investigated the uptake and incorporation of labeled leucine and cytidine as well as enzyme induction, for suitability as markers of viability.

^{3}H-leucine incorporation into protein. Chart 2 depicts the results of a study of the method we routinely use to evaluate the retention of TCA-soluble dpm from ^{3}H-leucine by cultured liver parenchymal cells. Normal cultures and cultures subjected to three episodes of rapid freezing and thawing (not shown) were incubated in medium containing 1 µCi/ml ^{3}H-leucine (0.38 mM final concentration) for 6 hours. The cultures were then washed 2 or 4 times with medium containing unlabeled leucine, and the TCA-soluble dpm observed in the freeze-thawed cultures (not shown) subtracted from the TCA-soluble dpm of the normal cultures (the freeze-thawed cultures contained 25% of the total dpm observed in the normal cultures after 2 rinses with medium, but less than 5% after 4 rinses). The corrected TCA-soluble values (dpm from freeze-thawed cultures subtracted) for the normal cultures are shown in Chart 2, and it can be readily seen that greater than 13,000 dpm were retained in the TCA-soluble fraction after 2 or 4 rinses with unlabeled medium. The TCA-insoluble fraction contained about 46,000 dpm (Chart 2).

Based on the results of this study, cultures were routinely washed 4 times with medium containing unlabelled leucine prior to extraction of TCA-soluble fractions. Freeze-thawed cultures were usually included as controls and handled in a similar manner.

Chart 3 shows the incorporation of ^{3}H-leucine into TCA-insoluble material on the 1st and 2nd days in culture. Incorporation into intracellular material was essentially linear on both days, but the rate of incorporation on the 2nd day was about twice the rate of the 1st day (Chart 3) suggesting repair of functional properties during the 1st day of culture. The level of labeled leucine in extra-cellular TCA-insoluble material also increased linearly, after a lag, on both days. The presence of this material in the culture medium suggests the synthesis and export of serum proteins, a

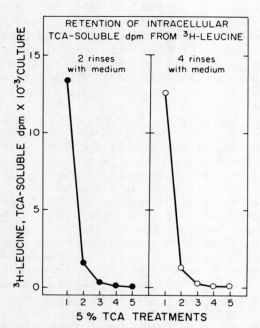

CHART 2:

Retention of intracellular TCA-soluble label from ^{3}H-leucine following 2 or 4 rinses with medium containing unlabeled leucine. Intact cultures and cultures subject to 3 episodes of rapid freezing and thawing were incubated in medium containing 1 µCi/ml ^{3}H-leucine (0.38 mM) for 6 hours and then washed 2 or 4 times with medium containing unlabeled leucine (0.38 mM) prior to TCA-treatments. Label (TCA-soluble) retained by the cultures subjected to freeze-thaw after 2 rinses (4000 dpm) or 4 rinses (600 dpm) were subtracted from TCA-soluble label retained by the intact cultures, and the results plotted above.

Label incorporated into TCA-insoluble material was 44,000 dpm (solid circles) and 48,000 (open circles). Cultures subjected to freeze-thaw contained no TCA-insoluble label.

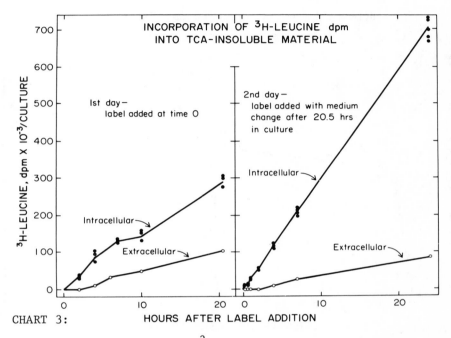

CHART 3: Incorporation of dpm from ^{3}H-leucine into TCA-insoluble material. Medium contained 1 µCi ^{3}H-leucine/ml (0.38 mM).

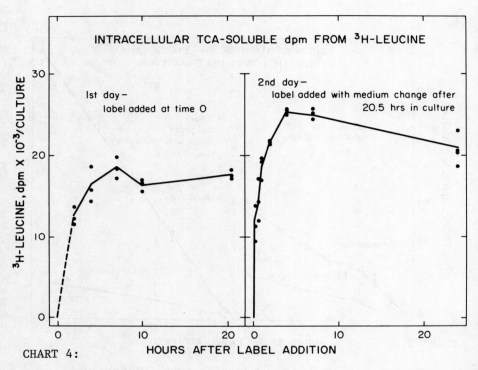

CHART 4: Intracellular TCA-soluble dpm from ^{3}H-leucine in cultures shown in previous chart.

hypothesis currently under study.

Chart 4 shows the levels of TCA-soluble ^{3}H-leucine in the same cultures, and the data correlate well with those of Chart 3. During the 2nd day in culture, cells achieved higher levels of TCA-soluble label than during the 1st day, thus supporting the hypothesis that functional properties are not optimal immediately after cell isolation, but that they improve with time in culture.

The experiment depicted in Charts 3 and 4 was performed under conditions where the medium was changed only once: at 20.5 hours after cell inoculation. However, the finding that approximately 50% of the cells do not attach (Chart 1) and appear to rapidly degenerate led us to hypothesize that a medium change at an early time point might remove inhibitory substances and thereby accelerate repair of functional properties by the cells which do attach. Table 1 summarizes the findings of two experiments designed to address this question. Medium was changed at 4 hours and/or at 24 hours, label was added at 4 hours or 24 hours, and cells were harvested at 24 or 48 hours depending upon when label was added (Table 1). Data are calculated on a <u>per culture basis</u> for comparison with Charts 3 and 4, and on a <u>per μg DNA basis</u> in order to assess the effect, if any, of variability in cell number between cultures. Taken either way, the results are unequivocal: changing the medium at 4 hours resulted in 50% increase of ^{3}H-leucine incorporation into TCA-insoluble material over not changing the medium when cultures were examined 20 hours after label addition (Table 1, Experiment 2). In contrast, TCA-soluble material did not appear to be so affected. During the 2nd 24 hour period in culture, cells exposed to medium changes at 4 and 24 hours also appeared to fare better than cells exposed to a medium change at 24 hours only: ^{3}H-leucine incorporation into TCA-insoluble material was about 10% greater (on a <u>per μg DNA basis</u>) in cultures exposed to the two medium changes (Table 1, Experiment 2). In both instances, levels of TCA-insoluble and TCA-soluble label were higher at the end of the 2nd 24-hour period than at the end of the 1st 24-hour period. Table 1, Experiment 3, shows the results of assessing incorporation of ^{3}H-leucine into TCA-insoluble and TCA-soluble fractions at the end of the 2nd 24-hour period in cultures exposed to a 4-hour medium change only or 4- and 24-hour medium changes. The incorporation of ^{3}H-leucine into both fractions was the same, indicating that the critical medium change was the 4-hour medium change.

In this experiment, it was assumed that the unlabeled leucine concentration in the medium did not change appreci-

EFFECT OF MEDIUM CHANGE ON TCA-INSOLUBLE AND TCA-SOLUBLE RADIOACTIVITY FROM ^{3}H-LEUCINE

	HOURS IN CULTURE					^{3}H-LEUCINE		dpm/µg DNA x 10^{-3}	
						dpm/CULTURE x 10^{-3}			
EXPERIMENT	MEDIUM CHANGE 4 HOURS	24 HOURS	LABEL ADDED (HOURS)	CULTURE HARVEST (HOURS)		TCA-INSOLUBLE	TCA-SOLUBLE	TCA-INSOLUBLE	TCA-SOLUBLE
2	NO	NO	4	24		320.4	18.8	20.7	1.2
	YES	NO	4	24		465.4	19.0	31.8	1.3
	NO	YES	24	48		728.4	27.6	60.1	2.3
	YES	YES	24	48		847.6	24.7	67.8	2.0
3	YES	YES	24	48		619.5	22.9	41.6	1.5
	YES	NO	24	48		628.5	26.5	41.5	1.7

TABLE 1

ably between 4 and 24 hours in cells exposed only to a 4-hour medium change. Consistent with this assumption is the finding that cells exposed to a 4-hour medium change and harvested at 24 hours incorporated only 5% of added labeled leucine into TCA-soluble and -insoluble intracellular fractions (calculated from the data of Table 1 and Pariza, M. W., unpublished.)

The data thus far presented describe the incorporation of ^{3}H-leucine into TCA-soluble and -insoluble fractions on a <u>whole culture basis</u> but they do not tell us what individual cells are doing. This question can only be approached with autoradiography.

<u>Autoradiography with ^{3}H-leucine</u>. Figure 1 is a photomicrograph of a hematoxylin-eosin stained autoradiogram of cells given a medium change after 4 hours in culture, and then exposed continuously to ^{3}H-leucine for 20 hours. The cells are from Experiment 1, Table 2.

Most of the cells (Figure 1) are well-flattened, contain nuclei with distinct nucleoli, and had incorporated label. There are a few cells with eosinophilic cytoplasm lacking morphological detail and condensed, pyknotic nuclei. These cells had not incorporated label during the periods of exposure, and appeared to be attached to other cells rather than to the collagen-coated surface.

When the medium is not changed at 4 hours, the cells appear considerably more rounded at 24 hours (not shown). However, 48 hour cultures (not shown) contain many islands of contiguous, well flattened labeled cells (as in Fig. 1) whether exposed to medium changes at 4 and 24 hours, 24 hours <u>only</u> or 4 hours <u>only</u> (^{3}H-leucine added at 24 hours). Unlabeled cells possessing abnormal staining characteristics were present in all cultures, regardless of protocol.

The presence of cells in the cultures which did not incorporate label and which appeared to be dying or dead was investigated with reference to medium change. Table 2 shows the results of three separate experiments in which the percentages of labeled cells and unlabeled or very lightly labeled cells were evaluated at 24 hours and 48 hours following media changes at 24 hours, 4 and 24 hours, or 4 hours only. Experiments 2 and 3 in Table 2 are from Experiments 2 and 3, respectively shown in Table 1.

The data indicate that the percent of labeled cells at 24 hours was not appreciably affected by a medium change 4 hours after cell inoculation. However, the percent of labeled cells at 48 hours was <u>higher</u> in cultures subjected to a 4-hour medium change <u>in all three experiments</u>. Additionally, cultures subjected to a 4-hour medium change only, labeled at

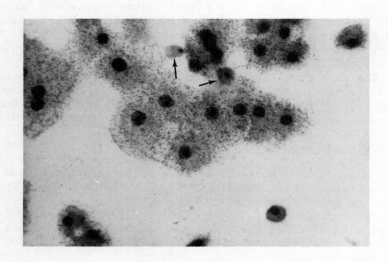

FIG. 1:

Photomicrographs of hematoxylin-eosin stained autoradiogram of cells given a medium change after 4 hours in culture and then exposed continuously to ^{3}H-leucine (1 µCi/ml, 0.38 mM) for 20 hours. Cells from Experiment 1, Table 2. Most of the cells are well-flattened, contain prominent nuclei with distinct nucleoli, and have many grains above cytoplasm. Note the four small, degenerating cells containing very few or no grains above cytoplasm (arrows). X 250.

AUTORADIOGRAPHIC STUDY OF THE EFFECT OF MEDIUM CHANGE ON CELL LABELING WITH ^{3}H-LEUCINE

EXPERIMENT	HOURS IN CULTURE AT TIME OF:			TOTAL CELLS COUNTED	PERCENT	
	MEDIUM CHANGES	LABEL ADDITION	CULTURE HARVEST		LABELED CELLS	UNLABELED CELLS
1	-*	4	24	1043	96	4
	4	4	24	1009	96	4
	24	24	48	1587	79	21
	4,24	24	48	1112	88	12
2	*	4	24	2088	91	9
	4	4	24	2679	93	7
	24	24	48	548	81	19
	4,24	24	48	551	93	7

TABLE 2 (Continued on the following page)

AUTORADIOGRAPHIC STUDY OF THE EFFECT OF MEDIUM CHANGE ON CELL LABELING WITH ^{3}H-LEUCINE

EXPERIMENT	HOURS IN CULTURE AT TIME OF:			TOTAL CELLS COUNTED	PERCENT	
	MEDIUM CHANGES	LABEL ADDITION	CULTURE HARVEST		LABELED CELLS	UNLABELED CELLS
3	—	4	24	1185	84	16
	4	4	24	1043	81	19
	24	24	48	1528	66	34
	4, 24	24	48	1041	78	22
	4	24	48	1294	77	23

*NO MEDIUM CHANGE

TABLE 2 (Continuation)

24 hours and harvested at 48 hours had no more unlabeled cells than cultures subjected to medium changes at 4 and 24 hours (Table 2, Experiment 3), supporting the conclusion that the critical medium change is the 4 hour medium change.

Double-labeling with ^{3}H-leucine and ^{14}C-cytidine

Parameters of viability for adult liver parenchymal cells during the first 48 hours in primary monolayer culture have been presented in some detail, and it is now pertinent to ask how well the cells survive beyond the first 48 hours in culture. The experiment shown in Chart 5 was designed to address this question. Under a protocol where medium was changed at 4, 24, 48, and 72 hours, ^{3}H-leucine incorporation into TCA-insoluble material and ^{14}C-cytidine incorporation into TCA-insoluble, KOH hydrolyzable material was determined with 20 minute double-label pulses, and ornithine decarboxylase (ODC) activity was assessed in sister cultures maintained in the same medium. Additionally, these cells were studied with scanning and transmission electron microscopy.

Prior to this experiment, it was determined that ^{14}C-cytidine and ^{3}H-leucine at the levels employed in the experiment in Chart 4 (3.5 µM and 0.38 mM, respectively) were linearly incorporated into TCA-insoluble material for one hour by 48 hour cultures simultaneously exposed to both labels (data not shown).

In Chart 5, the levels of incorporation of ^{3}H-leucine during 20 minute pulses increased slowly during the 1st 24 hours, appeared to "dip" after the 24 hour medium change, and then increased with a relatively steep slope through the medium change at 48 hours. At 68 hours in culture, the levels of label incorporated during the pulse period reached a plateau and remained constant through 96.5 hours.

The increase in the levels of ^{3}H-leucine incorporated between 24 and 48 hours was somewhat unexpected, as the data shown in Chart 3 suggested that the rate of incorporation was constant during this time period. However, a 20 minute pulse is likely to be more sensitive to metabolic shifts than is a cumulative value reflecting incorporation under conditions of synthetic, degradative and pool equilibria.

The incorporation of ^{14}C-cytidine into TCA-insoluble, KOH-hydrolyzable material during the 20 minute pulse periods exhibited somewhat different characteristics. During the 1st 24 hours, the levels incorporated were lower, on a relative basis, than the levels of ^{3}H-leucine incorporated, and the patterns of incorporation for the two precursors appeared similar. However, following the 24 hour medium change the incorporation of ^{14}C-cytidine during the pulse periods rapid-

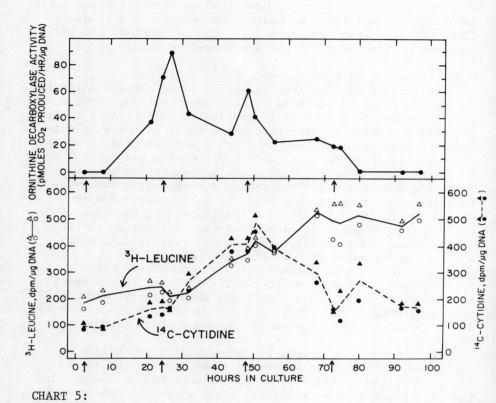

CHART 5:

Top: Ornithine decarboxylase (ODC) activity, one culture/point. Bottom: ^{3}H-leucine incorporation into TCA-insoluble, KOH-hydrolyzable fractions during 20 minute exposures. ^{3}H-leucine at 1 µCi/ml (0.38 mM), ^{14}C-cytidine at 0.1 µCi/ml (3.5 µM). Triangles (open and closed) at a time point refer to the same culture, and circles (open and closed) at a time point refer to the same culture. Arrows indicate medium changes.

ly increased and surpassed, on a relative basis, the levels of incorporation of ^{3}H-leucine. The levels of incorporation of ^{14}C-cytidine peaked at 50.5 hours, 2 hours after the third medium change, and then declined. The last medium change, at 72 hours, may have stimulated incorporation during the following 6 hours but the data are quite variable.

In addition to the results shown in Chart 5, two cultures treated with dexamethasone (1 µM) at 93 hours in culture exhibited a 240% increase in ^{14}C-cytidine incorporation at 100 hours in culture over untreated cells harvested at 93 and 96.5 hours. Incorporation of ^{3}H-leucine was unaffected (data not shown). This stimulation of ^{14}C-cytidine incorporation by dexamethasone suggests that the cells retained glucocorticoid responsiveness after 4-5 days in culture. Further investigation of this phenomenon is in progress.

Ornithine Decarboxylase Activity

Chart 5 also shows the oscillation of ornithine decarboxylase activity in sister cultures maintained in the same medium. As we previously reported (8), ODC activity was undetectable during the first few hours of culture. However, by 21.5 hours ODC activity had risen to detectable levels, and continued to rise until 26.5 hours in culture. Lower levels of activity were assayed at 30 and 44 hours, and then the ODC activity appeared to rise and peak a second time at 48.5 hours. Ornithine decarboxylase activity then declined and appeared to reach a plateau at 56 hours which lasted through 74.5 hours. No activity was detectable at 80, 92 or 96.5 hours.

The relationship between ODC activity and the incorporation of ^{3}H-leucine and ^{14}C-cytidine is not clear, but at least one conclusion is indicated: the optimal time for studying ornithine decarboxylase activity is not during the 1st day in culture, and this may also not be the best time for investigating protein and RNA synthesis. This conclusion is supported by data presented earlier (Charts 3 and 4, Tables 1 and 2). Whether conditions are indeed optimal or just better on day 2 cannot yet be ascertained, but our system as presently employed appears to respond best (with respect to the parameters discussed) during the time period of approximately 20 to 60 hours in culture. This is also the time period in which we have observed 2-2.5 fold inductions of tyrosine aminotransferase activity by glucocorticoids (4,5,8; Gurr, J. A., unpublished).

Ultrastructural Studies

Sister cultures from the experiment depicted in Chart 5

were studied with scanning electron microscopy and transmission microscopy. Figure 2 is a transmission electron micrograph of cells immediately after isolation. The plasma membrane appears intact, but the rough endoplasmic reticulum (ER) appears focally dilated. Additionally, mitochondria are dense and contracted, indicating an energy depleted state.

After four hours in culture (Fig. 3), cell morphology appears considerably improved. The rough ER has assumed a linear arrangement and mitochondria are no longer dense. Additionally, glycogen is appearing in areas of the cell periphery in association with the smooth ER.

Figure 4 is a scanning electron micrograph of cells immediately after isolation. Many rounded cells of the type shown are evident, but irregularly shaped cells are also observed, many with apparent defects resembling small holes (not shown).

After 4 hours in culture (Fig. 5), many cells have adhered to the collagen and appear to be spreading out by extension of peripheral veils of cytoplasm. Clefts between cells are often observed, and while probably artifactual, nonetheless permit evaluation of abutting faces of cells. These areas are relatively smooth in comparison with the outer cell surfaces which appear to be covered with distinct, delicate microvilli. Delicate fibers, possibly fibrin, connect cells to debris, to the collagen-coated culture surface, and to each other. We have preliminary evidence that the extension of peripheral cytoplasmic veils may be positively correlated with functional ability under the present conditions of culture (Goldfarb, S., and Pariza, M. W., unpublished).

Berry and Friend (1) reported that isolating hepatocytes by perfusion of liver with collagenase resulted in few alterations in ultrastructure detectable with transmission microscopy, whereas Schreiber and Schreiber (2) observed many of the same ultrastructural changes in freshly isolated hepatocytes which we report (Fig. 2). However, there is little doubt that by 4 hours in culture (Fig. 3) our cells have regained many of the ultrastructural properties of normal liver cells in vivo including aggregates of glycogen absent from the cells described by Berry and Friend (1). Additionally, the 4 hour cells appear similar in many respects to cells observed by Chapman et al. (25) after one and three days in monolayer culture.

Scanning electron microscopic studies (Figs. 4 and 5) reveal surface changes associated with the first 4 hours in culture. The most notable of these are the appearance of peripheral cytoplasmic veils, apparently involved in cell spread-

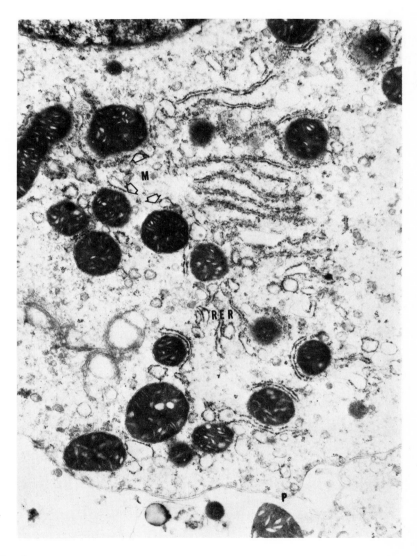

FIG. 2:

<u>Transmission electron micrograph of freshly isolated hepatocyte</u>. Mitochondria (M) appear dense and contracted. Rough endoplasmic reticulum (RER) is focally dilated. The plasma membrane (P) is devoid of microvilli. X 15,500.

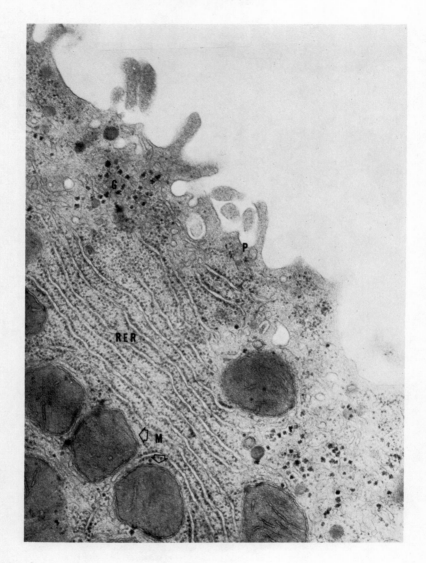

FIG. 3:

<u>Transmission electron micrograph of hepatoctye after 4 hours in culture</u>. Mitochondria (M) are less dense than in Fig. 3, and rough endoplasmic reticulum (RER) is no longer dilated. Glycogen aggregates (G) present in cell periphery in association with smooth endoplasmic reticulum. Plasma membrane (P) shows microvilli. X 36,000.

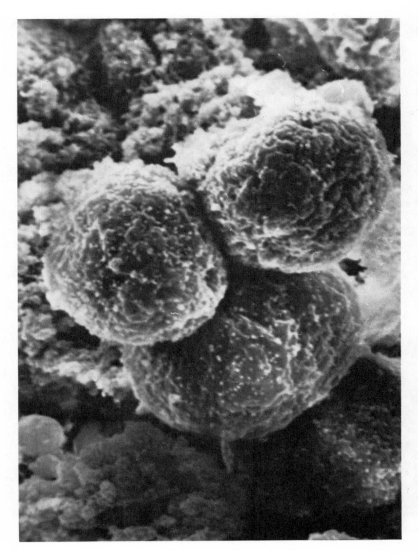

FIG. 4:

<u>Scanning electron micrograph of freshly isolated hepatocytes</u>.
The three rounded cells in the foreground have clumped together. Their surfaces show broad folds and only rare blunted microvilli. X 3,900.

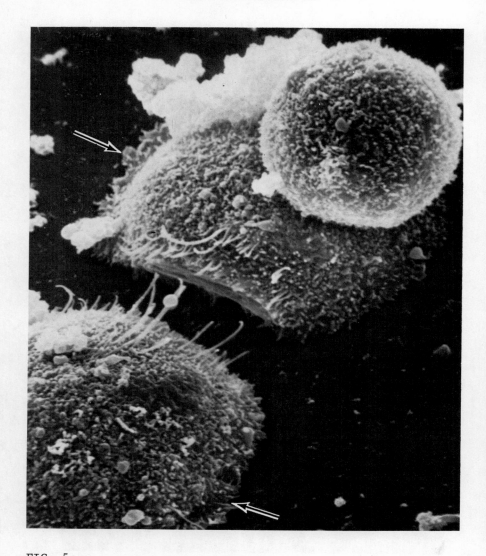

FIG. 5:
<u>Scanning electron micrograph of hepatocytes after 4 hours in culture</u>. Cells adherent to collagen coat have flattened, and are extending veils of cytoplasm (arrows). Cell surfaces are covered with numerous delicate microvilli. One hepatocyte has settled on adherent cell and remains rounded. The interface between the adherent cells appears exposed and relatively free of microvilli. Fibers that bridge the gap between the cells appear to have snapped apart. Debris appears on surface of hepatocytes. X 3.800.

ing, and microvilli.

Taken together, the ultrastructural studies strongly support the proposition that the hepatocytes repair functional deficiencies in culture. Because cells from the same preparation were studied in the experiments of Chart 5 and Figs. 2-5, we conclude that improvement in the ability to incorporate leucine and cytidine, and the appearance of ODC activity, are positively correlated with morphological amelioration in such important cell structures as mitochondria, endoplasmic reticulum, and the plasma membrane.

Relationship Between ^{3}H-leucine and ^{14}C-cytidine Incorporation and Staining with Trypan Blue

The ability to exclude the vital dye, trypan blue (TB), is commonly used to determine cell "viability". We routinely treat the initial cell suspensions with trypan blue, and generally find that less than 30% of the cells are stained. However, occasional suspensions with a high TB index (Table 3) are encountered, and we have undertaken studies to determine the suitability of such preparations for culture inocula.

Table 3 shows the results of one such study. A cell suspension with a TB index of 53% was inoculated into culture dishes under usual conditions, and subjected to medium changes at 4, 21, 45, and 69 hours in culture. At various times, cultures were pulsed for 20 minute intervals with ^{3}H-leucine plus ^{14}C-cytidine, and the levels of radioactivity in TCA-soluble and -insoluble fractions determined. Table 3 summarizes the results of these experiments and includes similarly tabulted data from the experiment shown in Chart 4 for comparison.

The data show that at 24, 33, 70 and 79 hours the cultures inoculated from the suspensions showing a TB index of 53% incorporated ^{3}H-leucine into TCA-insoluble material to levels comparable with those of cultures from Chart 4 (TB index of 29%). However, incorporation of ^{14}C-cytidine was much lower, possibly indicating an impaired ability to synthesize RNA. That both ^{3}H-leucine and ^{14}C-cytidine were entering and being retained by these cells during washing is apparent from the TCA-soluble fractions, which contained label. Thus, it appears from these data that the incorporation of labeled cytidine may be a more sensitive indicator of functional ability than the incorporation of labeled leucine.

Retention of Adult Liver Characteristics

The maintenance of adult liver properties by liver parenchymal cells in vitro is of considerable importance if one is to be reasonably certain that adult liver cells are being

COMPARISON OF ^{3}H-LEUCINE AND ^{14}C-CYTIDINE INCORPORATION INTO TCA-INSOLUBLE MATERIAL

dpm/ug DNA

TB INDEX*	HOURS IN CULTURE	TCA-INSOLUBLE		TCA-SOLUBLE	
		^{3}H	^{14}C	^{3}H	^{14}C
53%	24	347	38	1803	182
	33	539	32	2960	207
	70	967	17	3564	105
	79	210	18	5225	161
29%	24	210	170		
	32	225	268		
	68	522	300		
	80	300	263		

*TB INDEX REFERS TO TRYPAN BLUE STAINING
(PERCENT STAINED CELLS IN THE INOCULUM)

TABLE 3

studied. We have begun to approach this problem biochemically by studying glycolytic isoenzymes present in the cultured cells.

Total pyruvate kinase (PK) activity in the cells after one day in culture appears to be similar to that in normal rat liver on a per mg protein basis. For example, in one experiment a liver lobe removed prior to collagenase perfusion contained 0.262 units (mole substrate utilized per minute) per mg protein, whereas the isolated cells after one day in culture had 0.329 units/mg protein. Additionally, the majority of the PK activity migrated as PK-I (26,27) when subjected to polyacrylamide gel electrophoresis (Yanagi, S., unpublished). This pattern is consistent with that observed for normal adult liver where approximately 90% of the PK activity is present as PK-I (26,27).

Glucokinase activity has also been investigated. Unlike PK activity, glucokinase activity appears to be rapidly lost in culture (Grossman, S. H., unpublished), a finding which is in agreement with the report of Bonney et al. (4,5). Ways to preserve glucokinase activity, or to induce it in the cultured cells, are currently being investigated.

Isolated cell suspensions have been examined for glucokinase activity with fluorescent microscopy using rabbit antisera against pure rat liver glucokinase (28). Figure 6 is a photomicrograph of acetone-fixed hepatocytes which have been treated with rabbit antisera against rat liver glucokinase, and then with fluorescein-conjugated anti-rabbit serum from goat (29). Lack of fluorescence in the nucleus indicates that glucokinase is a cytoplasmic enzyme. Additionally, variability in the intensity of the staining suggests the possibility of diversity among liver parenchymal cells with respect to glucokinase concentration.

DNA Synthesis and Mitosis in Cultured Cells from Rats Subjected to Partial Hepatectomy

Our long-term goal is to establish conditions whereby liver parenchymal cells in primary culture can be made to modulate between the fetal and adult states thereby allowing for the in vitro study of normal liver cell ontogeny and liver carcinogenesis. However, with respect to DNA synthesis and mitosis, the differentiated state appears to be remarkably stable in cultured hepatocytes from normal adult rats, and attempts to stimulate division have not been successful (8; Yager, J. D., Jr., unpublished). Because of this, the problem has been approached from another direction: will cells isolated from regenerating liver (following surgical partial hepatectomy) synthesize DNA and divide in culture?

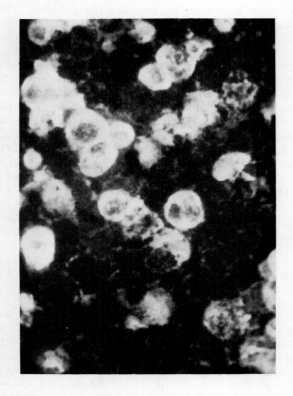

FIG. 6:

<u>Fluorescence photomicrograph of hepatocytes immediately after isolation</u>. Cells were fixed with acetone, treated with rabbit antiserum against rat liver glucokinase, and then with fluorescein-conjugated anti-rabbit serum from goat by the method of Nowinski et al. (29). X 250.

To facilitate in this study, use has been made of controlled feeding and lighting schedules previously developed in this laboratory (30) for the purpose of synchronzing metabolic activity in vivo. Hopkins et al. (31) described the pattern of ^{3}H-TdR incorporation into liver DNA in vivo in rats entrained to a controlled feeding and lighting protocol and subjected to partial hepatectomy at various times during a 24 hour period. Our intention was to determine whether the cells would continue in vitro along the path taken by liver parenchymal cells in vivo.

Initial observations, which have already been published (8,32), indicated that cells isolated 12 hours post-partial hepatectomy did not synthesize DNA in culture, but that cells isolated 15 hours post operation did synthesize DNA in vitro, and the pattern of ^{3}H-TdR incorporation paralleled that expected in vivo where the 1st peak of DNA synthesis occurred 24 hours after the operation under the protocol employed (31). However, we also reported that the proportion of cells synthesizing DNA was only 1-2%, which is considerably lower than that observed in vivo where nearly 60% of the liver parenchymal cell nuclei are labeled during the 1st period of DNA synthesis (33).

In more recent experiments, the schedule for perfusion and culture was moved to a later time in the protocol. Table 4 shows the incorporation of ^{3}H-TdR and the percent labeled nuclei observed when cells were isolated from rats 22 hours after partial hepatectomy and placed in culture by 24 hours post-operation. During the time period 26-28 hours after partial hepatectomy, 43.5% of the nuclei were labeled by a continuous pulse of ^{3}H-TdR, while in the time period 38-48 hours post-operation 23.2% of the nuclei were labeled. When the time period was broken down into 4 hour increments and cells exposed to label during the first 2 hours of each increment, the percent labeled nuclei and the TCA-insoluble radioactivity (on a per g DNA basis) began at an elevated level and slowly declined, and the second DNA peak, observed in vivo at 48 hours after partial hepatectomy, was not seen in culture (Table 4).

Mitotic figures were also found in the cultures described above. Their frequency has not yet been determined but they are considerably more numerous than in cultures of liver parenchymal cells obtained from intact livers where they are almost never seen.

The data of Table 4 together with those reported previously (8) indicate that if the parenchymal cells are to synthesize DNA in vitro under the conditions employed in this study, they must be in the S phase of the cell cycle at

^{3}H TdR INCORPORATION (2 hr.) AND PERCENT LABELED NUCLEI IN PRIMARY CULTURES OF PARENCHYMAL CELLS ISOLATED FROM REGENERATING RAT LIVER 22 HOURS AFTER PARTIAL HEPATECTOMY

HOURS AFTER OPERATION	SPECIFIC ACTIVITY dpm/μg DNA (10^{-3})	% LABELED NUCLEI
26-38	—	45.3
38-48	—	23.2
26-28	20.8 ± 0.3	20.0
30-32	13.4 ± 1.0	18.4
34-36	13.8 ± 0.4	12.9
38-40	7.87	9.9
42-44	3.03 ± 0.18	8.0
46-48	1.98 ± 0.20	2.3

TABLE 4

the time of isolation. If they are not, the culture conditions do not permit the cells to enter S. However, if cells are in S <u>at the time of isolation</u>, they apparently can complete this phase of the cell cycle and possibly even move through G_2 into mitosis.

Discussion

It is becoming increasingly popular to study biochemical phenomena in rat liver parenchymal cells immediately after isolation. While there may be several technical advantages to this system (2), it is nonetheless clear from data presented in this paper and other published reports (3,4,5,8) that freshly isolated hepatocytes do not possess optimal functional properties, and furthermore that they tend to degenerate rather than undergo repair during incubation in suspension (10). These disadvantages, which tend to preclude the use of liver cell suspensions for many kinds of studies, may be overcome by placing the cells in primary monolayer culture. The results presented in this paper support the hypothesis that cells will repair functional deficiencies <u>in vitro</u> given the proper environment, although we do not wish to imply that the "best" conditions for cell recovery have been found. However, several conclusions are evident from the results which have been reported:

1. Functional properties are not optimal immediately after cell isolation, but improve with time in culture.

2. A medium change after a few hours in culture is preferable to changing the medium only at 24 hours.

3. The optimal time for studying protein and RNA synthesis and the activities of at least some enzymes (notably ODC) is not during the 1st day in culture, but that rather during the period 20-60 hours in culture.

4. The incorporation of RNA precursors into TCA-insoluble, KOH-hydrolyzable material may be a more sensitive indicator of functional recovery than the incorporation of protein precursors (into TCA-insoluble material).

The retention of differentiated liver properties by the cultured cells is of particular importance, and our initial studies with PK activity suggest that PK-I may be retained by the cells at normal (<u>in vivo</u>) levels. In contrast, the rapid loss of glucokinase is somewhat puzzling, but studies with perfused liver (34) indicate that the regulation of glucokinase activity may be considerably more complex than <u>in vivo</u> studies have indicated (34).

Cells can also be isolated from regenerating liver, and by using controlled feeding and lighting protocols (31) developed in this laboratory (30) it is possible to ask the question: when are parenchymal cells in a regenerating liver *irreversibly* committed to DNA synthesis? The answer to this question appears to be that if the cells are to synthesize DNA in culture (and possibly divide) under present conditions they *must* be in the S phase of the cell cycle at the time of isolation. We are pursuing these studies in an effort to define the conditions necessary to initiate S phase *in vitro*.

References

1. Berry, M.N. and Friend, D.S. J. Cell Biol. **43** (1969) 506.
2. Schreiber, G. and Schreiber, M. Subcell. Biochem. **2** (1973) 321.
3. Bissell, D.M., Hammaker, L.E. and Meyer, U.A. J. Cell Biol. **59** (1973) 722.
4. Bonney, R.J. In Vitro, in press.
5. Bonney, R.J., Becker, J.E., Walker, P.R. and Potter, V.R. In Vitro, in press.
6. Bonney, R.J., Walker, P.R. and Potter, V.R. Biochem. J. **136** (1973) 947.
7. Johnson, M.E.M., Das, N.M., Butcher, F.R. and Fain, J.N. J. Biol. Chem. **247** (1971) 3229.
8. Pariza, M.W., Becker, J.E., Yager, J.D.,Jr., Bonney, R.J. and Potter, V.R. Princess Takamatsu Symposium on Cancer Res., in press.
9. Zahlten, R.N., Stratman, F.W. and Lardy, H.H. Proc. Nat. Acad. Sci., U.S.A. **70** (1973) 3213.
10. East, A.G., Louis, L.N. and Hoffenberg, R. Exptl. Cell Res. **76** (1973) 41.
11. Quistorff, B., Bondensen, S. and Grunnet, N. Biochim. Biophys. Acta **330** (1973) 503.
12. Iype, P.T. J. Cell Physiol. **78** (1971) 281.
13. Lambiotte, M., Susor, W.A. and Cahn, R.D. Biochemie **54** (1972) 1179.
14. Perske, W.F., Parks, R.E. Jr. and Walker, D.L. Science **125** (1957) 1290.
15. Potter, V.R. Cancer Res. **32** (1972) 1998.
16. Walker, P.R., Bonney, R.J., Becker, J.E. and Potter, V.R. In Vitro **8** (1972) 107.
17. Walker, P.R. and Potter, V.R. J. Biol. Chem. **248** (1973) 4610

18. Oshiro, Y., Gerschenson, L.E. and DiPaolo, J.A. Cancer Res. 32 (1972) 877.
19. Borek, C. In Vitro 7 (1972) 264.
20. Bausher, J. and Schaeffer, W.I. In Vitro 9 (1974) 286.
21. Bausher, J. and Schaeffer, W.I. In Vitro 9 (1974) 294.
22. Ham, R.G. Proc. Nat. Acad. Sci., U.S.A. 53 (1965) 288.
23. Waymouth, C. J. Nat. Cancer Inst. 22 (1959) 1003.
24. Morrison, S.J. and Jenkin, H.M. In Vitro 8 (1972) 94.
25. Chapman, G.S., Jones, A.L., Meyer, U.A. and Bissell, D.M. J. Cell Biol. 59 (1973) 735.
26. Potter, V.R., Walker, P.R. and Goodman, J.I. Gann Monograph on Cancer Res. 13 (1972) 121.
27. Yanagi, S., Makiura, S., Kamamoto, Y., Matsumura, K., Hirao, K., Ito, N. and Tanaka, T. Cancer Res., in press.
28. Grossman, S.H., Dorn, C.G. and Potter, V.R. J. Biol. Chem., in press.
29. Nowinski, R.C., Sarkar, N.H. and Fleissner, E. Methods in Cancer Research VIII (1973) 237.
30. Watanabe, M., Potter, V.R. and Pitot, H.C. J. Nutrition 95 (1968) 207.
31. Hopkins, H.A., Campbell, H.A., Barbiroli, B. and Potter, V.R. Biochem. J. 135 (1973) 955.
32. Yager, J.D. Jr., Pariza, M.W., Becker, J.E. and Potter, V.R. Proc. Third International Workshop on Exptl. Injury: Liver Regeneration after Exptl. Injury, in press.
33. Yager, J.D. Jr., Hopkins, H.A., Campbell, H.A. and Potter, V.R. Proc. Third International Workshop on Exptl. Injury: Liver Regeneration after Exptl. Injury, in press.
34. Hodderik, G. and Spydevoid, O. Acta Physiol. Scand. 90 (1974) 41.

The authors wish to express their gratitude to Dr. Harold Campbell for providing the methodology for the assay of ornithine decarboxylase, to Mr. Richard Hibma for assistance in the electron microscopic studies, and to Mrs. Henryka Brania for help in maintaining the cell cultures.

Note: We would like to thank the University of Tokyo Press, publisher of "Topics in Chemical Carcinogenesis," for permission to reproduce Figure 1.

DIFFERENTIATED FUNCTIONS IN CLONAL STRAINS OF HEPATOMA CELLS

Armen H. Tashjian, Jr., U. Ingrid Richardson, Robert Strunk and Peter Ofner

Introduction

The purpose of this report is to describe the preservation of several highly differentiated functions in clonal strains of cells derived from transplantable rat hepatomas. A number of the preceeding papers in this Symposium have served to document the difficulties of establishing in culture permanent lines of "normal" hepatocytes. Until it is possible to accomplish this at will, we have chosen to establish in culture cells derived from hepatomas which perform, in the host animal, the particular function we are interested in investigating. The technical advantages of using neoplastic cells derived from transplantable tumors to initiate primary cultures are now well established. From such primary lines we have cloned strains which preserve differentiated or organ-specific functions. The initial goal of this strategy is to learn how to handle such neoplastic cells in culture, to exploit them as model systems for studies of cellular control mechanisms, and eventually to use this experience to establish functional cultures of nonneoplastic hepatic cells. In this communication we describe hepatoma cell strains that preserve in culture the 5 enzymes of the urea cycle, that produce several components of the complement system, and that metabolize testosterone by pathways similar to those in normal liver.

Results and Discussion

Because this paper is meant to be a summary of results obtained in the 3 areas mentioned, we have not emphasized methodological details. The reader is referred to appropriate references for these methods.

Urea Cycle Enzymes. It has been our experience that explants, organ cultures, or primary dispersed cell cultures of adult, newborn or fetal rat liver rapidly lose the activities of the urea cycle enzymes. The half-times of disappearance being measured in hours for carbamoyl phosphate synthetase (CPS) and argininosuccinate synthetase (AS), and in days for argininosuccinate lyase (AL), ornithine carbamoyl transferase (OCT) and arginase (ARG).

A search of the literature revealed several transplantable rat hepatomas which were reported to contain the 5 enzymes. We confirmed that Morris hepatomas 7800, 7793, 7316A and 5123 each had all 5 enzymes, and that the specific enzyme activities were reasonably close to those of normal rat liver (1). Because of the especially high activities of the enzymes in the 7800 tumor (the percentages of the activities of the 5 enzymes compared to normal rat liver were CPS 240%, OCT 70%, AS 140%, AL 130%, and ARG 75%), we elected to establish in culture and clone cells from this tumor. This was successfully accomplished (1), and the epithelial cells have been in culture for 29 months. They have a population doubling time of about 3.5 days in Eagle's minimal essential medium supplemented with 5% horse serum and 5% fetal calf serum. The clone, $7800C_1$, does not produce measurable quantities of rat serum albumin, but it does have tyrosine aminotransferase (3 mU/mg, basally) which is increased about 2-fold by treatment with hydrocortisone (10^{-6}M for 24 hr).

The urea cycle enzyme activities of normal rat liver, the 7800 tumor, the $7800C_1$, clone in culture, and a tumor derived from injection of 7800 C_1 cells into the Buffalo rat host are given in Table 1. The activities of AL, CPS and AS do not change during the growth cycle of the cells (1).

Several factors which have been reported to change (usually to increase) urea cycle enzyme activities in the animal have not altered the activites in $7800C_1$ cells. These stimuli include glucagon (5 g/ml), hydrocortisone (10^{-6}M) and dibutyryl cyclic AMP (5 x 10^{-4}M). They have been tested separately and together for periods of up to one week. Whether these negative results mean that the $7800C_1$ cells are not responsive to these signals or that the results observed in the animal are indirect effects remains to be explained by further studies. Some preliminary experiments on the adaptive value of the urea cycle pathway in $7800C_1$ cells have been initiated. The cells have been placed in minimal essential medium containing dialyzed serum in which the arginine concentration has been reduced from 0.60 to 0.02 mM and which is supplemented with 0.60 mM ornithine. After two weeks in this medium several control hepatoma cells strains (MH_1C_1 and Li) and fibroblasts die, but the $7800C_1$ strain survived although the cells did not proliferate. When given complete medium containing 0.60 mM arginine they grew well whereas there were no survivors in control cell strains. This cycle to low arginine, ornithine-supplemented, medium and then to complete medium could be repeated again indicating that $7800C_1$ cells were able to

Sample	Enzyme activity (Units/mg protein)				
	CPS	OCT	AS	AL	ARG
Normal rat liver*	2.5 ± 0.44	52 ± 13	1.0 ± 0.27	1.4 ± 0.21	650 ± 220
7800 tumor	6.1	37	1.4	1.8	500
7800C$_1$ clone**	0.70	0.63	0.41	1.2	39
7800C$_1$ tumor	1.5	1.3	0.38	1.4	16

*Liver from 150 g Sprague-Dwwley rats maintained on a 23% protein diet. Values are means ± SD, 20 livers per group

**Results are mean values of 12 experiments (1).

TABLE 1: Activities of urea cycle enzymes

survive substantial arginine deprivation. However, whether this survival is due to the presence of the urea cycle pathway is not yet clear, for a clone of hepatoma cells, H_4, derived from Reuber hepatoma H35 by Pitot et al. (2), that did not contain significant activities of OCT, AS or ARG, did give rise to a subset of cells in low arginine medium which were able not only to survive but to grow in this medium. These proliferating cells have not yet been assayed for urea cycle enzyme activities.

From these results we conclude that it is possible to establish and maintain in culture cells of hepatic origin which have all 5 of the urea cycle enzymes. However, the low levels OCT and ARG in the cultured cells in comparison with the tumor from which they were derived and with normal rat liver suggests to us that these enzymes may be subject to control in the host which is difficult to reproduce <u>in vitro</u> and that it will be difficult indeed with our present understanding, to maintain these enzymes in nonneoplastic cells in culture.

<u>Production of Serum Complement Components.</u> Several reports have described the production of complement (C) components by hepatoma cells in culture. These results have been summarized by Strunk et al. (3) and have included the ninth component (C9) as well as the second (C2), third (C3) and sixth (C6) components. Our goal in these studies was to identify and characterize cell strains which produced substantial quantities of the early C components for use in somatic cell hybridization experiments with cells derived from patients with genetic C deficiencies. We examined 4 clonal strains (H_4, $7800C_1$, MH_1C_1 and $7316AC_1$). None of these strains synthesized measurable amounts of Cl and only $7800C_1$ produced C4; however, each strain synthesized biologically active C2 and C3. Three of the four synthesized C5 and the natural inhibitor of Cl (Cl INH). We summarize here results obtained with the $7800C_1$ and H_4 strains (3,4).

In these experiments, cells were grown in serum-containing complete medium until the beginning of a C experiment. They were then washed free of C components with chemically defined serum-free medium and maintained in this medium (Neuman and Tytell) for the remainder of the experiment. Components were measured by functional hemolytic assays using appropriately sensitized erythrocytes (3).

The production of C2, C3, C4, C5 and Cl INH by $7800C_1$ and H_4 and two control strains are given in Table 2. These measurements were made in medium harvested after 48 hr: synthesis was shown to be linear for at least 72 hr (3).

Strain	Complement production*				
	C2	C3	C4	C5	Cl INH
	(effective molecules x 10^{-8}/µg cell protein)				
$7800C_1$	136	3.6	4.5	13.8	27.2
H_4	4.5	9.0	<0.05	19.1	8.7
R_5	0.4	<0.05	<0.05	<0.05	<0.05
GH_3	0.5	<0.05	<0.05	<0.05	<0.05

Mean values of at least duplicate cultures

TABLE 2: Production of complement components by two rat hepatoma cell strains ($7800C_1$ and H_4) and by rat fibroblasts (R_5) and rat pituitary cells (GH_3).

Labeled amino acids present	Complement production*				Labeled protein*
	C2	C3	C5	Cl INH	
	(effective molecules x 10^{-8}/µg cell protein)				(cpm/µg)
Throughout	50	4.8	2.4	104	6350
Added at end	51	5.2	2.7	100	5

*
Mean values, duplicate cultures

TABLE 3: Accumulation of labeled protein in the medium of $7800C_1$ cells incubated for 72 hr with 6.0 µCi of $\{^{14}C\}$ amino acids.

Synthesis of all components was inhibited by the addition of cycloheximide (2 µg/ml) to the medium, and this effect was reversible (3).

Several sorts of experiments were performed to demonstrate that the hepatoma cells were synthesizing C-complement components in culture. Medium was supplemented with {^{14}C} amino acids, dialyzed, concentrated, and the newly synthesized labeled products examined by radioimmunoelectrophoresis, double diffusion and by binding to appropriately sensitized erythrocytes (3). The results in Table 3 show that when labeled amino acids were present throughout the incubation period there was no effect on C production, but labeled macromolecules accumulated in the medium. When labeled amino acids were added to medium just prior to collection, dialysis and concentration, there was no nonspecific accumulation of label found. The concentrated labeled medium was examined by double diffusion and autoradiography with rabbit antirat C3 and C1 INH. Precipitation lines of "identity" developed with anti-C3 and both purified rat C3 and 7800C_1 medium (Fig. 1). The radioautograph showed that the C3 produced by 7800C_1 was labeled. Immunodiffusion with anti-C1 INH gave a visible precipitation line only with partially purified rat C1 INH; however, a labeled arc was seen with 7800C_1 medium (Fig. 1). The labeled protein and partially purified C1 INH gave a reaction of partial identity. A similar experiment with anti-C1 INH diffusing against C1 INH in whole serum and partially purified C1 INH also showed a line of partial identity suggesting the loss of an antigenic determinant in the rat C1 INH during purification as an explanation for the lack of complete identity.

Additional evidence, in the case of C2, that the newly synthesized labeled component was functional was shown as follows. Concentrated, dialyzed, labeled 7800C_1 medium was incubated under conditions in which C2 binds to EAC_4 cells and the radioactivity associated with the complex was determined (3). In another experiment, any EAC_{42} that formed was allowed to decay and again the radioactivity associated with the erythrocytes was measured. The results given in Table 4, show that labeled 7800C_1 medium bound to EAC_4 cells 15 times more extensively than to control EA cells alone, and that in the second experiment during decay conditions about 40% of the radioactivity adsorbed to EAC_4 was lost, demonstrating that at least 40% of the adsorbed labeled protein was functional C2. Further evidence that the C proteins synthesized in culture were similar to the corresponding rat serum C components was

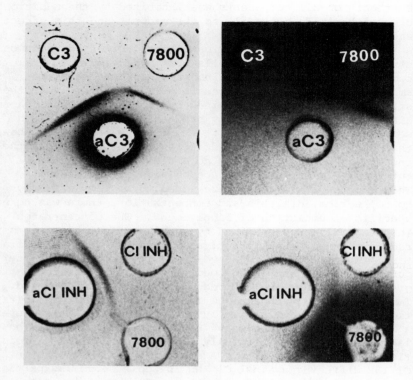

FIG. 1:

Immunodiffusion and autoradiography of culture medium from 7800C$_1$ cells incubated with {^{14}C} amino acids. Upper left. Immunodiffusion with anti-C3 (aC3), purified rat C3 (C3), and concentrated dialyzed medium from 7800C$_1$ cells. Upper right. Autoradiography of A. Lower left. Immunodiffusion with anti-Cl INH (aCl INH), partially purified Cl INH (Cl INH), and concentrated dialyzed medium from 7800C$_1$ cells. Lower right. Autoradiography of C. (Reproduced from ref.3).

Binding Conditions*			Decay conditions*		
+EAC4	+EA	Blank	Before	After	Blank
(cpm/10^9 cells)**			(cpm/10^9 cells)**		
300 ± 8	22 ± 6	21 ± 5	557 ± 10	330 ± 8	21 ± 2

*See ref. 3 for details of methodology

**Mean value of duplicate determinations ± range

TABLE 4: Incubation of medium from 7800Cl grown in the presence of {^{14}C} amino acids with C4-sensitized erythrocytes: Evidence for synthesis of labeled C2.

obtained by comparing the gel filtration elution positions of the components obtained from culture medium with those observed in rat serum. For those components examined, C2, C3 and Cl INH, the elution positions for the components synthesized in culture were the same as those in rat serum (3).

In the course of examining the control of C production we made an unexpected observation with hydrocortisone (Fig. 2). At a concentration of 4×10^{-7}M there was a large stimulation of the production of C3 but not of C2, C4 (not shown) or C5 by H_4 cells. In studies of the dose-response relationship of this effect, the maximum response was observed at hydrocortisone concentrations between 10^{-7} and 10^{-6}M (3).

We conclude that $7800C_1$ and H_4 cells synthesize and secrete into the medium authentic rat serum C components including C2, C3, C4, C5 and Cl INH. These cell strains should be useful for studies of the control of C production and for somatic cell hybridization experiments.

<u>Metabolism of Testosterone</u>. The slowly-growing MH_1C_1 clonal strain (5) derived from Morris hepatoma 7795 in a male Buffalo rat host maintains in culture a number of differentiated functions (6). These include the synthesis and secretion of albumin with response to hydrocortisone (5,7), the production of C9 (8), and increase in tyrosine aminotransferase activity by treatment with hydrocortisone (6), an increase in phosphoenolpyruvate carboxykinase by treatment with dibutyryl cyclic AMP (9), and the conjugation (10,11) and production (12) of bilirubin. In addition, these cells retain in culture several of the enzymes known to be involved in C_{19}-steroid metabolism by rat liver.

To study C_{19}-steroid metabolism we incubated 6 μM singly- or doubly-labeled testosterone or 5α-dihydrotestosterone with MH_1C_1 cells (at a protein/steroid ratio of 93:1), control R_5 fibroblasts (of rat thyroid origin) and with serum-free Ham's F 10 medium alone. Enzyme activity was terminated at intervals from 15 min to 24 hr by shaking with ethyl acetate and the aqueous and solvent fractions were analyzed by several methods. Free radiosteroids were examined by thin-layer chromatography in several solvent systems followed by autoradiography, for loss of tritium originally present in the 7-$\{^3H\}$ -labeled substrates, by oxidation with chromium trioxide in acetone to form ketonic derivatives, and by crystallization to constant specific activity. The conjugates were fractionated into glucuronide and

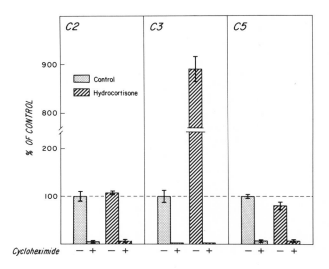

FIG. 2:

Effects of hydrocortisone on C production by H_4 cells.
Cultures were incubated for 48 hr without or with hydrocortisone (4×10^{-7}M) in presence or absence of cycloheximide (2 µg/ml), and the medium collected and assayed for C components.

sulfate constituents by the procedure of Baulieu and
Mauvais-Jarvis (6,13). The fractions were studied per se
by partition-column and thin-layer chromatography and their
free-steroid components after incubation with β-glucuronidase
and solvolysis.

The results showed a particularly high rate of conjugation of testosterone: 88% in 24 hr. Of the 12% remaining
as free steroid, 2% was unchanged testosterone and 6.5% and
3.5% consisted of $C_{19}O_2$- and $C_{19}O_3$-radiometabolites, respectively (6,14). Based on these analyses, we have either
preliminary chromatographic or definitive evidence for the
following C_{19}-steroid metabolizing enzymes in MH_1C_1 cells:
definitive evidence for 17β-hydroxysteroid oxidoreductase,
Δ^4-3-keto-C_{19}-steroid 5 α-reductase, 3 α-hydroxysteroid oxidoreductase, 7α (and/orβ)-hydroxylase(s) (5α-androstane-3, 7,17-
trione carrier was crystallized to constant SA with CrO_3-
oxidation product of hydroxylated 5α-dihydrotestosterone
radiometabolites), and UDP glucuronyl transferase; preliminary evidence for 3β-hydroxysteroid oxidoreductase,
6β-, 7α- and 16α-hydroxylases and 17β-hydroxysteroid sulfotransferase.

The metabolic pattern is characteristic of the hepatic
transformation of testosterone. All enzymes looked for,
except Δ^4-3-ketosteroid 5β-reductase, were found in MH_1C_1
cells. With this one exception, there were no important
qualitative differences in the metabolic patterns derived
from incubation of testosterone with normal and host rat
liver, with hepatoma tissue derived from injection of
MH_1C_1 cells, and with MH_1C_1 cells grown in monolayer culture.
We conclude that these cells are a valid model system in
which to study hepatic C_{19}-steroid transformation and should
prove useful for studies of steroid uptake and receptor
binding.

Summary

Several differentiated functions have been studied in 4
strains of rat hepatoma cells (MH_1C_1, $7800C_1$, $7316AC_1$ and
H_4). The $7800C_1$ strain maintains measurable activities of
the 5 enzymes of the urea cycle. The activity of argininosuccinate lyase is approximately that found in normal rat
liver, while argininosuccinate synthetase, carbamoyl transferase activities are respectively, 40%, 28%, 6%, and 1% of
normal values. The synthesis of serum complement (C)
components has been examined. None of the strains synthesized detectable amounts of Cl and only one strain ($7800C_1$)

produced C4. However, each strain synthesized biologically active C2 and C3 and three of the four strains produced C5 and the natural inhibitor of C1 (C1 INH). Studies with $7800C_1$ and H_4 revealed reversible inhibition of C synthesis by cycloheximide, incorporation of labeled amino acids into C2, C3 and C1 INH, and chromatographic profiles of the components which were the same as those in normal rat serum. Cortisol (10^{-6} to 10^{-7}M) markedly stimulated the production of C3 but not C2 or C5. Lastly, we studied testosterone metabolism by MH_1C_1 cells using singly-or doubly-labeled testosterone and 5α-dihydrotestosterone. Steroid radiometabolite patterns gave evidence for the following enzymes: 17β-hydroxy-C_{19}-steroid oxidoreductase, Δ^4-3-keto-C_{19}-steroid 5α-reductase, 3α- and 3β- hydroxysteroid oxidoreductases, C19-steroid 6β-, 7α-, and 16α-hydroxylases, 17β-hydroxysteroid sulfotransferase, and UDP glucuronyl transferase. We conclude that appropriately chosen or newly established strains of hepatoma cells can maintain in culture a wide spectrum of differentiated functions.

This investigation was supported in part by research grants from the USPHS (AM11011, HD05916 and FR00128). The authors thank A.J. Der Marderosian, J. McDonough, Y.E. Santo and R. Vena for expert assistance.

References

1. Richardson, U.I., Sondgrass, P.J., Nuzum, C.T. and Tashjian, A.H., Jr. J. Cell Physiol. 83 (1974) 141.
2. Pitot, H.C., Peraino, C., Morse, P.A., Jr., and Potter, V.R. Natl. Cancer Inst. Monograph 13 (1964) 229.
3. Strunk, R.C., Tashjian, A.H., Jr. and Colten, H.R. submitted for publication (J. Immunol.).
4. Strunk, R.C., Tashjian, A.H., Jr. and Colten, H.R. Fed. Proc. 33 (1974) 795.
5. Richardson, U.I., Tashjian, A.H., Jr. and Levine, L. J. Cell Biol. 40 (1969) 236.
6. Tashjian, A.H., Jr., Bancroft, F.C., Richardson, U.I. Goldlust, M.B., Rommel, F.A. and Ofner, P. In Vitro 6 (1970) 32.
7. Bancroft, F.C., Levine, L. and Tashjian, A.H., Jr. Biochem. Biophys. Res. Commun. 37 (1969) 1028.
8. Rommel, F.A., Goldlust, M.B., Bancroft, F.C., Mayer, M.M. and Tashjian, A.H., Jr. J. Immunol. 105 (1970) 396.
9. Wicks, W.D., Barnett, C.A. and McKibbin, J.B. Fed. Proc.

33 (1974) 1105.
10. Rugstad, H.E., Robinson, Yannoni, C. and Tashjian, Jr. J. Cell Biol. 47 (1970) 703.
11. Rugstad, H.E., Robinson, S.H., Yannoni, C. and Tashjian, A.H., Jr. Science 170 (1970) 553.
12. Robinson, S.H., Rugstad, H.E., Yannoni, C. and Tashjian, A.H., Jr. Proc. Soc. Exp. Biol. Med. 136 (1971) 684.
13. Baulieu, E.E. and Mauvais-Jarvis, P. J. Biol. Chem. 239 (1964) 1569.
14. Ofner, P. and Tashjian, A.H., Jr. In Vitro 6, (171) 385.

EFFECT OF GLUCOCORTICOIDS ON THE ULTRASTRUCTURE OF
CULTURED LIVER CELLS

Judith A. Berliner

Introduction

For a number of years it has been recognized that the injection of corticoid hormones into animals causes alterations in liver cell morphology (1,2). It was shown that when cortisone was injected into rats liver mitochondria enlarged to several times the normal size with a concommitant decrease in mitochondrial number (3). It has subsequently been shown by Kimberg and Loeb (4) that these enlarged mitochondria, which I will refer to as megamitochondria, are formed by fusion of smaller mitochondria. Megamitochondria exhibit altered membrane properties (5) and there are suggestions from other *in vivo* studies that outer cell membranes are also altered (6). However, the question of whether cortisone was the direct causal agent and exactly how the hormone could bring about these changes were not answered due to the complexity of studying these questions in whole animal models.

We have now developed an *in vitro* system in which both mitochondrial and outer membrane morphology are altered by a synthetic glucocorticoid, dexamethasone sodium phosphate. For these studies we used a cell line, RLC-GAI, originally isolated from rat liver, which grows in defined medium (7). This line has become spontaneously transformed in culture (8). We have studied these morphological changes using both transmission and scanning electron microscopy and have employed the technique of electron microscopy stereology in analyzing our data.

Materials and Methods

For transmission microscopy cells were grown in modified Ham's F-12 (7) in Falcon plastic petri dishes at $37°C$. In treated cultures $10^{-5}M$ dexamethasone sodium phosphate (dex) was added to cultures 3 days after plating and left in the medium for 5 days. Then cells were fixed in 2% glutaraldehyde, post-osmicated and embedded on the dishes using a method described elsewhere (9). Mitochondrial dimensions were determined in transmission micrographs using methods described by Loud *et al* (10). Data for each treatment described represent the results of five separate experiments taken from 10 blocks and 50 micrographs.

For scanning electron microscopy cells were grown on coverslips (11) as described in detail elsewhere, fixed, dehydrated with alcohol and critical point dried in liquid CO_2. They were photographed in a Cambridge Stereoscan at magnifications of 600-6000 times.

For autoradiography studies cells were treated for 5 days with dex or 2 days with ethidium bromide. Then H^3-thymidine at 20 µc/ml was added for 4 hours. Cells were fixed, embedded in epon and autoradiography performed on 5µ sections. The number of grains was determined using the method of Saltpeter (11).

Results

The appearance of mitochondria in cells cultured for 5 days with dex is represented in Figure 1. Enlargement appears to occur mainly by mitochondrial lengthening and there is always an increase in the number of branched mitochondria from 0.1% in controls to 5% in treated cells. Our statistical data which are presented in much more detail elsewhere (9), show that dex caused an increase of approximately 2.5x in mitochondrial volume and an approximate halving of mitochondrial number. The volume fraction of mitochondria in the cell was unchanged by the corticoid. We found that the mitochondrial effects of dex were reversible upon removal of the hormone from the medium for 2 days. So we investigated the signal that causes mitochondria to return to normal size. We looked at the ability of drugs to inhibit the division of the megamitochondria. Two "non-mitochondrial", hydroxyurea and cycloheximide, and two "mitochondrial", ethidium bromide and chloramphenicol, inhibitors were utilized. These were added to the culture medium during the two day reversal period. Hydroxyurea at 50 µg/ml, cycloheximide at 2.8 µg/ml, and chloramphenicol at 10 µg/ml had no effect on mitochondrial morphology when added alone to the medium for 2 days. They did not cause enlargement nor a decrease in number as determined statistically. Surprisingly ethidium bromide at 0.1 µg/ml caused megamitochondria formation and these appeared similar to enlarged mitochondria formed by dex treatment. When included in the reversal medium cycloheximide and hydroxyurea had little effect in blocking the return to normal size, but both chloramphenicol and ethidium bromide completely blocked reversal. The size of mitochondria seen after 5 days of dex treatment followed by 2 days of ethidium bromide treatment was not the sum of the two treatments separately. This perhaps suggests that they act on a similar regulatory

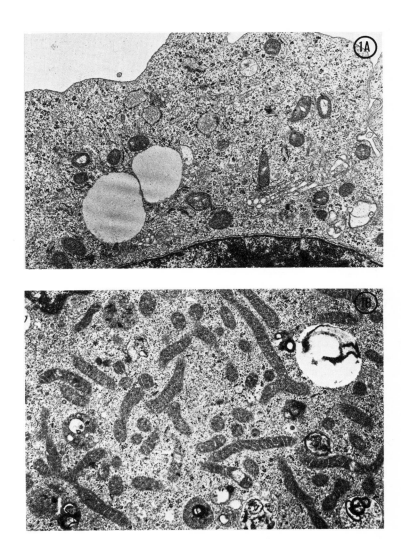

FIG. 1:

Transmission electron micrographs of control cells (a) and cells after 5 days of dex treatment to show mitochondrial changes. X 24,000

process in the cell.

Since ethidium bromide in some systems is a known inhibitor of mitochondrial DNA synthesis (12) and since cortisone is known to inhibit mitochondrial DNA synthesis in vivo (3), it seemed possible that this common inhibition could lead to megamitochondria formation. We examined the effects of these agents on DNA synthesis in our cells in the nucleus and cytoplasm by autoradiography. These results are reported in Table I. Ethidium bromide in our cells completely inhibited "cytoplasmic", i.e. mitochondrial incorporation of H^3-thymidine and only slightly effected incorporation into the nucleus, while dex inhibited nuclear and cytoplasmic DNA synthesis to the same degree. Part of the "in hibition of DNA synthesis" measured in this way, i.e. labeling after hormone treatment was probably due to permeability differences since prelabeling experiments showed about a 30% inhibition of DNA synthesis by dex after 5 days of treatment.

In addition to the mitochondrial changes observed with dex, it was noted previously that the general cell shape was altered with dex treatment (15), so we examined these changes in more detail with phase contrast and scanning electron microscopy. Within 24 hours after addition of dex, the overall cell shape changed from a more bipolar to a more epithelioid form (Fig. 2). With dex the cells appeared to be more "contact inhibited", When examined in the scanning electron microscope treated cells had a strikingly different appearance from untreated cells (Fig. 3). The number of cells possessing blebs was reduced by about 50% and the % of the total upper surface covered with projections was reduced by about 30% as determined by a modification of Weibel's methods (13). These data are analyzed in more detail elsewhere (11).

Discussion

We have developed an in vitro system for megamitochondria formation using a synthetic glucocorticoid. The mitochondria in untreated cells are much smaller than those seen in liver in vivo and so the megamitochondria are also smaller and have a different shape than those seen in vivo. For these reasons we attempted to obtain mitochondrial enlargement in primary cultures but were unsuccessful in this or in RLC cells in serum containing medium. Since the inhibition of mitochondrial and nuclear DNA synthesis by dex was similar, it seems unlikely that this sort of inhibition could cause mitochondria fusion. From our studies it appears that the signal for mitochondria to return to normal size after dex

EFFECT OF GLUCOCORTICOIDS

	Nucleus	Cytoplasm
Control	12.32 ± 0.7	4.61 ± 0.3
Dex	6.41 ± 0.6	2.41 ± 0.3
EBr	11.90 ± 0.7	0.17 ± 0.3

TABLE I. EFFECTS ON DNA SYNTHESIS OF DEX AND ETHIDIUM BROMIDE.
Values are in grains/unit area and are the average (± S.D.) of 15 fields, with approximately 20 cells/field.

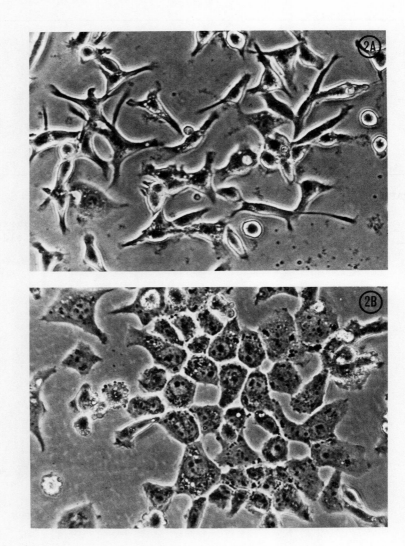

FIG. 2:

Phase contrast photographs of control cells (a) and cells treated 1 day with dex (b). X 400

EFFECT OF GLUCOCORTICOIDS

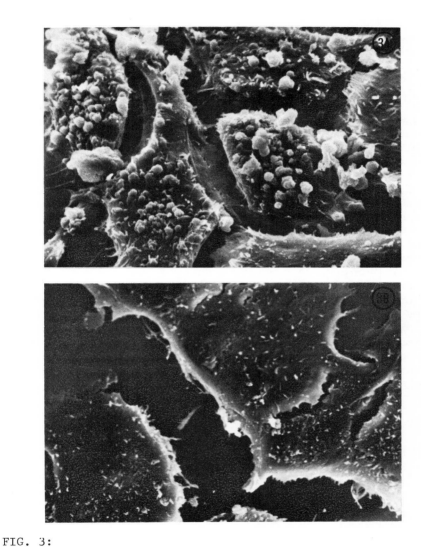

FIG. 3:

Scanning electron micrographs of (a) control cells and (b) cells treated 1 day with dex (b). X 2,160

is withdrawn from the medium can be propagated through the mitochondria since neither hydroxyurea or cycloheximide blocked reversal very effectively whereas ethidium bromide and chloramphenicol did. So although we have not really found the signal for mitochondrial fusion we have perhaps excluded some possibilities.

Our observations with the scanning electron microscope suggest gross changes in the outer cell membrane with dex treatment. Most tumor cells in culture, such as we examined, exhibit large numbers of blebs and microvilli while normal cells have considerably smaller numbers. The surface differences we observed in our cells with dex might be thought of as a change to a more normal surface configuration and thus this system could provide a model for such changes. These surface changes could also provide an explanation for the decreases in permeability that others have observed in hepatoma cells with dex (14) and further studies are in progress on this question.

In summary, we have found that dex induces 3 distinct morphological changes in our cells all of which may involve membrane alterations. We have not completed studies on the time course of the mitochondrial alterations, but the outer membrane changes begin within a day after addition of the hormone. We now have a good model to study the effect of dex on cell morphology and we are investigating more of the parameters of the signal which passes from the hormone to the membrane.

References

1. Khandekar, B., Garg, D., Tuchweber, B. and Kovacs, K. Revue Canad. Biol. 32 (1973) 121.
2. Kodama, T. and Kodama, M. Cancer Res. 32 (1972) 208.
3. Weiner, J., Loud, A., Kimberg, D. and Spiro, D. J. Cell Biol. 37 (1968) 47.
4. Kimberg, D. and Loeb, J. J. Cell Biol. 55 (1972) 635.
5. Kimberg, D., Loeb, J. and Weiner, J. J. Cell Biol. 37 (1968) 63.
6. Birchmeiser, B. Austr. Biol. Sci. 22 (1969) 965.
7. Gerschenson, L.E., Davidson, M.B. and Andersson, M. Eur. J. Biochem. 41 (1974) 139.
8. Oshiro, Y., Gerschenson, L.E. and Di Paolo, J. Cancer Res. 32 (1972) 877.
9. Berliner, J. Manuscript submitted for publication.
10. Loud, A., Barany, W. and Pack, B. Lab. Invest. 14 (1965) 996.

11. Berliner, J., de Vellis, J. and Gerschenson, L.E. Manuscript submitted for publication.
12. Smith, A., Jordan, J. and Vinograd, J. J. Mol. Biol. $\underline{59}$ (1971) 255.
13. Weible, E., Kistler, G. and Scherle, W. J. Cell Biol. $\underline{30}$ (1966) 23.
14. Risser, W. and Gelehrter, T. J. Biol. Chem. $\underline{248}$ (1973) 1248.
15. Gerschenson, L.E. Unpublished observation.

HORMONAL REGULATION OF AMINO ACID TRANSPORT IN RAT HEPATOMA CELLS IN TISSUE CULTURE

Thomas D. Gelehrter, William L. Risser and Samuel B. Reichberg

HTC cells are an established line of rat hepatoma cells in tissue culture which provide a simple system in which to study hormonal effects on hepatic protein metabolism. Incubation of these cells in a chemically-defined serum-free medium with 0.1 µM dexamethasone results in a 5-10 fold increase in the rate of synthesis of tyrosine aminotransferase without any increase in the rate of total protein synthesis (19). The addition of insulin (4 µg/ml) to HTC cells previously induced with dexamethasone causes a rapid further 2-fold increase in tyrosine aminotransferase activity. Insulin also stimulates a modest increase in the rate of total protein synthesis (7). Since insulin is known to stimulate amino acid uptake into muscle and other tissues (15), we have studied the role of amino acid transport in the effects of dexamethasone and insulin on specific and general protein synthesis in hepatoma cells.

Amino acid transport and the effects of hormones on transport have not previously been studied in HTC cells. Furthermore, although amino acid transport has been studied extensively in a variety of tissues (4), to our knowledge relatively few detailed kinetic studies of such transport in liver cells, either in vivo or in vitro, have been reported (2,5,10,12,17,18). In our initial studies we have investigated the transport of the model amino acid, α-aminoisobutyric acid (AIB) by hepatoma cells in suspension culture. AIB is transported in a number of tissues by the so-called alanine preferring system for neutral amino acids (13). Following uptake, AIB is neither incorporated into protein nor catabolized; hence its transport can be studied independently of the complications introduced by such intracellular metabolism. Subsequently we have extended our studies to the hormonal modulation of transport of the natural amino acid glycine.

Methods

Cells and Media

HTC cells were grown in spinner culture in Eagle's minimal essential medium supplemented with 5% bovine and 5%

fetal calf serum (8). Glucose content of the medium was 1 gram per liter, and amino acid content was: L-arginine, 0.60 mM; L-cystine, 0.13 mM; L-lysine, 0.40 mM; L-methionine, 0.10 mM; L-phenylalanine, 0.20 mM; L-threonine, 0.40 mM: L-tryptophan, 0.05 mM; L-tyrosine, 0.20 mM; L-histidine, 0.15 mM; L-valine, 0.40 mM; L-isoleucine, 0.40 mM; and L-leucine, 0.40 mM. Induction medium was the same as the growth medium except that it lacked serum. Cells were harvested by centrifugation while in the logarithmic phase of growth, washed with induction medium and resuspended in fresh induction medium. All experiments were carried in suspension culture at $37°$ in a gyrotory water-bath shaker. In all experiments the cells were preincubated in induction medium for 18 to 20 hours with or without hormones prior to the transport studies. Cell viability as measured by trypan blue exclusion was 70-85% after such a preincubation.

AIB Uptake

Experimental details have been described previously (16). Briefly, cell suspensions were incubated with ^{14}C - AIB as described in the legends to the figures, and samples transferred into tared, chilled tubes and centrifuged to pellet the cells. The pellets were washed to remove medium from the tube and surface of the pellet; the tube was carefully dried and reweighed after warming to room termperature. The pellets were resuspended in Protosol tissue solubilizer and counted in a triton X-toluene scintillation solution. The amount of intracellular and extracellular water in each pellet was calculated using the results of inulin space (trapped water) experiments as described previously (16). The pellet counts were corrected for trapped supernatant counts, and the CPM/ml intracellular water was calculated. The distribution ration (CPM/ml intracellular water divided by the CPM/ml extracellular water) was then determined for each sample.

Glycine Uptake

Glycine, unlike AIB, can be metabolized as well as incorporated into cellular protein; therefore different techniques were devised to study the initial rates of uptake of glycine. Following a 1 to 5 min incubation of the cells with ^{14}C -glycine, small aliquots were centrifuged rapidly through a layer of silicone oil to separate the cells from the supernatant medium. Following removal of the supernatant medium and silicone oil, the pellet was suspended in distilled water and lysed by freezing and thawing. Macromolecules were precipitated with trichloroacetic acid and

the radioactivity in the acid-soluble supernatant was
counted in a xylene-based scintillation solution (Reichberg
& Gelehrter, unpublished observations). Correction for
trapped water was made from simultaneous measurement of
[^{3}H] -methoxyinulin. The acid-recipitable pellet was dissolved in alkalai and the protein content determined by the
method of Lowry et al. (11). Glycine uptake was expressed as
pmoles/min/mg protein, or per 10^6 viable cells.

Results

Characterization of AIB Transport

The time course of AIB uptake by hepatoma cells in
suspension culture is shown in Figure 1 in which the distribution ratio is plotted against time of incubation with AIB.
Uptake of AIB was linear for 30 min and reached equilibrium
at approximately 175 min. At an AIB concentration of 0.133
mM, the distribution ratio at equilibrium was 21:1, signifying active transport against a concentration gradient.

The uptake of AIB is saturable at concentrations above
40 mM, and demonstrates Michaelis-Menten kinetics. Over a
concentration range of 0.05 mM to 40 mM, a plot of the initial velocity of AIB uptake, V, against V/S, where S is the
AIB concentration in the medium, yields a single straight
line without break, suggesting that only one transport system is involved in AIB uptake. The maximum transport velocity, V_{max}, was 1.42 $\pm$ 0.06 mMoles/l/min, and the apparent
affinity constant, K_m, was 2.18 + 0.13 mM (mean $\pm$ one SEM
of 4 experiments) (Table I).

The transport of AIB is an energy dependent process as
indicated by an approximately 40% inhibition by 5 mM sodium
cyanide, an inhibitor of oxidative phosphorylation. Glycine,
known to be a competitive inhibitor of AIB transport in several tissues, inhibited initial uptake of AIB by 50% at a
concentration 100 times that of AIB.

Effects of Dexamethasone on AIB Transport

Incubation of HTC cells with 0.1 µM dexamethasone markedly reduced the uptake of AIB. Figure 2 shows the time
course of AIB uptake in cells exposed to dexamethasone for
18 hours and then incubated with AIB in the continued presence of the steroid. At all times tested, AIB uptake was
reduced; at equilibrium the distribution ratio in dexamethasone-treated cells was 17% of that in untreated cells. Efflux of AIB was not increased from dexamethazone-treated
cells, suggesting that dexamethasone decreased AIB uptake

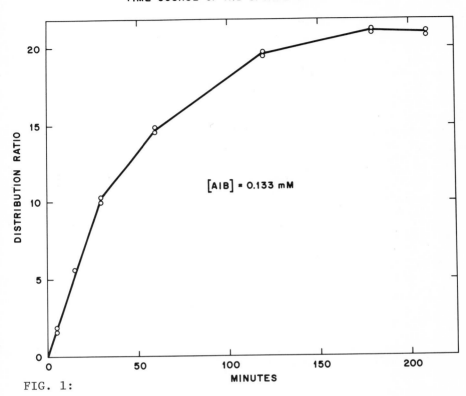

FIG. 1:

Time course of AIB uptake in HTC cells. A suspension culture of HTC cells at 1.5 x 10^6 cells/ml in induction medium was incubated with 0.133 mM AIB including 0.15 μCi/ml[^{14}C]AIB. Duplicate samples were removed at the times indicated and analyzed as described in Methods to obtain distribution ratios (Fig. 1 from ref.16).

Hormone treatment	V_{max}	K_m
	mmoles/liter/min	mM
None (control) (4)	1.42 ± 0.06[a]	2.18 ± 0.13[a]
Dexamethasone, 2 hours (1)	0.36	2.60
Dexamethasone, 18-20 hours (3)	0.28 ± 0.03[a]	7.04 ± 0.30[a]
Dexamethasone, 18 hours, then Dexamethasone + Insulin, 2 hours (2)	0.90; 1.10	4.21; 3.74

[a] mean $\pm$ one S.E.

HTC cells at 1.5×10^6 cells/ml in induction medium were incubated for 16 to 20 hours, following which they were incubated with or without 0.1 μM dexamethasone and/ or 4 μg/ml insulin for the times indicated. Aliquots of the cell suspensions were transferred to flasks containing 0.15 μCi/ml (^{14}C)- AIB and unlabeled AIB to give final concentrations of 0.132 mM, 1.32 mM, 5.03 mM, and 10 mM. After 18 minutes incubation, duplicate samples were taken and analyzed as described in Methods. Kinetic constants were obtained as described previously. The number of experiments is indicated in parentheses. (Table I from Ref. 16).

TABLE I. EFFECTS OF DEXAMETHASONE AND INSULIN ON THE KINETICS OF AIB TRANSPORT.

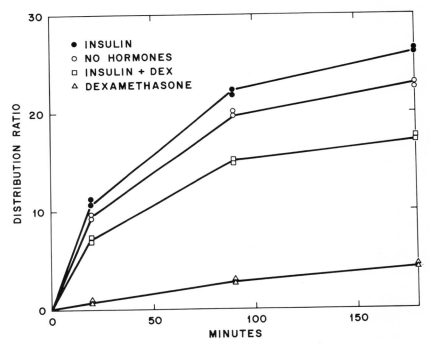

FIG. 2:

Time course of AIB uptake under various hormonal conditions. HTC cells at 1.4×10^6 cells/ml in induction medium were incubated for 18 hours with or without 0.1 μM dexamethasone. Half the dexamethasone-treated and half the untreated cells were incubated with 4 μg/ml insulin for a further two hours. AIB was then added to all cells to a final concentration of 0.5 mM including 0.15 μCi/ml [^{14}C]-AIB, and duplicate samples were taken at the indicated times and analyzed as described in Methods (Fig. 3 from ref.16).

by decreasing influx rather than by increasing AIB efflux. Cell viability, measured by the ability of the cells to exclude the dye trypan blue, was the same in cells exposed to dexamethasone and in untreated cells; therefore the decreased influx was not due to a greater proportion of dead cells after incubation with dexamethasone.

The time course of the inhibition of AIB transport by dexamethasone is shown in Figure 3, in which the distribution ratio for AIB achieved after an 18 min incubation is plotted against the time of exposure of cells to dexamethasone. After a lag of 30 min during which there was no effect of dexamethasone on AIB transport, AIB uptake decreased exponentially; the decrease in transport was half-maximal in one hour.

Kinetic constants for AIB transport were obtained after 2 hours and after 18 to 20 hours of exposure of cells to dexamethasone. As shown in Table I, in cells exposed to dexamethasone for 2 hours the capacity of the tranport system for AIB, given by V_{max}, declined to 25% of the control values, while the apparent affinity of the system for substrate, given by K_m, was essentially unchanged. After 18 to 20 hours exposure to dexamethasone, transport capacity declined still further and the apparent affinity also declined to approximately 1/3 of the control value. Thus dexamethasone rapidly decreased the capacity of the transport system for AIB and after a longer time also decreased the affinity of the transport system for substrate.

Effect of Insulin on AIB Transport

Insulin (4 µg/ml) increased the transport of AIB consistently, but only slightly in untreated cells. Figure 2 shows the time course of AIB uptake in cells exposed to insulin for 2 hours before AIB was added. The distribution ratio at equilibrium was 10 to 15% greater in insulin-treated cells than in untreated cells. Similarly, V_{max} was consistently slightly greater in insulin-treated cells than in untreated cells; K_m was unaffected.

In contrast, in cells incubated with dexamethasone for 18 hours, and then with dexamethasone plus insulin for a further 2 hours before AIB was added, insulin markedly stimulated transport, restoring uptake toward normal. The equilibrium distribution ratio increased to 5 times that in dexamethasone-treated cells, to 75% of the control value (Figure 2).

The time course of the enhancement of AIB transport by insulin in cells incubated for 23 hours with dexamethasone before insulin was added is shown in Fig. 3. Within 30

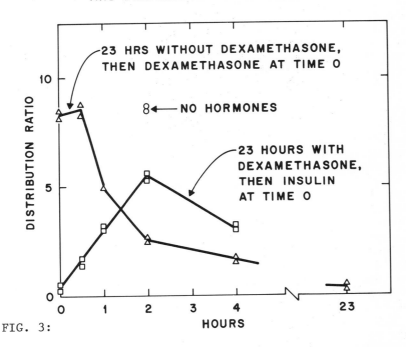

FIG. 3:

Time course of the effects of insulin and dexamethasone on AIB uptake. HTC cells at 1.5 x 10⁶ cells/ml in induction medium were incubated for 23 hours. Dexamethasone (0.1 µM) was added, and at the indicated times, aliquots of the cel suspension were transferred to 25 ml Erlenmeyer flasks containing 0.5 mM AIB with 0.15 µCi/ml [^{14}C]-AIB. After incubation for 18 minutes, duplicate 2.5 ml samples were analyzed as described in Methods. A duplicate culture of HTC cells was incubated for 23 hours with 0.1 µM dexamethasone. Insulin (µg/ml) was added, and at the indicated times, aliquots of cell suspensions were transferred to flasks containing 0.5 mM AIB with 0.15 µCi[^{14}C] -AIB. After incubation for 18 minutes, duplicate samples were analyzed as described in Methods. (Modified from figs. 4 and 6 from ref.16).

minutes insulin stimulated an apparent increase in the initial velocity of AIB transport that was maximal in 2 hours and which had again declined by 4 hours.

Kinetic constants were determined after exposure of cells to insulin for 2 hours following incubation with dexamethasone for 18 hours (Table I). The capacity of the transport system for AIB had increased to nearly 4 times that observed in the dexamethasone-treated cells, with $V_{max} = 1.00$ mMoles/l/min; and the apparent affinity of the tranport system for substrate had increased to 2 times its value in dexamethasone-treated cells, with $K_m = 3.98$ mM (average of 2 experiments).

Role of Protein and RNA Synthesis

Inhibitors of RNA and protein synthesis were used to characterize further these hormonal effects. The results are shown in Figure 4, in which the heights of the bars represent the distribution ratio for AIB after an 18 minute incubation. The first bar shows uptake of AIB in untreated cells; the distribution ratio for AIB was 10. Incubation for 2.5 hours with 0.4 µg/ml actinomycin D, a concentration that inhibits RNA synthesis by 90-95% in HTC cells, inhibited uptake of AIB by only 20%. Similarly, incubation for 2.5 hours with 0.1 mM cycloheximide, a concentration which inhibits protein synthesis by more than 98%, also inhibited AIB uptake by only 25%. These results imply that the protein or proteins involved in the AIB transport system are relatively stable under these conditions.

Exposure of cells to dexamethasone for 2 hours decreased uptake by 75%. Inhibition of protein synthesis by cycloheximide added 30 minutes prior to dexamethasone completely prevented the decrease in AIB transport caused by this hormone, implying that the effect of dexamethasone on AIB transport is dependent on protein synthesis. Inhibition of RNA synthesis by actinomycin D added 30 minutes prior to dexamethasone partially blocked the effect of dexamethasone on AIB transport (Figure 4A).

In cells exposed to dexamethasone for 18 hours (Figure 4B), AIB transport decreased in this experiment to 7% of the control value. Incubation with insulin for a further 2 hours caused a 7-fold incrase in AIB transport. Inhibition of protein synthesis by cycloheximide added 30 minutes prior to insulin prevented the increase in AIB transport caused by this hormone; again, the hormonal effect on transport appears to be dependent on protein synthesis. Inhibition of RNA synthesis by actinomycin D added 30 minutes prior to insulin partially blocked the effect of insulin on AIB transport;

EFFECT OF INHIBITORS OF RNA AND PROTEIN SYNTHESIS ON HORMONAL ALTERATIONS OF AIB TRANSPORT

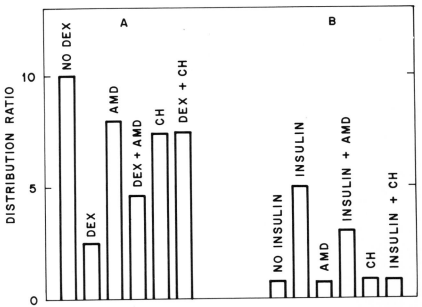

FIG. 4: <u>Effect of inhibitors of RNA and protein synthesis on hormonal alterations of AIB transport.</u> HTC cells at 1.5×10^6 per ml in induction medium were incubated 18 hr with or without 0.1 µM dexamethasone   A. Six aliquots of the untreated cells were incubated for an additional 2.5 hr with the following additions: no additions (NO DEX); no addition for 0.5 hr, then 0.1 µM dexamethasone for 2 hr (DEX); actinomycin D, 0.4 µg/ml (AMD); 0.4 µg/ml actinomycin D for 0.5 hr, then actinomycin + 0.1 µM dexamethasone for 2 hr (DEX + AMD); 0.1 mM cycloheximide (CH); 0.1 mM cycloheximide for 0.5 hr, then cycloheximide plus 0.1 µM dexamethasone for 2 hr (DEX + CH). After 2.5 hr, aliquots of the 6 cell suspensions were placed in flasks containing 0.5 mM AIB + 0.15 µCi/ml (^{14}C)-AIB & incubated 18 min. Duplicate samples were analyzed as described in Methods. B. 6 aliquots of the dexamethasone-treated cells were incubated 2.5 hr as follows: no additions (NO INSULIN); nothing for 0.5 hr, then 4 µg/ml insulin for 2 hr (INSULIN); actinomycin D (AMD); actinomycin D for 0.5 hr, then actinomycin + insulin for 2 hr (INSULIN + AMD); cycloheximide (CH); cycloheximide for 0.5 hr, then cycloheximide + insulin for 2 hr (INSULIN + CH). AIB uptake studies were then performed as described in Part A (Fig. 7 from <u>ref.</u> 16.)

transport increased only 4-fold. The inhibitors alone after 2.5 hours incubation had no effect on AIB transport in dexamethasone-treated cells.

Effect of Dexamethasone on Natural Amino Acid Transport

In order to determine whether the inhibition of amino acid transport by dexamethasone is a general phenomenon, we have studied the effect of this hormone on the transport of natural amino acids as well as AIB. Preliminary studies indicated that dexamethasone decreased by approximately 50% the initial rate of uptake of alanine, glycine, serine, phenylalanine and leucine. Glycine was chosen for more detailed study. During short incubation periods (less than 5 min) uptake of labeled glycine is linear with time and there is no significant incorporation of the labeled amino acid into protein. Under the conditions of our experiments, dexamethasone does not effect either the intracellular metabolism of glycine or its rate of incorporation into protein during this experimental period. Approximately 95% of the radioactivity in the acid soluble fraction of the cell can be recovered as glycine, as indicated by high voltage electrophoresis.

Incubation of the cells for 18 hours with 0.1 µM dexamethasone resulted in an approximately 50% decrease in the initial rates of glycine uptake indicating that, as in the case of AIB, dexamethasone inhibits influx of glycine into the cells. The time course of dexamethasone inhibition of glycine transport, however, was quite different from that observed with AIB. The reduction in glycine influx was detectable only after 5 to 10 hours and maximal only after 18 to 24 hours incubation with the hormone.

Analysis of the effect of substrate concentration on the initial rate of glycine uptake revealed the presence of at least two components for the transport of this amino acid: a high affinity ($K_m \cong 0.6$ mM), low capacity system; and a low affinity ($K_m \cong 2-3$ mM), high capacity system. Dexamethasone appeared to inhibit both components of glycine transport.

Both components of glycine uptake were also inhibited by high concentrations of unlabeled AIB. Surprisingly, this inhibition was observed even in cells treated with dexamethasone, in which AIB uptake itself is depressed by more than 90%. This latter finding raised the possibility that the hormone did not alter the initial binding step in the transport process but affected some subsequent step in the flux of the amino acid across the membrane (Reichberg & Gelehrter, in preparation).

DISCUSSION

HTC cells in suspension culture provide a simple system in which to study the effects of hormones on amino acid transport in hepatocytes. In the intact animal, it is difficult to separate the effect of one hormone from that of other hormones whose concentrations are changed directly or indirectly. Cells in tissue culture, liver slices, and isolated perfused liver all permit study of the action of a single hormone. In addition, experiments with cultured cells allow one to study isolated intact parenchymal cells free of non-parenchymal cells, and to manipulate the experimental environment in a controlled fashion.

We have found that HTC cells transport AIB by a saturable, energy-dependent, active process. A single transport system has been demonstrated which exhibits Michaelis-Menten kinetics. Dexamethasone markedly decreases AIB influx in HTC cells, quickly decreasing the capacity of the transport system for AIB, and, after a longer time, also decreasing the apparent affinity of the system for substrate. Insulin has only a slightly stimulatory effect on AIB transport in untreated cells; in contrast, in cells previously exposed to dexamethasone, insulin restores transport toward normal, rapidly increasing both the capacity and the affinity of the transport system for AIB. Both hormonal effects require protein synthesis.

The mechanisms of action of dexamethasone and insulin on AIB transport are presently unknown; however, certain characteristics of these actions have been defined. Dexamethasone depresses AIB uptake by decreasing AIB influx rather than increasing its efflux; dexamethasone-treated cells lose AIB less rapidly than do untreated cells. Furthermore, dexamethasone does not decrease cell viability, and it has been demonstrated previously that dexamethasone does not change the overall rate of cellular protein or RNA synthesis (6,19).

The effect of dexamethasone on AIB transport appears to be dependent on concomitant protein synthesis, and is completely inhibited by cycloheximide. Thus it is unlikely that dexamethasone acts directly on the transport system in the cell membrane. Similarly, it seems unlikely that dexamethasone prevents the synthesis of a protein or proteins normally active in AIB transport; such a mechanism has been proposed by Kostyo and Redmond('66) to explain the identical and non-additive suppression of AIB uptake by corticosterone and puromycin in the isolated perfused rat diaphragm.

Our observations are consistent with there being a single transport system which mediates AIB transport in untreated HTC cells. Thus it seems unlikely that dexamethasone inactivates one or two or more transport systems for AIB, uncovering a low-capacity, low-affinity system. However, this possibility cannot be excluded, for heterogeneity of amino acid transport agencies is well known in other tissues, and may be quite difficult to demonstrate unless there are significant differences between the kinetic parameters of the various agencies (3).

Although insulin stimulates a marked increase in AIB transport in dexamethasone-treated cells, it has only a slight effect on untreated cells. Thus it seems unlikely that insulin induces the appearance of a second AIB transport system that is insensitive to dexamethasone. Again, such a possibility cannot be excluded for the reasons noted above.

The observations reported here are at least consistent with the hypothesis that dexamethasone induces the synthesis of a protein or proteins that act on the AIB transport system to decrease influx of this amino acid. Insulin might then stimulate the production of another protein that antagonizes the action of the protein induced by dexamethasone.

The observation that dexamethasone inhibits the transport of several natural amino acids indicates that this hormonal effect is a general one not limited to the non-metabolized amino acid analog, AIB. The reason for the difference in the time course of hormonal inhibition of amino acid transport between glycine and AIB, which both share some common transport systems, is not presently known.

Of considerable interest is the finding that AIB competitively inhibits the transport of glycine even in dexamethasone-treated cells in which AIB transport is markedly inhibited. These data suggest that competitive inhibition does not require simultaneous transport of the inhibiting species and that such an inhibition probably occurs at the initial binding step. Furthermore, it appears that dexamethasone inhibits a step in the transport process distal to binding itself, presumably acting on some aspect of the translocation of carrier-bound amino acid through the membrane. If this hypothesis is supported by detailed competition experiments currently in progress it would provide a unique example of the dissection of the components of a complex transport process using hormones. Such an approach could be of considerable value since transport mutants, which have proven so useful in the elucidation of amino acid transport in prokaryotic organisms (13) are not currently available in mammalian cells. Although the present studies were undertaken to explore the role of substrate transport in the effects of dexamethasone and insulin on specific and general protein synthesis, they have raised more questions than they have answered. On the other hand these studies may provide a fruitful approach to the understanding of hormonal regulation of the transport process per se.

Finally, these investigations demonstrate that dexamethasone alters amino acid transport, a function occurring at the cell membrane. Ballard and Tomkins (1) found that dexamethasone modified the surface of HTC in such a way that cells more readily adhered to glass and had different electrophoretic and antigenic properties. This effect required both protein and RNA synthesis. They concluded that dexamethasone induced the synthesis of a protein which either modified the cell surface or was incorporated into surface structures. The relation, if any, between these two regulatory effects of dexamethasone is unknown; it is significant however that there are now two known specific regulatory effects of dexamethasone on HTC cell membrane function.

References

1. Ballard, P.L. and Tomkins, G.M. J. Cell Biol. 47 (1970) 222.
2. Chambers, J.W., Georg, R.H. and Bass, A.D. Molec. Pharm. 1 (1965) 66.
3. Christensen, H.N. Fed. Proc. 25 (1966) 850.
4. Christensen, H.N. Perspect. Biol. Med. 10 (1967) 471.
5. Crawhall, J.C. and Segal, S. Biochim. Biophys. Acta 163 (1968) 163.
6. Gelehrter, T.D. and Tomkins, G.M. J. Molec. Biol. 29 (1967) 59.
7. Gelehrter, T.D. and Tomkins, G.M. Proc. Nat. Acad. Sci. USA 66 (1970) 390.
8. Gelehrter, T.D., Emanuel, J.R. and Spencer, C.J. J. Biol. Chem. 247 (1972) 6197.
9. Kostyo, J.L. and Redmond, A.F. Endocrinology 79 (1966) 531.
10. Kravitt, E.L., Baril, E.F., Becker, J.E. and Potter, V.R. Science 169 (1970) 294.

11. Lowry, O.H, Rosebrough, N.J., Farr, A.L. and Randall, R.J. J. Biol. Chem. 193 (1951) 265.
12. Mallette, L.E., Exton, J.H. and Park, C.R. J. Biol. Chem. 244 (1969) 5724.
13. Oxender, D.L. and Christensen, H.N. J. Biol. Chem. 238 (1963) 3686.
14. Oxender, D.L. Metabolic Pathways 6 (1972) 133.
15. Riggs, T.H. Hormones and transport across cell membranes, in Biochemical Actions of Hormones (ed. G. Litwack) Ac. Press, N.Y., vol. 1, p. 157.
16. Risser, W.L. and Gelehrter, T.D. J. Biol. Chem. 248 (1973) 1248.
17. Scott, D.F., Reynolds, R.D., Pitot, H.C. and Potter, V.R. Life Sci. 9 (1970) 1133.
18. Tews, J.K. and Harper, A.E. Biochem. Biophys. Acta 183 (1969) 601.
19. Tomkins, G.M., Thompson, E.B., Hayashi, S., Gelehrter, T., Granner, D. and Peterkofsky, B. Cold Sprg. Hrbr. Symp. Quant. Biol. 31 (1966) 349.

Note: We would like to thank the publishers of the J. Biol. Chem. for the permission to reproduce Table I and the figures indicated from ref. 16.

REGULATION OF SPECIFIC PROTEIN SYNTHESIS IN CULTURED HEPATOMA CELLS BY ANALOGS OF CYCLIC AMP

Wesley D. Wicks, Kay Wagner, Michael D. Roper, Ben H. Leichtling and Jayantha Wimalasena.

Abstract

A series of new 6- and 8- substituted analogs of cyclic AMP have been tested for their abilities to increase the activity of tyrosine aminotransferase in cultured Reuber H-35 hepatoma cells. The analogs were found to fall into two basic groups, those active and those inactive as inducers. The various changes in enzyme activity were shown to be paralleled by corresponding alterations in the relative rate of aminotransferase synthesis. Several of the analogs proved to be as potent as Bt_2cAMP and generated comparably rapid increases in the activity of the aminotransferase. An excellent correlation was found to exist between the ability of any analog to increase the activity of the aminotransferase and that of phosphoenolpyruvate carboxykinase. The analogs were essentially equipotent, including non-inducers, as activators of protein kinase in rat liver crude extracts. Preliminary studies have shown that, in contrast, only derivatives active as enzyme inducers are capable of activating protein kinase in whole cells. The suggestion is put forward that protein kinase mediates the effects of cAMP analogs on enzyme induction.

Introduction

Previous reports from this laboratory have documented the fact that the synthesis of two minor liver proteins is subject to regulation by cyclic AMP (cAMP) (1-4). Cultured Reuber H35 hepatoma cells have been chosen as the main biological system for most of the recent studies. Although the responsiveness of the adenylate cyclase system is defective in H-35 cells, the internal response to cAMP (in the form of various analogs) is identical, for all intents and purposes to that in rat liver (3-5). Thus, the activities and rates of synthesis of the two proteins, tyrosine aminotransferase (TAT) and phosphoenolpyruvate carboxykinase (PEPCK), are elevated 2-4 fold by $N^6, O^{2'}$-dibutyryl cAMP (Bt_2cAMP) over the same time intervals in both rat liver and H-35 cells (2-5).

In both systems the onset of the response is rapid and removal of the cyclic nucleotide prompts a rapid first order decay in both elevated enzyme activities (3-5). Actinomycin D does not block the early rise in both activities but is ultimately inhibitory (2,5). These results have been interpreted to suggest that Bt_2cAMP acts by overcoming some rate-limitation in enzyme synthesis at a post-transcriptional site, possibly at the polysomal level (1-5).

As one approach to elucidating the mechanism by which the synthesis of these two proteins is regulated by cAMP, we have asked the question whether protein kinase plays a mediating role in this process or not. Use has been made of a variety of new 6- and 8- substituted analogs of cAMP which possess diverse abilities to mimic the effects of cAMP in several biological systems (6-12). As will be seen, these properties have enabled us to demonstrate an excellent correlation between induction of TAT by any given analog with that of PEPCK. These results are consistent with the existence of a common intracellular mediator of the action of cAMP such as protein kinase. Studies with other hepatoma cell lines where one or more response to cAMP analogs is altered are also consistent with a role for protein kinase in enzyme induction. Finally, preliminary experiments have shown that protein kinase activity in <u>whole cells</u> (monitored by site-specific f_1 histone phosphorylation) is enhanced only by cAMP analogs that are enzyme inducers and not by non-inducers. These results support the conclusion that cAMP exerts its effects on TAT and PEPCK synthesis by virtue of activating protein kinase.

Materials and Methods

Reuber H-35 cells (H4-II-E-C-3) were maintained in monolayer culture in $25cm^2$ Falcon plastic flasks in a dry incubator at 35-37 C(1-4). Eagle's basal medium enriched 4-fold in vitamins and amino acids containing 10% calf serum and 5% fetal calf serum was used as the growth medium. Cells near confluency were used in most studies after overnight incubation in serum-free medium containing penicillin and streptomycin. The assays used for TAT and PEPCK have been described elsewhere (13). Protein was determined by the Lowry method (14). The immunochemical procedure used to measure the relative synthetic rate of TAT has been described in detail (1,3,5,13). Protein kinase activity was monitored by the incorporation of γ-^{32}P ATP into purified f_1 histone. Calf thymus f_1 histone was purified before use (15). The tryptic peptide of f_1 histone containing the serine-37 residue was isolated by paper electrophoresis and thin layer chromatography according to the techniques of Langan (16).

All the cyclic AMP analogs were purchased commercially or were generous gifts from Drs. R. K. Robbins, M. Stout, and L. N. Simon of the ICN Nucleic Acid Research Institute. The various phosphodiesterase inhibitors were obtained through commercial sources or were kindly given to us by Drs. F. F. Giarusso of the Squibb Institute (SQ 20006) or W. E. Scott of Hoffman-LaRoche Inc. (RO-20-1724). All isotopically labelled compounds were purchased from NEN Corp., Boston, Mass., except for $NaH^{14}CO_3$ which came from Amersham-Searle Corp., Chicago. All tissue culture components were obtained from Grand Island Biological Co., Grand Island, N. Y.

Results and Discussion

Although the $\underline{N}^6,\underline{O}^{2'}$-dibutyryl analog of cAMP causes a rapid 2-4 fold increase in TAT activity in H-35 cells, cAMP itself is only a very weak inducer even at exceedingly high concentrations (Table I). Theophylline has little effect when added alone but does enhance the effects of cAMP to a modest extent. $\underline{O}^{2'}$-monobutyryl cAMP is also a rather feeble inducer but $\underline{N}^6$-monobutyryl cAMP and Bt_2cAMP are both very effective even in the absence of phosphodiesterase inhibitors. The presumed basis for this difference appears to be two-fold; first, the butyrated analogs are reputed to penetrate cells more readily than cAMP. Secondly, $\underline{N}^6$-monobutyryl cAMP and Bt_2cAMP are resistant to attack by phosphodiesterase whereas $\underline{O}^{2'}$-monobutyryl cAMP and cAMP are not (17).

With respect to the first point, we have observed that Bt_2cAMP does enter H-35 more readily (in the sense that it remains intact) than cAMP (18). In preliminary studies we have found that more actual cpm are found inside the cells after 60 mins exposure to ^3H cAMP, but the bulk of these are present in ATP and ADP. Similar results have been reported by Michal et al (19) in perfused liver and by Kaukel and coworkers in HeLa cells (20-21).

In contrast to cAMP itself, a number of new 6- and 8-substituted analogs of cAMP, which are devoid of butyryl groups, are very effective in elevating TAT activity. The kinetics of their effects are similar to that observed with Bt_2cAMP (1-4) (Figure 1). The response is rapid and the maximal increase is obtained by 3-4 hours.

The analogs fall into two groups in terms of TAT induction: active and inactive compounds. The active analogs exhibited either high, intermediate, or moderate potencies (Figure 2). Although not shown, N^6-$CH_2C_6H_5$-cAMP is the most

Cyclic Nucleotide Added	Concentration	Tyrosine Aminotransferase (%increase in Activity)
$O^{2'}$-monobutyryl cyclic 3':5'-AMP	0.5 mM	0 (6)
$O^{2'}$-monobutyryl cyclic 3':5'-AMP	2.0 mM	85 (2)
Cyclic 3':5'-AMP	0.8 mM	0 (7)
Cyclic 3':5'-AMP	1.6 mM	8 (4)
Cyclic 3':5'-AMP	10.0 mM	21 (1)
Cyclic 3':5'-AMP	20.0 mM	2 (1)
Cyclic 3':5'-AMP + Theophylline	0.8 mM 1.0 mM	0 (3)
Cyclic 3':5'-AMP + Theophylline	10.0 mM 1.0 mM	55 (2)
Cyclic 3':5'-AMP + Theophylline	20.0 mM 1.0 mM	35 (2)

Confluent H35 cells were placed in serum-free for 16-18 hours. The various additions were made and 5 hours later the cells were harvested for assays except with 10 mM and 20 mM cyclic 3':5'-AMP and $O^{2'}$-monobutyryl cyclic AMP which were harvested at 3 hours. The values shown are the average percentage increase in aminotransferase activity observed in one or two experiments with the number of separate observations in parenthesis.

TABLE I: ABILITY OF VARIOUS CYCLIC NUCLEOTIDES TO INFLUENCE TYROSINE AMINOTRANSFERASE ACTIVITY IN H35 CELLS*

*from Wagner et al. (28) adapted

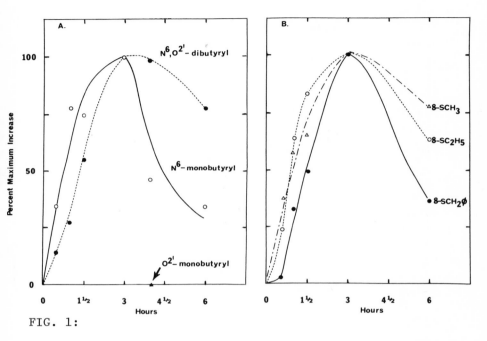

FIG. 1:

<u>Time course of effects of cyclic AMP analogs on tyrosine aminotransferase activity in H35 cells.</u> All analogs were added at 0.5 mM and cells were harvested for assay at the times indicated as described in Experimental Procedures. Each value represents the average of 4-6 observations with standard errors of 5-10%. The maximum fold increase (i.e. 100% of the maximum response) in aminotransferase activity observed with each analog is given in brackets.

A. $N^6,O^{2'}$-dibutyryl cyclic AMP [3.8] (●---●), N^6-monobutyryl cyclic AMP [2.7] (o---o), $O^{2'}$-monobutyryl cyclic AMP [0] (▲).

B. 8-SCH_3 cyclic AMP [2.6] (Δ---Δ), 8-SC_2H_5 cyclic AMP [2.9] (o---o) and 8-$SCH_2C_6H_5$ [3.5] (●---●). Taken from Wagner et al (28).

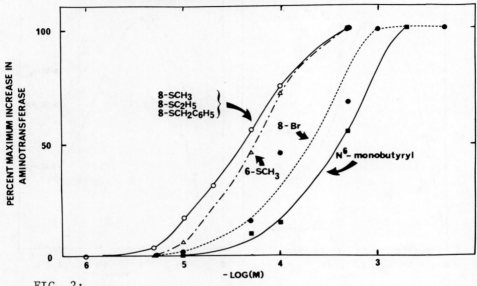

FIG. 2:

<u>Dose-response relationships among cyclic AMP analogs for effects on tyrosine aminotransferase activity in H35 cells.</u>
Analogs were added at the final concentrations indicated and cells were harvested 4 hr later for assay as described in Experimental Procedures. Each value represents the average of 6 separate observations with average standard errors of 5-10%. The maximum fold increase (i.e. 100% of the maximum response) in aminotransferase activity is given in brackets. The curves for the 8-SCH_3, 8-SC_2H_5 and 8-$SCH_2C_6H_5$ cyclic AMP analogs were indistinguishable and have been combined [8-SCH_3, 2.6; 8-SC_2H_5, 3.4; 8-$SCH_2C_6H_5$, 2.5] (o---o), N^6-monobutyryl cyclic AMP [4.3](■——■), 6-SCH_3 cyclic AMP [2.2] (Δ---Δ), 8-Br cyclic AMP [3.5] (●---●). Taken from Wagner et al (28).

potent analog we have tested with an ED_{50} of $\sim 10\mu M$, some 5 times lower than that for Bt_2cAMP (1). The non-inducers include the $8-NH_2-$, $8-NHC_2H_4OH-$ and $6-SH-cAMP$ analogs which actually depress basal TAT activity. The reason for the great discrepancy between the ED_{50} for Bt_2cAMP and that for N^6-monobutyryl cAMP is not understood at present.

A number of the analogs were tested for their ability to influence the relative rate of TAT synthesis (Table II). An excellent correlation was found to exist between the degree of increase in enzyme activity and that in relative rate of synthesis. A small but inconsistent increase in total soluble protein labelling was detected which may result from an increase in amino acid uptake (22). The non-inducing analogs actually inhibited general protein synthesis which may account for their inability to induce TAT.

Bt_2cAMP has been found to induce both TAT and PEPCK to similar extents in adult rat liver (5) and in H-35 cells (1). Glucagon and isoproterenol also produce comparable increases in the activity of both enzymes in fetal and adult rat liver (13). These results have led to the suggestion that the synthesis of these two proteins is regulated by a common intracellular component. Considerably stronger support for this suggestion has resulted from studies with the new cAMP analogs. Using a fixed concentration and time period, an excellent correlation has been found to exist between the ability of any given analog to increase TAT activity and that of PEPCK (Fig. 3).

At least two possible explanations can be proposed to explain the observed correlation.

1) The analogs inhibit phosphodiesterase activity to varying extents which leads to increases in the intracellular cAMP concentration thereby generating the observed differing degrees of response.

2) The analogs directly mimic the action of cAMP by interacting to varying degrees with some intracellular component which generates the observed biological responses.

Three pieces of evidence argue against the first suggestion. First, although many of the present analogs are potent phosphodiesterase inhibitors, one, 8-OH-cAMP is essentially inactive in this regard and yet is a potent enzyme inducer (10,11). $\underline{O}^{2'}$-monobutyryl cAMP is a more effective phosphodiesterase inhibitor than $\underline{N}^6$-monobutyryl cAMP (17) and yet is a very weak inducer. Second, well-known inhibitors of phosphodiesterase exhibit little, if any, ability to elevate TAT activity (Table III). Third, in preliminary experiments, we have observed no elevation in the low intracellular concentration of cAMP after incubation of H-35 cells with an

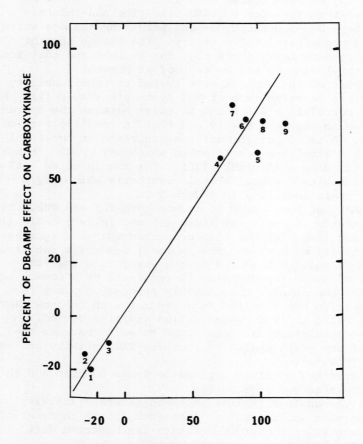

FIG. 3: <u>Correlation of effects of cyclic AMP analogs on tyrosine aminotransferase and PEP carboxykinase activities in H35 cells</u>. The analogs were all added at 0.5 mM & cells were harvested for assay at 5 hr as described in Experimental Procedures. Each value is the mean of at least 3 observations, & on the average 8, with standard errors of 10-20%. The average increase in activity generated by dibutyryl cyclic AMP (DBcAMP) was 2.9-fold for the aminotransferase & 2.0-fold for the carboxykinase. The values for analogs #1,2 & 3 are given as % <u>decrease</u> in control activity. 1. 8-NHC_2H_4OH cAMP; 2. 8-NH_2 cAMP; 3. 6-SH cAMP; 4. 6-SCH_3 cAMP; 5. 8-SC_2H_4OH cAMP; 6. 8-Br cAMP; 7. 8-$SCH_2C_6H_5$ cAMP; 8. 8-SC_2H_5 cAMP; 9. 9-SCH_3 cAMP. Taken from Wagner, et al (28).

TABLE 2: EFFECTS OF ANALOGS OF CYCLIC AMP ON THE RELATIVE RATE OF SYNTHESIS OF TYROSINE AMINOTRANSFERASE IN CULTURED H35 CELLS*

Cyclic Nucleotide Added	Tyrosine Aminotransferase Activity	Radioactivity in:		Relative Synthetic Rate
		Tyrosine Aminotransferase	Total Soluble Protein	
	(% of control)	(cpm) A	(cpm x 10^{-3}) B	A/B x 10^3
None	100± 9 (5)	1609±112 (5)	532±22 (5)	3.0±0.3 (5)
$N^6,O^{2'}$-Dibutyryl cyclic AMP	358±69 (2)	6681±379 (2)	732±11 (2)	9.1±0.7 (2)
8-SCH_3 cyclic AMP	288±10 (2)	4503± 68 (2)	712±14 (2)	6.3±0.2 (2)
N^6-Monobutyryl cyclic AMP	219±17 (2)	2883±435 (2)	547±47 (2)	5.3±0.3 (2)
8-$SCH_2C_6H_5$ cyclic AMP	190± 7 (2)	3196±174 (2)	699±28 (2)	4.6±0.1 (2)
8-NH_2 cyclic AMP	97±11 (2)	878±115 (2)	470± 6 (2)	1.8±0.2 (2)
8-NHC_2H_4OH cyclic AMP	63± 2 (2)	736± 6 (2)	456± 4 (2)	1.6±0 (2)

Confluent H35 cells were placed in serum-free medium for 14 hours and then into serum-free medium containing 1/4 the usual leucine concentration. 2 hours later the various analogs listed were added at 0.5 mM. 3 1/2 hours later 25 μCi of [^3H]-leucine was added and 30 minutes later the cells were harvested. After lysis, tyrosine aminotransferase was partially purified and incubated with anti-aminotransferase antibodies. The resulting immunoprecipitate was washed 3 times with cold saline and then counted. The values shown are the mean ± derivation. Each value was that obtained with 4 pooled flasks.

*Taken from Wagner et al. (28)

Inhibitor Added	Concentration	Tyrosine Aminotransferase
		(% Increase in Activity)
Theophylline	1.0 mM	6 (6)
Papaverine	0.1 mM	0 (9)
Methylisobutylxanthine	0.1 mM	0 (9)
RO-20-1724	0.5 mM	32 (9)
SQ-20006	0.5 mM	0* (9)

Confluent H35 cells were placed in serum-free medium 16-18 hours before additions. Cells were harvested for assays 4 hours later in one group of experiments (6 flasks) and 24 hours later in the other (3 flasks except for theophylline). The values from the two time points were combined since the results were similar

*Actually caused a decrease in aminotransferase activity

TABLE 3: ABILITY OF INHIBITORS OF PHOSPHODIESTERASE TO INFLUENCE TYROSINE AMINOTRANSFERASE ACTIVITY IN H35 CELLS*

*Taken from Wagner et al. (28)

analog active as an inducer.

Although none of these considerations can eliminate the possibility that phosphodiesterase inhibition may occur as a result of exposure of H-35 cells to these analogs, they make it unlikely that this is a primary mechanism by which these derivatives act.

The second possibility, that of direct action, is supported by the fact that many of the analogs tested are potent activators of protein kinase (7,8,11,12). We have also found that most of the analogs, including non-inducers, are able to activate rat liver protein kinase with f_1 histone as substrate (Table IV). It is of interest to note that Bt_2cAMP is a very poor activator of protein kinase in vitro as has also been reported by others (11,17,20). These results would appear to argue against a role for protein kinase in enzyme induction. However, a possible resolution of this apparent inconsistency has been provided by preliminary analyses of site-specific f_1 histone phosphorylation in intact H-35 cells. All the analogs tested which are active as inducers have been found to increase phosphorylation of the serine-37 residue of endogenous f_1 histone. In marked contrast, the three non-inducing analogs (8-NH_2-, 8-NHC_2H_4OH-and 6-SH-cAMP) are totally unable to activate protein kinase in vivo. (This analysis can be considered as a functional assay of the degree of protein kinase activation based on the extensive studies of Langan (16,23). These results provide the strongest and most direct evidence for the involvement of protein kinase in enzyme induction. Bourne, Tomkins, and Lion (24) have also recently obtained evidence consistent with the participation of protein kinase in the regulation of protein synthesis.

The reason for the effectiveness of Bt_2cAMP in whole cells as an activator of protein kinase and inducer appears to result from its conversion to N^6-monobutyryl cAMP (and or cAMP itself). This analog is an effective protein kinase activator both in vivo and in vitro and is a good inducer. The removal of the $O^{2'}$-butyryl group from Bt_2cAMP, which appears to severely hinder or abolish binding to protein kinase (11,17,20), has been shown to be catalyzed by an esterase present in rat liver (25). The reason that the non-inducing analogs only activate protein kinase in vitro may be that they are converted (by some process as yet unknown) to metabolites which are no longer able to activate the kinase. We are in the process of preparing radioactive analogs in order to examine this possibility.

The hypothesis that protein kinase mediates all the effects of cAMP analogs predicts that cellular variants ought

Cyclic Nucleotide Added	ATP^{32} Incorporated Into f_1 Histone	Fold Increase in Kinase Activity
	(pmoles/min/mg/protein)	
None	428 ± 24 (7)	
Cyclic AMP	1630 ± 42 (7)	3.8
N^6-Monobutyryl cyclic AMP	1660 ± 22 (4)	3.9
$N^6, O^{2'}$-Dibutyryl cyclic AMP	1028 ± 26 (2)	2.4
8-SCH_3 cyclic AMP	1654 ± 66 (7)	3.9
8-SC_2H_5 cyclic AMP	1598 ± 100 (2)	3.7
8-$SCH_2C_6H_5$ cyclic AMP	1969 ± 94 (2)	4.0
8-Br cyclic AMP	1718 ± 12 (2)	4.0
8-NH_2 cyclic AMP	1662 ± 26 (2)	3.9

50 µl of a 1:50 dilution of a rat liver postmitochondrial supernantant preparation was used as the enzyme source (25 mg protein per ml undiluted). Purified calf thymus f_1 histone (0.25 mg per tube) was used as the protein substrate. All the nucleotides were added at 1 µM. Incubation was continued for 20 minutes with ATP-γ-^{32}P (10, 325 cpm/nmole; 1.29 x 10^6 cpm total), 50 mM Tris pH 7.5, 5 mM $MgCl_2$ and 1 mM dithiothreitol at 37°C. Each value is the mean ± SEM with the number of observations in parenthesis.

TABLE 4: ACTIVATION OF RAT LIVER PROTEIN KINASE BY ANALOGS OF CYCLIC AMP*

*Taken from Wagner et al. (28)

Cell Line	Additions	Concentration	Tyrosine Aminotransferase	PEP Carboxykinase
			(units/mg protein)	
H35	None		36.3 (13)	51.4 (13)
	Bt_2 cAMP	0.5 mM	107.3 (12)	95.4 (13)
	Dexamethasone	0.2 uM	171.4 (8)	94.9 (8)
MH_1C_1	None		46.1 (16)*	36.6 (11)
	Bt_2 cAMP	0.5 mM	57.8 (18)*	72.7 (13)
	Dexamethasone	0.2 μM	113.8 (10)	112.4 (10)
HTC	None		19.3 (5)	4.5 (5)
	Bt_2 cAMP	0.5 mM	26.0 (6)	4.7 (6)
	Dexamethasone	0.1 μM	42.7 (6)	4.9 (6)

Cells near confluency (6-8 days except 12-14 days with MH_1C_1 cells) were placed in serum-free medium 16-20 hours prior to addition of inducers. Cells were harvested 5 hours after additions for assays as described in materials and methods. Each value represents the average of the number of observations shown in parentheses with standard errors ranging from 5 to 15%.

*The disparity in the number of observations between the two enzymes is due to the fact that the results of one experiment in which the carboxykinase was not assayed are included.

TABLE 5: ENZYME INDUCTION BY Bt_2 cAMP AND DEXAMETHASONE IN VARIOUS HEPATOMA CELL CULTURES*

* Taken from Van Rijn et al. (18)

to be found in which one or more of the responses to cAMP ought to be altered or missing altogether. Two existing hepatoma cell lines have in fact been found to exhibit such behavior. MH_1C_1 cells possess normal levels of both TAT and PEPCK and each is inducible by glucocorticoids (Table V). In contrast, only PEPCK exhibits significant induction by Bt_2-cAMP. In HTC cells the levels of both enzymes are much lower and only TAT exhibits any increase in activity. Bt_2cAMP and other analogs active as inducers have been shown to inhibit the growth of H-35 cells by reducing the rate of DNA replication (26,27). MH_1C_1 cells exhibit a similar growth response to Bt_2cAMP and other analogs but the growth of HTC cells is totally unaffected by exposure to these cAMP derivatives (18).

The data described in this paper lend support to the suggestion that protein kinase plays a crucial role in the regulation of specific enzyme synthesis by cAMP analogs. However, other theories can be proposed which also accomodate these results. The involvement of protein kinase in enzyme induction cannot be accepted as certain until such time as a key phosphorylated intermediate is isolated whose function is altered by phosphorylation in a manner consistent with its proposed role.

References

1. Barnett, C.A. and Wicks, W.D. J. Biol. Chem. 246 (1971) 7201.
2. Wicks, W.D. and McKibbin, J.B. Biochem. Biophys. Res. Commun. 48 (1972) 205.
3. Wicks, W.D., Van Wijk, R. and McKibbin, J.B. Adv. Enzyme Regul. 11 (1973) 117.
4. Wicks, W.D., Barnett, C.A. and McKibbin, J.B. Fed. Proc. 33 (1974) 1105.
5. Wicks, W.D., Kenney, F.T. and Lee, K-L J. Biol. Chem. 244 (1969) 6008.
6. Posternak, T. and Cehovic, G. Amer. N.Y. Acad. Sci. 185 (1971) 42.
7. Muneyama, K., Bauer, R.J., Shuman, D.A., Robins, R.K. and Simon, L.N. Biochemistry 10 (1971) 2390.
8. Meyer, R.B., Shuman, D.A., Robbins, R.K., Bauer, R.J., Dimmitt, M.K. and Simon, L.N. Biochemistry 11 (1972) 2704.
9. Michal, G., DuPlooy, M., Woschee, M., Nelbock, M. and Weimann, G. Anal. Chem. 252 (1970) 183.
10. Free, C.A., Chasin, M., Paik, V.S. and Hess, S.M. Biochemistry 10 (1971) 3585.

11. Simon, L.N., Shuman, D.A. and Robbins, R.K. Adv. Cyclic Nucleotide Res. 3 (1973) 225.
12. Bauer, R.J., Swiatek, K.R., Robbins, R.K. and Simon, L.N. Biochem. Biophys. Res. Communs. 45 (1971) 526.
13. Wicks, W.D. J. Biol. Chem. 244 (1969) 394.
14. Lowry, O.H., Rosebrough, N.J., Farr, A.L. and Randall, R.J. J. Biol. Chem. 193 (1951) 265.
15. Johns, E.W. Biochem. J. 92 (1964) 55.
16. Langan, T.A. J. Biol. Chem 244 (1969) 5763.
17. Miller, J.P., Shuman, D.A., Dimmitt, M.K., Stewart, C.M., Khwaja, T.A., Robbins, R.K. and Simon, L.N. Biochemistry 12 (1973) 1010.
18. Van Rijn, H., Bevers, M.M., Van Wijk, R. and Wicks, W.D. J. Cell Biol. 60 (1974) 181.
19. Michal, G., Woschee, M. and Wanderwald, P. Life Sci. 10 (1971) Part II, 841.
20. Kaukel, E., Mundhenk, K. and Hilz, H. Eur. J. Biochem. 27 (1972) 197.
21. Hilz, H. and Kaukel, E. Mol. and Cell. Biochem. 1 (1973) 229.
22. Chambers, J.W., George, R.H. and Bass, A.D. Endocrinology 87 (1970) 366.
23. Langan, T.A. Amer. N.Y. Acad. Sci. 185 (1971) 166.
24. Bourne, H.R., Tomkins, G.M. and Lion, S. Science 181 (1973) 952.
25. Blecher, M. and Hunt, N.H. J. Biol. Chem. 247 (1972) 7479.
26. Van Wijk, R. Clay, K. and Wicks, W.D. Cancer Res. 32 (1972) 1905.
27. Van Wijk, R., Wicks, W.D., Bevers, M.M., Van Rijn, H. Cancer Res. 33 (1973) 1331.
28. Wagner, K., Roper, M.D., Leichtling, B.H., Wimalasena, J. and Wicks, W.D. J. Biol. Chem. in press.

This work was supported in part by NIH Grants AM 16753, AM 53197 and GM 01983 and by a grant from the American Diabetes Association. We wish to thank the group at ICN Nucleic Acid Research Institute for their invaluable assistance in providing us with cAMP analogs.

Note: We would like to thank the publishers of the J. Biol. Chem. for the permission to reproduce Table I.

HORMONAL EFFECTS ON TWO RAT LIVER CELL LINES CULTURED IN
CHEMICALLY DEFINED MEDIUM

L.E. Gerschenson

The establishment of a spectrum of cell lines derived from liver or hepatomas, differing in their retention of liver-specific functions, should provide us with useful models for genetics and differentiation studies not only in liver but in all mammalian cells.
In my own laboratory, we have established several cell lines derived from normal rat liver and determined their response to several hormones with the hope that the experimental system(s) could be used for hormonal mechanism of action and/or recombinational studies for the analysis of the regulation of the gene expression in mammalian cells. This presentation concerns the description of the multiple hormonal regulation of the enzyme tyrosine α-ketoglutarate transaminase (TKT) in 2 epithelioid rat liver cell lines cultured in chemically defined medium.

Materials and Methods

RLC-GAI and BRLC-GAI, two aneuploid and epithelioid tissue culture lines derived from rat liver using different techniques (1,2) and adapted to grow in chemically defined medium by a described technique (3) were grown in monolayer culture by established methods (3). The enzyme activity, hydroxyapatite chromatography and heat stability of TKT were assayed as previously described (4). Protein content was determined by means of Lowry's technique (5).
The hormones dexamethasone sodium phosphate (Dex), insulin (I), glucagon (G) and epinephrine (E) were dissolved in distilled water (Dex), 0.003 N HCl (I and G) or 0.003N NaOH(E). Dibutyril cyclic AMP (cAMP) was also diluted in distilled water. Concentrated solutions were prepared shortly before addition to the culture medium in aliquots of 10-50 µl per 5 ml of medium

Results and Discussion

The enzyme TKT was chosen as a marker for hormonal response, since it has been found that many hormones (glucocorticoids, insulin, glucagon and epineprhine (6)) induced, in liver, an increase in its specific activity, which appeared to be due to an increase in de novo synthesis

CELL LINES CULTURED IN CHEMICALLY DEFINED MEDIUM

CELL LINE		TKT LEVELS ($mMoles \times min^{-1} \times mg\ protein^{-1}$)	
RLC-GAI	BASAL:	5 - 10	
	DEX. INDUCED:	30 - 100	(3-10 FOLD INCREASE)
BRLC-GAI	BASAL:	50 - 70	
	DEX. INDUCED	120 - 170	(2-3 FOLD INCREASE)

TABLE 1: Basal and Dex-induced levels of TKT (After 18 hours of incubation with $10^{-5}M$ Dex) in 2 different cell lines. Results indicate ranges among 15 different experiments.

CELL LINE	I	E	G	CAMP
RLC-GAI	0	0	0	0
BRLC-GAI	80% (4H)	60% (4H)	80% (4H)	200% (2H)
	DX + I	DX + E	DX + G	DX + CAMP
RLC-GAI	60% (1H)	49% (1H)	40% (1H)	82% (1H)
BRLC-GAI	195% (2H)	240% (2H)	140% (2H)	210% (2 H)

TABLE 2: MULTIPLE HORMONAL REGULATION OF TKT IN TWO DIFFERENT CELL LINES

Results expressed as percentage of increase above controls (No hormones added above. Dex added below) after incubation with the hormones at times indicated by number of hours in parentheses, which also showed the time at which the enzyme activity reached a plateau. When indicated (Dx+) cells were incubated with $10^{-5}M$ Dex for 18 hours prior to $5 \times 10^{-8}M$ I, $10^{-4}M$ G, $10^{-3}M$ E or $10^{-4}M$ CAMP additions.

(6). However, the mechanisms by which these changes were elicited as well as their physiological significance are still under investigation.

The synthetic steroid Dex was found to induce the enzyme TKT in the two cell lines studied. However, the basal enzyme activity as well as the induced levels were different (Table I). Table 2 shows that peptide hormones, the catecholamine and cAMP did not change the specific activity of TKT of RLC-GAI cells unless they were pretreated with Dex. All the compounds assayed above induced an increase in TKT of BRLC-GAI cells without a need for incubation with Dex. However, pretreatment with the steroid induced a quantitative and qualitative change in the response. These effects of the glucocorticoid in enhancing the other hormones' effects are interpreted as similar to the described "permissive" roles of glucocorticoids in hepatic carbohydrate metabolism (7) and other experimental systems (8).

Using RLC-GAI cells it was found (4) that the lower and ineffective concentration of Dex were, however, effective "per se" in making the cells insulin-sensitive. Optimal concentrations of the glucocorticoid increased the enzyme activity 3 to 4-fold during the first 8-9 h of incubation, however, the cells were not insulin-sensitive until the 10th hour of incubation (Fig. 1, Ref. 4). These findings suggested that Dex induced a factor(s) necessary for the insulin effect(s). Our actual research hypothesis on this "permissive" effect is the interaction between steroid and other hormones through qualitative or quantitative regulation of the other hormone receptors.

Berliner (9) described changes in the surface of RLC-GAI cells, which are induced by Dex. This finding supports the possibility of glucocorticoid regulation of hormonal receptors located in the cell membrane.

Since the existence of multiple forms of TKT might complicate the interpretation of our results, we used a modification (4) of an established technique (10) to separate by chromatography different forms to TKT. Two forms of this enzyme were at least observed in RLC-GAI cells. Both forms were inducible by Dex (Fig. 2 Ref.4). One of the forms (I) was extremely heat-labile (65°C) while the other form (II) was completely heat-resistant. Incubation of the cells with Dex increased significantly the resistance of form I to heat treatment but did not change the heat stability of form II (Fig. 3, Ref. 4). The kinetics of the heat stability curve of form I suggested the existence of more than one form after Dex treatment of the cells (Fig. 3-A).

In summary, we believe that the use of the above

described cell lines cultured in chemically defined medium provides us with helpful models a) to study mechanism of hormonal action and b) to use as parents for cell hybridization, assuming that their lack of growth-dependence on serum would make it possible to prepare hybrids with serum-dependent cell lines and then eliminate them by plating in chemically defined media.

This work was supported by Contract AT (04-1) GEN 12 between the Atomic Energy Commission and the University of California and by Grants-in-Aid from the Diabetes Association of Southern California. The excellent assistance of Mrs. M. Andersson and J. Yang is gratefully acknowledged.

References

1. Gerschenson, L.E., Andersson, M., Molson, J. and Okigaki, T. Science 170 (1970) 859.
2. Gerschenson, L.E., Berliner, J. and Davidson, M. in Fleischer, S., Packer, L. and Estabrook, R.W. (Eds), "Methods in Enzymology" Ac. Press, New York, 32 1974 pp. 733-740.
3. Gerschenson, L.E., Okigaki, T., Andersson, M., Molson, J. and Davidson, M.B., Expt. Cell Res. 71 (1972) 49.
4. Gerschenson, L.E., Davidson, M.B. and Andersson, M. Eur. J. Biochem. 41 (1974) 139.
5. Lowry, O.H., Rosebrough, N.T., Farr, A.L. and Randall, J.T. J. Biol. Chem. 193 (1951) 265.
6. Kenney, F.T. in H.T. Munro (Ed.) "Mammalian Protein Metabolism," Ac. Press, New York, 1970 pp. 131-150.
7. Friedmann, N., Exton, J.H. and Park, C.R. Biochem. Biophys. Res. Commun. 29 (1967) 113.
8. Robison, G.A., Butcher, R.W. and Sutherland, E.W. "Cyclic AMP" Ac. Press, New York, 1971 pp. 392-396.
9. Berliner, J. (this book).
10. Iwasaki, Y., Lamar, C., Danenberg, K., and Pitot, H.C. Eur. J. Biochem. 34 (1973) 347.

Note: *We would like to thank the publishers of the Eur. J. Biochem. for the permission to reproduce Figs. 1, 2 and 3.*

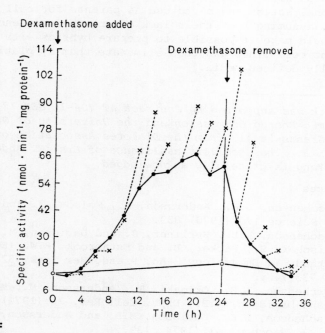

FIG. 1:

<u>Dexamethasone and Insulin Effects on TKT Specific Activity</u>
Specific activity is expressed as nmol p-hydroxyphenylpyruvate/min per mg protein. Cultured cells grown as monolayers in petri dishes of 5 cm diameter were given a change of medium (5 ml each) with or without 10 μM dexamethasone at zero time. After 24 h of incubation this culture medium was removed and the cells were washed 4 times with 5 ml culture medium devoid of the steroid. Insulin was added in 25μl volumes to each petri dish indicated by the black dot and the enzyme measured at the time indicated by (x). Each point is the average of enzyme determinations performed by triplicate in cells from three different petri dishes containing about 10^6 cells each. (o———o) Control, no additions; (●———●) dexamethasone added; (x———x) insulin added to dexamethasone-pretreated cells.

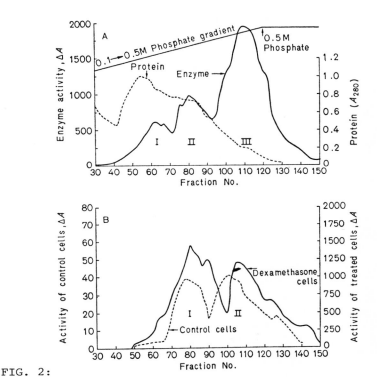

FIG. 2:

<u>Hydroxyapatite Chromatography Patterns of Tyrosine Transaminase.</u> The activity of the transaminase is plotted as change of absorbance at 335 nm in 20 min per 0.1 ml eluate and protein is plotted as absorbance at 280 nm. The linear phosphate gradient was added to the columns at the same time that fraction number 1 was collected. The patterns of eluted proteins were always similar, therefore, only one is illustrated and applies to both (A) and (B). Enzyme = normal rat liver cytosol; control cells = cells incubated in medium without additions; dexamethasone cells = cells incubated with 10µM dexamethasone for 18 h.

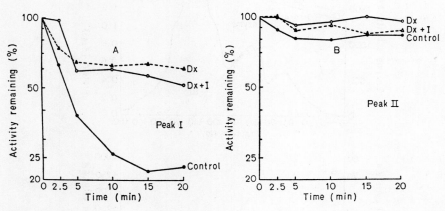

FIG. 3:

Heat Stability of the Two Forms of Tyrosine Transaminase from Culture Cells.   Cells were incubated as described before with 10 M dexamethasone for 18 h and then with 5 nM insulin for 2.5 h. Then cell extracts (see Methods) were first dialysed overnight at 4°C, versus 5 to 10,000 volumes of the homogenization medium and then chromatographed in hydroyapatite columns as in Fig. 2. Five fractions from the center of each peak were pooled and the heat stability determined at 65°C for different periods of time. Specific activity of the enzyme was determined as before. Each point is the average of triplicate determinations. (o) Control, no additions; (Δ), Dx, dexamethasone was added; (0) Dex. + I, dexamethasone plus insulin added.

ON THE SPECIFICITY OF THE INDUCTION OF TYROSINE AMINOTRANSFERASE

H. Kroger

It is known that steroid-hormones can induce a number of enzymes (1-3). In this connection we are interested in the question of specificity and therefore investigating the regulation of some enzymes.

We have been looking lately on the induction of tyrosine aminotransferase *in vivo* and *in vitro*. It turned out that the substrate L-tyrosine shows a remarkable effect on the regulation of tyrosine-aminotransferase.

1. Experiments in vivo

Previous studies have indicated that the induction of tyrosine-aminotransferase in adrenalectomized rats requires small doses of cortisone and also the substrate L-tyrosine (4). Fig. 1 gives a survey on the decrease of this form of induction if cortisone is injected prior to the substrate. 45 min time between the application of cortisone and L-tyrosine leads to values which are almost identical to the controls.

The induction of tyrosine-aminotransferase by cortisone and L-tyrosine simultaneously applied is strongly inhibited in the presence of actinomycin (6). Tab. 1 shows that this inhibition also takes place if L-tyrosine is given 15 or 30 min after cortisone.

A remarkable prolongation of the induction time is possible by adding a virostatic compound from Asta, Brackwede, Germany (Tab. 2). This substance inhibits DNA and RNA viruses, the biochemical mechanism, however, is unknown so far (7). At this stage of our experiments we like to conclude that the Asta-compound influences the cortisone substrate induction by an effect on the chromatin.

2. Experiments in vitro

For the purpose of investigation whether the results of the *in vivo* experiments are also true of the ones *in vitro* we worked with tissue culture. Dr. Gerschenson, Los Angeles, USA, kindly made RLC cells available to us (8). The tyrosine-aminotransferase can be induced in these cells by dexamethasone phosphate (9,10). This induction considerably

---

Enzyme Nomenclature:

Tyrosine-aminotransferase (EC 2.6.1.5)

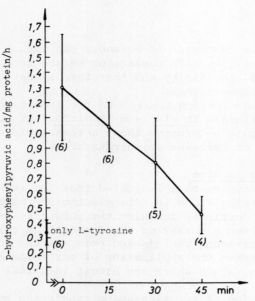

FIG. 1: L-tyrosine application after cortisone

Induction of tyrosine aminotransferase by cortisone and L-tyrosine in rat liver in dependence on time. Number of animals in brackets. (from ref. 5).

CELL LINES CULTURED IN CHEMICALLY DEFINED MEDIUM

min after cortisone	L-tyrosine	L-tyrosine + actinomycin D
0	1.3 ± 0.35 (6)	0.45 ± 0.10 (6)
15	1.16 ± 0.11 (4)	0.57 ± 0.14 (4)
30	0.66 ± 0.12 (3)	0.44 ± 0.09 (5)
without cortisone	0.34 ± 0.05 (6)	

TABLE 1. Influence of actinomycin D on the induction of tyrosine-aminotransferase by cortisone and L-tyrosine in liver of adrenalectomized rats. Number of animals in bracket (From Ref. 5).

compound (mg/kg)	μ mole p-hydroxyphenylpyruvic acid/g liver	
--	0.553	(2)
50	0.693 ± 0.027	(3)
100	0.804 ± 0.087	(3)
200	1.061	(2)

TABLE 2. Influence of the Asta-compound on the induction of tyrosine-aminotransferase by cortisone and L-tyrosine in rat liver. The compound and 2.5 mg/kg cortisone were applied 4 hours prior to 500 mg/kg L-tyrosine. Number of animals in brackets. Treatment 2 hrs. (From Ref. 6).

decreases if L-tyrosine is omitted from the experiment or added in rather small amounts (Tab. 3). Half the quantity of L-tyrosine only allows about 2/3 of the enzyme activity to be induced. This effect proved to be clearly specific, for varying the amount of L-leucine does not bring about a change in the induction process (Tab. 4).

Our results obviously indicate that the induction of tyrosine-aminotransferase needs both: the hormone and the substrate. We assume the synthesis of the specific m-RNA for the enzyme only takes place if hormone and substrate are present in an appropriate amount.

This work was supported in part by grants from the Deutsche Forschungsgemeinschaft and Stiftung Volkswagenwerk.

References

1. Civen, M., and Knox, W.E.   J. Biol. Chem. 234 (1959) 1787.
2. Knox, W.E. and Metzler, A.H.   Science 113 (1951) 237.
3. Feigelson, P. and Greengard, O.   J. Biol. Chem. 237 (1962) 3714.
4. Kroger, H. and Greuer, B.   Nature 210 (1966) 200.
5. Kroger, H., Philipp, J. and Wicke, A.   Biochem. Z. 344 (1966) 227.
6. Kroger, H. and Greuer, B.   Biochem. Z. 341 (1965) 190.
7. Kroger, H., Donner, I. and Skiello, in press.
8. Gerschenson, L.E., Okigaki, T., Andersson, M., Molson, J. and Davidson, M.B.   Exp. Cell Res. 71 (1972) 49.
9. Gerschenson, L.E., Davidson, M.B. and Andersson, M. Eur. J. Biochem. 41 (1974) 139.
10. Kroger, H., Donner, I., Voss, H. and Plotze, G. in preparation.

amount of L-tyrosine	p-hydroxyphenylpyruvic acid (μ mole/mg protein/h x 10^{-3})
--	44
3×10^{-5} M	91
1.5×10^{-5} M	64
3×10^{-6} M	42
3×10^{-7} M	41
3×10^{-8} M	38

TABLE 3. Influence of different amounts of L-tyrosine on the induction of tyrosine-aminotransferase by dexamethasone-phosphate in tissue culture. Incubation time: 16 hrs. (From Ref. 10).

amount of L-leucine	p-hydroxyphenylpyruvic acid (μ mole/mg protein/h x 10^{-3})
1×10^{-4} M	87
0.5×10^{-4} M	87
1×10^{-5} M	106
1×10^{-6} M	85
1×10^{-7} M	81

TABLE 4. Influence of different amounts of L-leucine on the induction of tyrosine-aminotransferase by dexamethasone-phosphate in tissue culture. Incubation time: 16 hrs. (From Ref. 10)

ISOZYME PATTERNS OF BRANCHED CHAIN AMINO ACID TRANSAMINASE IN CULTURED RAT LIVER CELLS

Akira Ichihara, Jiro Sato and Masayoshi Kumegawa

Introduction

Recent biochemical studies on hepatocellular differentiation and carcinogenesis have led to several basic biological findings. One example is the discovery of tissue specific proteins, particularly isozymes. The molecular basis of cellular differentiation can be defined as the acquisition of luxury molecules, and for this reason studies on isozymes have proved a very useful unique method (1-4). Other valuable findings are on slow growing differentiated hepatomas, such a as Morris or Reuber hepatomas. These tumors show a wide variety of characters, which correlate fairly well with their growth rates (5). Thirdly, progress in culture of differentiated hepatocytes has proved of immeasurable value in studies of not only liver cells per se, but also of the mechanism of carcinogenesis (6). Based on these studies it has become evident that there is a close relationship between hepatocellular differentiation and carcinogenesis. Since Greenstein's proposal, there have been a number of biochemical views on cancer, and it has been regarded as a state of retro-, de- or dis-differentiation, disease of differentiation, blocked ontogeny, fatalism or molecular correlation (1,3-5,7). Irrespective of how cancer as a disease is considered, it is evidently very important to pursue the mechanism of cellular differentiation, not only per se but also because of its possible relevance to cancer research.

We have found three forms of branched chain amino acid transaminase in various rat tissues and the changes of the isozyme pattern of the transaminase in hepatocytes during development and carcinogenesis of the liver are quite unique.

Enzymology of Isozymes of Branched Chain Amino Acid Transaminase

In 1966 we and Taylor and Jenkins found a specific transaminase for branched chain amino acids in hog heart (8,9). Subsequently, studies on the tissue distribution of this enzyme showed that heart, kidney and lactating mammary gland have the highest activity of the enzyme, while muscle and brain have moderate activity and liver and adipose tissue lower activity (Table 1) (10). Moreover, DEAE cellulose column chromatography of various tissues revealed that there

Tissue	Activity for leucine (nmole/hr/g w.w.)		Distribution of isozymes in supernatant (%)		
	Total[1]	Supernatant	I	II	III
Liver	6.4	2.0	25	75	0
Kidney	139.2	46.8	100	0	0
Muscle	50.4	14.4	100	0	0
Brain	74.4	37.2	5	0	95
Heart	147.6	86.4	100	0	0
Lung	17.6	4.6	100	0	0
Spleen	21.8	7.3	100	0	0
Gut	7.4	3.0	100	0	0
Ovary	50.4	22.8	70	0	30
Placenta			45	0	55
Testis			100	0	0
Adipose Tissue	7.8		100	0	0
Lactating Mammary gland	100.0		100	0	0

[1]*There are considerable activities in both the supernatant and mitochondrial fractions and their enzymatic properties may be different (11, 12).*

TABLE 1. DISTRIBUTION OF ISOZYMES OF BRANCHED CHAIN AMINO ACID TRANSAMINASE IN VARIOUS RAT TISSUE (10).

are three forms of this transaminase. Most tissues examined contain enzyme I, while liver has enzyme II and brain, ovary and placenta have a third form (enzyme III), as shown in Table 1. The three forms were isolated and purified and their general characters are shown in Table 2. Enzymes I and III are very similar in their kinetical properties, but their chromatographic and immunochemical properties are quite different. Enzyme II has unique characters and is specific for leucine and its Km is very high. The molecular weight of enzyme I of hog heart is 75,000 (9), while those of enzyme III of hog brain and of rat hepatoma are 39,000 and 42,000, respectively (14,15). The facts that enzyme I of hog heart contains one mole of pyridoxal phosphate per mole of enzyme and that there is no immunochemical cross reaction between enzymes I and III suggest that they are not in the relation of hybrids of subunits or interrelated by aggregation-disaggregation. Therefore, it is very likely that these three forms are controlled by independent genes.

Physiology of Branched Chain Amino Acid Transaminases

The physiological roles of branched chain amino acids are uniquely different; valine is glycogenic, leucine is ketogenic and isoleucine has both properties. Krebs discussed the importance of these amino acids as energy sources (16). Leucine particularly, has very interesting features: it forms β-hydroxy- β- methylglutaryl Co A, a precursor of acetoacetate and cholesterol. Therefore, it has been found that this amino acid is a very efficient substrate for lipogenesis as well as being an energy source (17-23). It also regulates urea formation (24-26) and is a specific inducer of tyrosine transaminase in cultured hepatoma cells (27). However, the physiological roles of the isozymes are not clear at present.

Enzyme II is induced readily by cortisol, while diabetes and hypophysectomy induce enzyme I in kidney (28,29).

Developmental Aspects of Branched Chain Amino Acid Transaminases

Fetal rat liver contains only enzyme I, but enzyme II appears and increases rapidly after birth (30). There is no difference in the isozyme patterns of muscle and brain of fetal and adult rats. From these results it seems reasonable to assume that in liver enzyme I is a proliferating type and enzyme II a differentiated type, as suggested by Richter (31). To study the mechanism of increase of enzyme II during development we used Rose's circumfusion

	Isozyme		
	I	II	III
Concentration of phosphate buffer for elution from DEAE cellulose column (M)	0.02	0.18	0.20
Mobility on polyacrylamide gel disc electrophoresis (Rm)	0.32	0.46	0.57
Inhibition of activity by antiserum(%)			
against enzyme I	57	0	0
against enzyme III	0	0	94
Km for substrate (mM)			
valine	4.3		2.5
leucine	0.8	25	0.6
isoleucine	0.8		0.5
α-ketoglutarate	1.0	0.07	0.5
pyridoxal phosphate	0.03	0.004	
Optimal pH	8.2	8.7	8.4

TABLE 2. PROPERTIES OF ISOZYMES OF BRANCHED CHAIN AMINO ACID TRANSAMINASE IN RATS (13,14).

Conditions	Isozymes	References
Adult liver	I + II	13
Fetal liver	I	30
Regenerating liver	I + II	30
Morris hepatomas		39
7794A, 7316A	I + II	
5123 tc, 7795, 7793	I + II + III	
7777	I	
Primary hepatomas induced by 3'-methyl DAB	I + III	39
Benign adenomas induced by 3'-methyl DAB	I + II	39
Yoshida ascites hepatomas	I + III	14,39
Brain, ovary, placenta	I + III	10

TABLE 3. CHANGES OF ISOZYME PATTERN OF BRANCHED CHAIN AMINO ACID TRANSAMINASE UNDER VARIOUS CONDITIONS.

chambers, since one of the authors showed that this apparatus is very useful for studies of cellular differentiation in vitro (32,33). It was found that some key glycolytic enzymes can be induced without addition of hormones during culture of fetal mouse liver and that a variety of immature tissues differentiate morphologically. Fig. 1 shows that tyrosine transaminase in fetal mouse liver can be induced either by glucocorticoid or glucagon. Similarly enzyme II can be induced by both hormones, although the effects of these hormones are not so great as with tyrosine transaminase (Fig. 2). Enzyme I was not induced by these hormones under these conditions. These results are compatible with those of in vivo experiments (28,29). Wicks also reported induction of tyrosine transaminase by various hormones in organ cultures of fetal rat liver (34,35). It is very interesting to find in Figs. 1 and 2 that in early periods of culture these hormones are not effective and rather inhibitory for enzyme induction. Greengard proposed that during development sensitivity to hormones may change and cyclic AMP may be effective before cortisol and glucagon become effective in early fetal liver (36), or a receptor protein for hormone may be acquired during culture of fetal liver (37). It is still uncertain whether adult enzymes can be induced in cultured fetal liver or whether some hormonal stimulations are necessary. The present experiments support the latter possibility and suggest that hormones may be important, if not essential. The mechanism of cellular differentiation was discussed recently by Rutter et al. (38).

Carcinogenesis and Branched Chain Amino Acid Transaminases

It is interesting and unexpected that a rapidly growing hepatoma (Yoshida ascites hepatoma 130) contained no enzyme II, but did contain enzyme III. The enzyme III in the hepatoma was found to be indistinguishable from that of brain enzymologically or immunochemically (14). Primary hepatomas induced in rats fed 3'-methyl DAB for different lengths of time also showed similar patterns to various Yoshida hepatomas, that is enzymes I and III, but no enzyme II (39). Benign adenomas showed a similar pattern to normal adult liver. It is interesting that various Morris hepatomas showed various patterns: enzyme I only (fetal pattern), enzyme I and II (normal adult pattern) or all three isozymes. It should be mentioned that regenerating liver after partial hepatectomy had the same isozyme pattern as normal adult liver. The changes of the isozyme pattern under various conditions are summarized in Table 3.

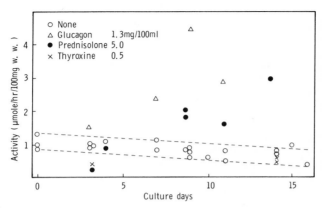

FIG. 1:

Induction of tyrosine transaminase of fetal mouse liver cultured in Rose circumfusion system. Culture conditions were described previously (32). Various hormones were added two days before the enzyme assay.

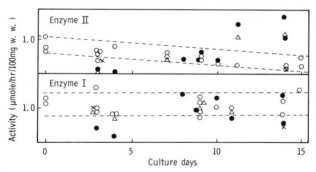

FIG. 2:

Induction of isozymes of branched chain amino acid transaminase of fetal mouse liver in Rose circumfusion system. Experimental conditions were as shown in Fig. 1.

It is still uncertain whether these changes of the isozyme pattern are due to aberration of gene expression or to mere selection of a cell population. In order to answer this question use of cloned cultured hepatocytes is a very powerful tool. When cultured rat hepatocytes, cloned from a single cell, PC-2 strain, which contained only enzyme I, were treated with chemical carcinogens, the cells were transformed and acquired tumorigenicity (40). Cultured cells from tumors, induced by intraperitoneal injection of transformed cells, showed a significant content of enzyme III (Table 4). These results strongly suggest that the acquisition of enzyme III during carcinogenesis is due to aberration of gene expression, although it is just possible also that cloned cells produced a variant population during long term culture, containing enzyme III and that this population was selected by chemical carcinogens. However, this possibility is very unlikely judging by the results shown in Table 5. There is a good correlation between the appearance of enzyme III and deviation of diploid chromosomal numbers. Cells with a lower percentage of diploid chromosomes contained more enzyme III. Spontaneously transformed cells also gave similar results (Table 6). When hepatocytes are cultured for long periods the cells are transformed (41-47), with deviation of chromosomal numbers and the appearance of enzyme III, though in small amount. Cells isolated from tumors, induced by injection of transformed cells into rats, contained more enzyme III. This can be explained as due to selection of tranformed cells only in vivo, while the cell population in cultures may be heterogeneous, containing both nontransformed and transformed cells. Therefore, enzyme III in transformed cells, if present, may be diluted in vitro, since non-transformed cells contain only enzyme 1.

It is still controversial whether hepatocytes in culture are "normal adult liver cells" or dedifferentiated cells or whether they are derived from stem cells (6). The established cell lines, used in this experiment had no enzyme II nor other liver type enzymes as seen in PC-2. However, some cell lines with high diploid percentage, such as RLN-B-2, retained relatively similar characters of liver as shown in Table 7. More pronounced expression of differentiated characters of liver cells was seen in cultured cells of Morris hepatoma 7316A, which is known to contain enzyme II in the tumor state (39). In culture these cells retain enzyme II and tyrosine transaminase and both are inducible by addition of glucocorticoid (Table 8). Tyrosine transaminase found in RLN-B-2 is not inducible by this

Cells	Culture days	Specific activity[1]	Isozyme pattern (%)		
			I	II	III
PC-2[2]	950	6.1	100	0	0
PCQT-2[3]	1141	13.0	81	0	19
PCDT-2[4]	874	10.0	71	0	29

[1] Activity for leucine: nmole/min/mg protein

[2] PC-2 cells were isolated and cloned from a single cell from the liver of 5 days old rat.

[3] PC-2 cells were treated with 4-NQO and the transformed cells were backtransplanted into rat. Ascites tumor cells formed were recultured.

[4] PC-2 cells were treated with DAB, backtransplanted and recultured as PCQT-2.

TABLE 4. EFFECT OF TRANSFORMATION OF PC-2 CELLS BY CHEMICAL CARCINOGENS ON THE ISOZYME PATTERNS (40).

Cells	Culture days	Specific activity[1]	Isozyme pattern (%)[2]		Diploid (%)
			I	III	
RLN-JC-1[3]	163	2.3	45	55	22
-13	174	14.0	100	0	58
-16	174	17.7	100	0	80
-20	161	3.4	66	34	14
RLN-B-2	278	23.1	100	0	82
dRLN-4[4]	272	8.4	100	0	70
-9	155		100	0	74
-53	451	10.0	77	23	2
-6	214	13.2	100	0	86
dRLh-84[5]	197	6.1	40	60	0
3'-mRLN-30[6]	172	17.0	42	58	0
-31	240		34	66	0

[1] as shown in Table 4.

[2] None of these cell lines contained enzyme II.

[3] These cells were isolated from a 7 days old rat, except RLN-B-2 which was isolated from another 7 days old rat.

[4] These cells were isolated from 75-125 days old rats which received DAB for 1-2 months.

[5] This cell line was isolated from a hepatoma of a 406 days old rat given DAB for 312 days.

[6] These cells were isolated from 116 and 125 days old rats which were fed 3'-methyl DAB for 2 months.

TABLE 5. COMPARISON OF ISOZYME PATTERNS AND CHROMOSOMAL NUMBERS OF VARIOUS CULTURED RAT HEPATOCYTES (40).

TABLE 6. SPONTANEOUS TRANSFORMATION OF CULTURED HEPATOCYTES AND THE APPEARANCE OF ENZYME III (for origins of cells, see 40, 43). For back transplantation $5 \times 10^6 - 1 \times 10^7$ cells were injected intraperitoneally. The cells dated in parenthesis of cell history were examined for isozyme pattern and chromosomal numbers.

```
                              1259 days        (1290)
RLN-8─────────────────────────────────┬─────────────────────────────── (9)
                                      ↓           155 Ascites
                                     Rat-------------
                                                     tumor

              (174)  (400)   579    (646)    (720)
RLN-JC-13─────────────────────┬─────────────────────────────────── (106)
                              ↓              165 Ascites    33    74 ┌Ascites\
                             Rat--------                       ↓    │tumor   │
                                             tumor            Rat---└        ┘

              (278)  (420)  (685)   796    (897)   (1129)
RLN-B-2───────────────────────────────┬───────────────────
                                      ↓                 296 ┌Ascites\
                                     Rat---------           │tumor   │
                                                            └        ┘
```

Cells	Culture days	Specific activity[1]	Isozyme pattern (%)[2]		Diploid (%)
			I	III	
RLN-8	1290	3.9	100	0	0
recultured	9	2.7	60	40	0
RLN-JC-13	174	14.0	100	0	58
	400				0
	646	15.3	100	0	
	720	3.0	90	10	
recultured	106	8.2	60	40	
tumor		0.5	43	57	
RLN-B-2	278	23.1	100	0	82
	420	20.6	100	0	40
	685	3.6	100	0	0
	897	2.4	80	20	
	1129	4.6	90	10	
tumor			40	60	

[1] as shown in Table 4.
[2] None of these cells contained enzyme II.

Tissue or cell	Aldolase[1]	Glucose-ATP phosphotransferase[2]	Tyrosine transaminase[3]	Tryptophan pyrrolase[3]
Muscle	60	0.5	0	0
Liver	1.0	6.0	17.5	10.5
PC-2[4]	50	0.5	0	
RLN-B-2[5]	14	1.4	1.2	3.3

[1] Activity ratio for fructose-1,6-diphosphate and fructose-1-phosphate.

[2] Activity ratio with high (1×10^{-1} M) and low (1×10^{-4} M) glucose concentration.

[3] Activities are expressed as nmoles of product formed /min/mg protein.

[4] Cultured for about 1000 days.

[5] Cultured for about 300 days. It was shown that the cells produced albumin (48).

TABLE 7. DIFFERENTIATED CHARACTERS OF CULTURED RAT HEPATOCYTE RLN-B-2 CELL LINE.

BRANCHED CHAIN AMINO ACID TRANSAMINASE

Cells	Culture	Enzyme II[1]	Tyrosine transaminase[1]
RLN-B-2	420	0	1.1
+ Cortisol[2]		0	1.1
Morris 7316A[3]	400	0.9	9.4
+ Cortisol[2]		2.5	52.7

[1] Activities are expressed as nmole product/min/mg protein

[2] $1 \times 10^{-5} M$ cortisol was added for 15 hrs.

[3] The modal chromosomal number is 43 as reported by Nowell et al (49).

TABLE 8. EXPRESSION OF DIFFERENTIATED CHARACTERS (LIVER SPECIFIC ENZYMES) IN CULTURED MORRIS HEPATOMA 7316A.

hormone, suggesting that the enzyme may be a non-specific transaminase or that the cells lost a receptor protein of glucocorticoid. It is still puzzling and paradoxical that some slow growing hepatoma cells retain more differentiated characters than cultured cells from normal liver. This suggests that once a phenotype is expressed in a hepatoma it may be fixed in a more stable way. The simplest explanation, which may be too simple, is that these cells have lost regulation of gene expression.

Based on these results on isozyme patterns we can draw the process of cellular differentiation and carcinogenesis and their relation to cultured cells, schematically, as shown in Fig. 3. In vivo matured cells may be derived from undifferentiated stem cells and between these two grades there must be several intermediate grades of differentiation. Although we could not identify the presence of enzyme III in fetal liver (30), it is still possible that very early embryonic liver may contain enzyme III. When these cells with different grades of differentiation are transformed, we can expect to obtain various hepatomas, varying between slow growing, highly differentiated hepatomas and fast growing, undifferentiated hepatomas. Similar processes can be applied to cultures of various liver cells. However, another possibility, which can not be excluded, is that these cells may be dedifferentiated when they are transformed or cultured. Studies on the isozyme patterns in various types of cells may provide more information to clarify these possibilities and indicate the process of differentiation and carcinogenesis more precisely.

Summary

Transaminase specific for branched chain amino acids in rat tissues were found in three forms (enzymes I - III). Liver contains enzymes I and II, while brain, ovary and placenta contain enzymes I and III. Other tissues examined contains enzyme I only and enzyme II appears after birth. Rapidly growing hepatomas lost enzyme II, but acquired enzyme III. Various Morris hepatomas showed a variety of the isozyme patterns. All established liver cell lines lost enzyme II, but retained enzyme I. Cells transformed by chemical carcinogens or spontaneously acquired enzyme III. There is a close relation between appearance of enzyme III, acquisition of tumorigeneicity and deviation of chromosomal numbers.

Expression of enzyme II in cultured liver cells were demonstrated by addition of hormones to mouse fetal liver

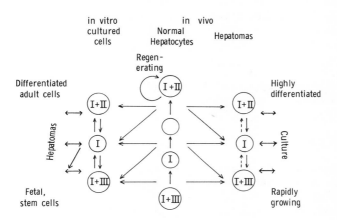

FIG. 3:

Hypothetical processes of hepatocellular differentiation, carcinogenesis and culture and expression of isozymes of branched chain amino acid transaminase in these cells.

or in Morris hepatoma 7316A. Tyrosine transaminase in these cells was also inducible by hormones.

From these findings a possible relation between hepatocellular differentiation, carcinogenesis and culture was discussed.

This work was supported by grants for cancer research from the Ministry of Education.

References

1. Weinhouse, S. Control Processes in Neoplasia (Mehlman, M.A. and Hanson, R.W. eds) Academic Press, New York (1974) p. 1.
2. Criss, W.E. Cancer Res. 31 (1970) 1523.
3. Weinhouse, S. and Ono, T. (eds) Gann monograph (1972) 13.
4. Walker, P.R. and Potter, V.R. Advances in Enzyme Regul. 10 (1972) 339.
5. Weber, G. and Lea, M.A. Advances in Enzyme Regul. 4 (1966) 115.
6. Potter, V.R. Cancer Res. 32 (1972) 1998.
7. Knox, W.E. Enzyme Patterns in Fetal, Adult and Neoplastic Rat Tissues, Karger, Basel (1972).
8. Ichihara, A. and Koyama, E. J. Biochem. 59 (1966) 160.
9. Taylor, R.T. and Jenkins, W.T. J. Biol. Chem. 241 (1966) 4396.
10. Ichihara, A., Yamasaki, Y., Masuji, H. and Sato, J. The 3rd International Conference of Isozymes, to be published by Academic Press, New York (1974).
11. Aki, K., Ogawa, K., Shirai, A. and Ichihara, A. J. Biochem. 62 (1967) 610.
12. Ito, T., Imai, K. and Ichihara, A. unpublished data.
13. Aki, K., Ogawa, K. and Ichihara, A. Biochim. Biophys. Acta 159 (1968) 276.
14. Ogawa, K., Yokojima, A. and Ichihara, A. J. Biochem. 68 (1970) 901.
15. Aki, K., Yokojima, A. and Ichihara, A. J. Biochem. 65 (1969) 539.
16. Krebs, H.A. Advances in Enzyme Regul. 10 (1972) 397.
17. Abraham, S., Madsen, J. and Chaikoff, I.L. J. Biol. Chem. 239 (1964) 855.
18. Ichihara, A. Enzyme 15 (1973) 210.
19. Ichihara, A., Noda, C. and Ogawa, K. Advances in Enzyme Regul. 11 (1973) 155.

20. Noda, C. and Ichihara, A. J. Biochem. in press.
21. Odessey, R. and Goldberg, A.L. Am. J. Physiol. 223 (1972) 1376.
22. Buse, M.G., Biggers, J.F., Drier, C. and Buse, J.F. J. Biol. Chem. 248 (1973) 697.
23. Meikle, A.W. and Klain, G.J. Am. J. Physiol. 222 (1972) 1246.
24. McGivan, J.D., Bradford, N.M., Crompton, M. and Chappel, J.B. Biochem. J. 134 (1973) 209.
25. Katunuma, N., Matsuda, Y. and Tomino, I. J. Biochem. 56 (1964) 499.
26. Strecker, H.J. and Eliasson, E. J. Biol. Chem. 241 (1966) 5750.
27. Lee, K.L. and Kenney, F.T., J. Biol. Chem. 246 (1971) 7595.
28. Ichihara, A., Takahashi, H., Aki, K. and Shirai, A. Biochem. Biophys. Res. Commun. 26 (1967) 674.
29. Shirai, A. and Ichihara, A. J. Biochem. 70 (1970) 741.
30. Ichihara, A. and Takahashi, H. Biochim. Biophys., Acta 167 (1968) 274.
31. Richter, D. Brit. Med. Bull. 17 (1961) 118.
32. Nakamura, T. and Kumegawa, M., Biochem. Biophys. Res. Commun. 51 (1973) 474.
33. Kumegawa, M., Nakamura, T. and Yamada, T. J. Cell Biol. in press.
34. Wicks, W.D. J. Biol. Chem. 243 (1968) 900.
35. Wicks, W.D. J. Biol. Chem. 244 (1969) 3941.
36. Greenhard, O. Biochemical Actions of Hormones (Litwack, G. ed.) Vol. I, Academic Press, New York (1970) p. 53.
37. Cake, M.H., Chisalberti, A.V. and Oliver, I.T. Biochem. Biophys. Res. Commun. 54 (1973) 983.
38. Rutter, W.J., Pictet, R.L. and Morris, P.W. Ann. Rev. Biochm. 42 (1973) 601.
39. Ogawa, K. and Ichihara, A. Cancer Res. 32 (1972) 1257.
40. Ogawa, K., Ichihara, A., Masuji, H. and Sato, J. Cancer Res. 33 (1973) 449.
41. Evans, V.J., Hawkins, N.M., Westfall, B.B. and Earle, W.R. Cancer Res. 18 (1958) 261.
42. Katsuta, H., Takaoka, T., Doida, Y. and Kuroki, T. Japan. J. Exp. Med. 35 (1965) 513.
43. Sato, J., Namba, M., Usui, K. and Nagano, D. Japan. J. Exp. Med. 38 (1968) 105.
44. Borek, C. Proc. Natl. Acad. Sci. U.S. 69 (1972) 956.
45. Oshiro, Y., Gerschenson, L.E. and DiPaolo, J.A. Cancer Res. 32 (1972) 877.

46. Breslow, J.L., Sloan, H.R., Ferrans, V.J., Anderson, J.L. and Levy, R.I. Exp. Cell Res. 78 (1973) 441.
47. Diamond, L., McFall, R., Tashiro, Y. and Sabatini, D. Cancer Res. 33 (1973) 2627.
48. Namba, M. Acta Med. Okayama 20 (1966) 251.
49. Nowell, P.C., Morris, H.P. and Potter, V.R. Cancer Res. 27 (1967) 1565.

THE PHOSPHORYLATION REGION OF LYSINE-RICH HISTONE IN DIVIDING HTC CELLS

Daryl Granner, David Sherod, Rod Balhorn, Vaughn Jackson and Roger Chalkley

Histones are highly basic proteins and are firmly bound to chromosomal DNA probably both by electrostatic and hydrophobic bonds. It was initially thought that there were a large number of different histone molecules and that these might play specific roles in the regulation of gene expression (1). More recently this diversity has convincingly been shown to be an artifact of histone isolation, due to contamination with other basic proteins or occurring as a result of active proteolytic enzymes (2). However, isolation of histone from chromatin in the presence of inhibitors of proteolysis such as bisulfite ions, reveals that essentially all eucaryotic cells contain the same five classes of histone molecules, namely F_1, F_{2a1}, F_{2a2}, F_{2b}, and F_3 (3,4). The primary structures of all these classes have been highly conserved, being nearly the same from pea plants to higher mammals. This, along with their intimate and extensive interaction with DNA, suggests that histones might play a role in chromatin structure rather than acting on specific gene controlling elements.

Recently it has been discovered that the diversity of histones can be increased by chemical modification. In most tissues histones F_3 and F_{2a1} are acetylated at specific lysine residues (5). Histones also contain N-methylated amino acid derivatives of lysine, arginine and histidine (6). A number of groups have shown that histones F_1 and F_{2a2}, in particular, are also modified by phosphorylation (7). This alteration involves serine and threonine residues. A variety of studies have been done in an effort to associate these modifications with the phenomena of gene activation (i.e. by hormones) or cell replication.

Modification of lysine-rich histone by phosphorylation has been implicated in both gene activation and cell replication. The concept of hormonal activation of selective portions of the genome is largely based on the finding that injected glucagon or cyclic AMP, or the latter added to <u>in vitro</u> preparations, increased the phosphorylation of a single, specific amino acid residue, serine 37, in rat liver F_1 histone (8). In large measure the concept of hormone action via cyclic AMP-mediated gene activation had its genesis in this series of experiments. Bulk levels of phosphorylation noted in this non-dividing tissue were quite low.

Under conditions in which liver cells, or tissue culture cells of related origin are induced to divide rapidly, F_1 histone is extensively phosphorylated. A review of this phenomenon, plus a study of the molecular localization of the phosphorylation sites, forms the basis of this paper.

Determination of histone phosphorylation

The addition of a phosphate group to the positively charged F_1 histone molecule retards the migration of the latter through a polyacrylamide gel at pH 2.8 under an electric field (9). In fact a series of equispaced bands are produced when multiple phosphate groups are present. However, because of the inherent sequence-induced microheterogeneity of F_1 histone which could complicate analyses based only upon gel migration, we have established the following criteria for phosphate-induced microheterogeneity: 1) Incubation of intact tissue or cells with ^{32}P should result in ^{32}P-labeling of the bands which migrate slower than the parent molecule. The extent of labeling should increase as the rate of migration decreases, indicating multiple levels of phosphorylation. 2) Treatment of such F_1 histone samples with alkaline phosphatase should (a) release ^{32}P radioactivity and (b) abolish the phosphate-induced microheterogeneity.

Figure 1 shows an acrylamide gel electrophoretic analysis of the F_1 histone of liver, regenerating liver and a hepatoma-derived permanent cell line, HTC cells. Liver has four forms of F_1 histone and phosphatase treatment results in no change in the pattern as confirmed by the gel scan analysis. Rapidly dividing tissues such as regenerating liver and HTC cells show an electrophoretic heterogeneity of F_1 histone, which is abolished by phosphatase treatment, thus affording preliminary indication of extensive F_1 phosphorylation.

Figure 2 shows that such electrophoretic heterogeneity is associated with increased labeling by ^{32}P. Thus this system satisfies the criteria outlined above. The extent of phosphorylation can be measured by slicing the gels and determining radioactivity, or by staining the gels, scanning them and quantitating the area under the curves with a Dupont curve analyzer.

Demonstration of cell division-associated phosphorylation

Using these techniques it was possible to directly test the hypothesis that histone phosphorylation was positively associated with cell division, as first proposed by Ord and Stocken (10). Our experimental approach has been to compare rapidly dividing cells with similar but non-dividing cells.

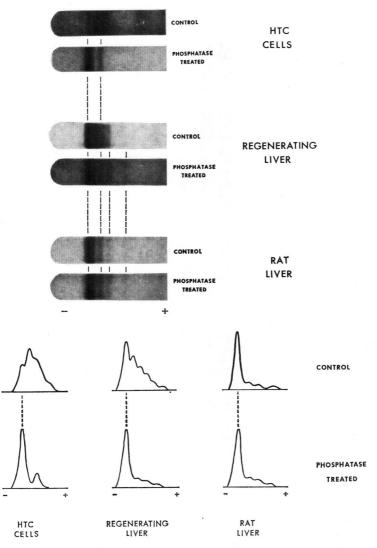

FIG. 1:

(Top) High-resolution gel electrophoretic analysis of normal and phosphatase-treated lysine-rich histone from adult rat liver, 29 hours regenerating liver and HTC cells in exponential growth. (Bottom) Microdensitometric scan of the electrophoretic patterns of the above gels plus and minus phosphatase.

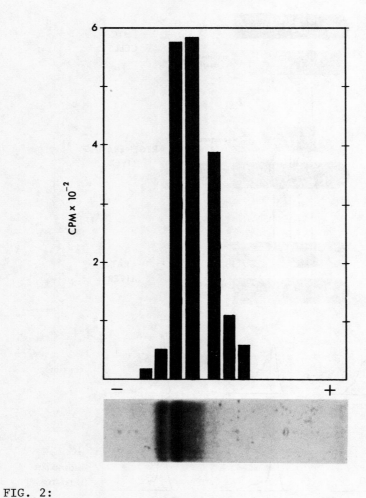

FIG. 2:

High resolution polyacrylamide gel electrophoresis of F_1 histone from HTC cells (Bottom) and the radioactivity associated with specific regions of the gel (Top).

It was noted that extensive phosphorylation was present 38 hours after partial hepatectomy as compared to a sham-operated control (11), that the extent of phosphorylation-induced heterogeneity was directly proportional to the rate of growth of a series of Morris minimal deviation hepatomas (12) and that fetal liver (which is undergoing rapid cell division) has extensive phosphorylation-induced heterogeneity that decreases the rate of cell division decreases (13).

These studies were most useful in establishing the relationship between histone phosphorylation and cell replication, however, it soon became apparent that further analysis of this problem would require a system more amenable to experimental manipulation. Established lines of cultured cells provide such flexibility in that it is relatively easy to alter the environmental composition by adding or subtracting from the culture medium, pulse-chase experiments can readily be employed to establish turn-over times of selected molecules and the growth rate of the cells can be manipulated by nutritional factors. We thus turned to HTC cells, a line derived from Morris hepatoma 7288c (14), and which grows readily in suspension culture (15) for our further studies.

Through a complex and poorly understood series of events suspension cultures of HTC cells enter stationary phase at about a cell density of 10^6/ml. At this stage the cells are capable of performing a variety of differentiated functions, however DNA synthesis is undetectable. Dilution of the cells into fresh medium stimulates cell division. One can thus use a single population of cells to test the hypothesis that histone phosphorylation is directly related to the rapidity of cell replication.

Figure 3 shows a high resolution electrophoretic analysis of lysine-rich histone throughout the growth curve of HTC cells. In essence it confirms the solid tissue studies in that stationary cells show little phosphorylation of F_1 histone and exponentially growing (log) cells have extensive phosphorylation of this molecule. In fact, during most of the exponential phase of growth, the bulk of the histone is in the second and third most rapidly migrating bands. As cells reenter stationary phase the F_1 histone re-assumes the non-phosphorylated pattern with the bulk of the material in the most rapidly migrating band.

Relationship of F_1 histone phosphorylation to the cell cycle
___
Stationary phase cells, which have low levels of F_1 phosphorylation, are considered to be blocked in the G1 phase of the cell cycle. Since histones are synthesized in S

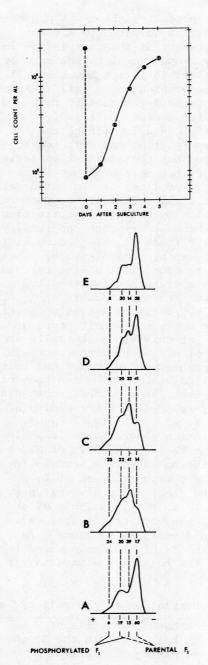

FIG. 3:

F_1 histone phosphorylation as a function of the HTC cell growth curve. Cells were subcultured into fresh medium from a stationary phase culture and an aliquot was taken immediately (A) and at various stages of growth (B-E). Microdensitometric scans of high-resolution electrophoretic analysis of lysine-rich histone from corresponding samples are shown below. The relative percentage of each fraction as determined by computer curve resolution is shown below each scan.

phase we reasoned that phosphorylation may primarily reflect an S-phase event.

Figure 4 suggests that this is indeed the case. HTC cells were synchronized in mitosis by colcemide block. During the block and at various times after release (resuspending cells in fresh, colcemid-free medium) 30 min pulses of ^{32}P and ^{3}H-thymidine were given to aliquots of the synchronized cells and isotopic incorporation into specific histone fractions and DNA was determined. It is apparent that, although there is a low rate of ^{32}P incorporation into F_1 histone (and F_{2a2}) in M and G1 phase, the greatest rate of phosphorylation occurs in concert with DNA synthesis, and in fact both peak at approximately the same time.

Other studies have served to further link F_1 phosphorylation to HTC cell replication. We have shown that both pre-existing and newly synthesized histone can be phosphorylated (16) although it is the latter which accounts for the rate increase in S phase (17). Phosphorylation occurs about 30-40 min after histone synthesis (16) and after histone has been deposited on the chromatin (18). Histone phosphorylation is not absolutely dependent upon DNA synthesis. Addition of 1 mM hydroxyurea to HTC cell cultures inhibits DNA synthesis by greater than 90% within 10 minutes, yet histone synthesis and subsequent phosphorylation are decreased by about 50% 6-8 hours after addition of the inhibitor (19).

Site of replication-associated F_1-phosphorylation

This strong association of histone phosphorylation with the replication of HTC cells led us to inquire as to whether the extensive modification of the F_1 molecule we noted was at a site different from that allegedly involved in gene activation. To facilitate these studies we cleaved F_1 histone by the procedure of Bustin and Cole (20). N-bromosuccinimide (NBS) cleaves proteins on the carboxy-terminal side of tyrosine. Since F_1 histone contains but a single tyrosine, cleavage results in two large fragments, which we have designated as B and C. The C fragment represents the amino-terminal portion of the molecule, has a molecular weight of 6000 daltons and contains serine 37. The B fragment is the larger (15,000 daltons) carboxy-terminal fragment.

HTC cells maintained in log growth were incubated with ^{32}P for 3 hours following which F_1 histone was isolated and purified as described previously (4). Following cleavage with NBS the mixture was separated by exclusion chromatography on Sephadex G-100. Figure 5 illustrates the results of such an experiment. It can be seen that three peaks of

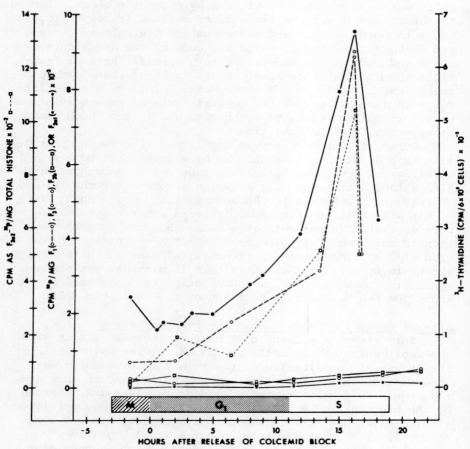

FIG. 4:

Correlation of ^{32}P incorporation into histones with DNA synthesis in synchronized HTC cells. HTC cells were synchronized in mitosis (M) by colcemid block. In M and at various times after release of the block, separate aliquots of cells were pulsed for 30 min with ^3H-thymidine (DNA) and ^{32}P (histone). Histones were fractionated by polyacrylamide gel electrophoresis and ^{32}P in various fractions was determined as described in the text. The mitotic index was > 95%.

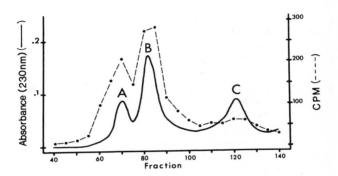

FIG. 5:

Sephadex G-100 column separation of the NBS fragments of HTC F_1 histone. ^{32}P-labeled F_1 histone was purified and cleaved with NBS (20). The mixture was applied to a 2.5 x 80 cm column equilibriated with 0.05 N HCl. Two ml fractions were collected and analyzed for absorbance at 230 nm (——) and ^{32}P radioactivity (•--•--•--•).

optical density are obtained. Peak A has been shown to be an intact F_1 molecule containing a spironolactone derivative of tyrosine, hence probably represents incomplete cleavage (21). Next to be eluted was fraction B followed by fraction C as defined above. It can be seen that ^{32}P radioactivity is associated primarily with A and B, that is the carboxy-terminal end of the F_1 histone.

Further analysis of the distribution of the label in the F_1 molecule was afforded by electrophoresis in 15% acrylamide containing 2.5 M urea as illustrated in Fig. 6. Although staining reveals the presence of fraction A plus the B and C fragments, ^{32}P radioactivity is again seen only in A and B, thus corroborating the chromatographic analysis.

As mentioned earlier, there is sequence-induced heterogeneity of F_1 histone and at least two F_1 molecules have been isolated from HTC cells (21). We next asked whether both were phosphorylated during cell replication. To answer this question we labeled, purified and cleaved the F_1 histone as usual then separated the fragments using G-100 chromatography. Analysis of the pooled ^{32}P-labeled fraction (A & B) was performed on sodium dodecyl sulfate (SDS) gels which separate on the basis of molecular weight. The results shown in Fig. 7 show again that both A and B consist of two molecular weight species, both of which contain ^{32}P radiolabel and both species of each apparently has the same specific activity. Thus both sequence and phosphate-induced heterogeneity of the carboxy-terminal region of F_1 histone appears to be involved in HTC cell replication.

F_1-phosphopeptide fingerprint analysis

In order to further characterize the replication-associated phosphorylation, ^{32}P-labeled F_1 histone was isolated and purified from HTC cells in log growth. After tryptic digestion the phosphopeptides were subjected first to descending chromatography then to high voltage electrophoresis. The left panel of the upper set of Fig. 8 shows the fingerprint analysis of log cell F_1 histone. Altogether 4 major and 3-5 minor phosphopeptides are seen. A rapidly migrating area consists of one major and two minor spots; an intermediate group consists of two major and a minor spot; and a slower major spot is also seen. In most analyses two faint spots are seen near the origin, a large spot is seen below the flow of chromatography and a considerable amount of radioactivity is seen at the origin. The latter does not stain with ninhydrin suggesting it is not peptide, and may well represent ^{32}P labeled nucleotides which have always vexed analysis of F_1 histone phosphorylation.

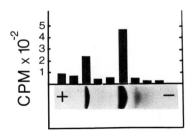

FIG. 6:

Electrophoretic analysis of ^{32}P distribution in the NBS cleavage products of HTC cell F_1 histone. Electrophoresis was in 15% acrylamide, 0.9 N acetic acid and 2.5 M urea at 130 V for 3.5 hours. The low bars represent the background radioactivity of gel slices.

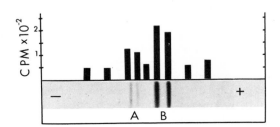

FIG. 7:

SDS gel analysis of ^{32}P-labeled A and B fragments from HTC cell F_1 histone. ^{32}P-F_1 histone was isolated from HTC cells, cleaved with NBS and fractionated by G-100 exclusion chromatography. The A and B fractions were then electrophoresed in acrylamide containing 0.05% sodium dodecyl sulfate and 0.2 M glycine buffer, pH 10.

By comparison, phosphopeptides generated by in vitro labeling of purified stationary phase HTC cell F_1 histone with HTC cell cytoplasmic protein kinase presents a different pattern as shown in the right, top panel. Two major and two minor spots are seen. The most prominent probably corresponds to one of the spots of intermediate migration seen in log phase (left, upper) cells, as might the adjoining minor spot. The other major spot probably corresponds to a faint, faster moving spot may be unique to the in vitro system, as the remaining phosphopeptides isolated from log phase cells appear to be specific to this stage.

It should be emphasized that log cells represent a mixed population. With a cell cycle of about 18 hours and an S phase of 10 hours, approximately 60% of the cells will be in S phase. Most of the rest will be in G1 with smaller numbers in G2 and M. The center panel of Fig. 8 compares the phosphopeptides of log cells with those of G1 cells, obtained after release of colcemid block. In G1, 5 phosphopeptides are detected rather than the 9 seen in the mixed population in log cells. It should be apparent that fewer cells and with a lower basal rate of ^{32}P incorporation into F_1 histone precluded quantitation of the density of the G1 spots. Nonetheless comparison of distance of migration suggests that the phosphopeptides seen in G1 are all apparent in the log cells, but 4 peptides are unique to the latter.

To further explore the difference between S and G1 we exploited the fact that the rate of phosphorylation is several times greater in S phase than in G1. Hence, shorter ^{32}P pulse times selectively enrich for S phase phosphorylation. The bottom panel of Fig. 8 compares a 2 min. pulse of log cells with the standard 60 min. pulse described earlier. Under these conditions 6 phosphopeptides are detected. Four appear to be common to both conditions, however there appear to be 2 phosphopeptides unique to the short pulse (S phase) and 4 seen only in the logner pulse. Again insufficient quantities precluded quantitation of the spots.

Recent studies indicate the presence of a nuclear phosphokinase which has high specificity for F_1 histone, is cyclic AMP-independent and which increases in activity in mitosis (22). Further analysis revealed that M phase Chinese hamster ovary cell F_1 histone had a different tryptic phosphopeptide map than interphase cells and this pattern could be mimicked by an in vitro system using isolated F_1 histone and the nuclear phosphokinase (23). These studies, those of Langan (8) and those reported in this paper make it clear that a variety of sites on the F_1 histone molecule can be phosphorylated, that agents such hormones and/or cyclic AMP can modify the rate of phosphorylation at a given site,

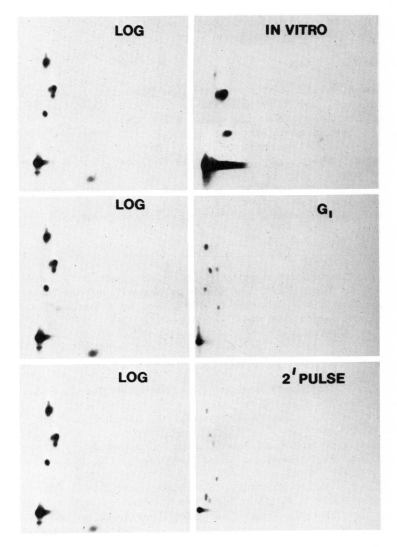

FIG. 8:

Autoradiograms of paper chromatographic-electrophoretic separation of the tryptic phosphopeptides obtained from HTC cell F_1 histone. (Top) Compares log phase HTC cells with HTC cell F_1 histone labeled in vitro with HTC cell cytoplasmic protein kinase; (Middle) compares log phase HTC cells with G1 cells; (Bottom) compares a 60 min pulse of log cells with a 2 min pulse.

and that rates and sites of phosphorylation vary throughout the cell generation cycle.

Conclusions

1. Phosphorylation of F_1 histone is positively correlated with cell replication. Three-fold increases in the rate of phosphorylation are seen in S phase.
2. Replication-associated phosphorylation involves the carboxy-terminal portion of the F_1 histone molecule; at least 6 phosphopeptides can be isolated from S phase cells.
3. Unique phosphopeptides are found in various phases of the cell cycle, or with in vitro labeling of F_1.
4. Replication-associated phosphorylation does not appear to involve serine 37.

Daryl Granner is supported by USPHS Grant # CA 12191 and by Veterans Administration Research Funds. Roger Chalkley is supported by USPHS Grant # CA 10871 and by a Research Career Development Award #GM 46410.

References

1. Huang, R.C.C. and Bonner, J. Proc. Nat. Acad. Sci. U.S. 48 (1962) 1216.
2. Bartley, J.A. and Chalkley, R. J. Biol. Chem. 247 (1972) 3647.
3. Philips, D.M.P. and Johns, E.W. Biochem J. 94 (1965) 127.
4. Panyim, S., Bilek, D. and Chalkley, R. J. Biol. Chem. 246 (1971) 4206.
5. Allfrey, V.G., Faulkner, R. and Mirsky, A.E. Proc. Nat. Acad. Sci. U.S. 51 (1964) 786. Candido, E.P.M. and Dixon, G.H. J. Biol. Chem. 246 (1971) 3182. Shepherd, G.R., Noland, B.J. and Hardin, J.M. Biochem. Biophys. Acta 228 (1971) 544.
6. Murray, K. Biochemistry 3 (1964) 10. Paik, W.K. and Kim, S. Biochem. Biophys. Res. Commun. 27 (1967) 479. Gershey, E.L., Haslett, G.W., Vadali, G. and Allfrey, V.G. J. Biol. Chem. 244 (1969) 4871.
7. Ord, M.G. and Stocken, L.A. Biochem. J. 98 (1966) 888. Langan, T.A. Science 169 (1968) 579. Sherod, D., Johnson, G. and Chalkley, R. Biochemistry 9 (1970) 4611.

8. Langan, T.A. J. Biol. Chem. 244 (1969) 5763. Langan, T.A. Proc. Nat. Acad. Sci. U.S. 64 (1969) 1267. Mallette, L.E., Neblett, M., Sexton, J.H. and Langan, T.A. J. Biol. Chem. 248 (1973) 6289.
9. Panyim, S. and Chalkley, R. Arch. Biochem. Biophys. 130 (1969) 337.
10. Ord, M.G. and Stocken, L.A. Biochem. J. 112 (1969) 81.
11. Balhorn, R., Rieke, W.O. and Chalkley, R. Biochemistry 10 (1971) 3952.
12. Balhorn, R., Balhorn, M., Morris, H.P. and Chalkley, R. Cancer Res. 32 (1972) 1775.
13. Balhorn, R., Balhorn, M. and Chalkley, R. Devel. Biol. 29 (1972) 199.
14. Thompson, E.B., Tomkins, G.M. and Curran, J.F. Proc. Nat. Acad. Sci. U.S. 56 (1966) 296.
15. Granner, D.K., Thompson, E.B. and Tomkins, G.M. J. Biol. Chem. 245 (1970) 1472.
16. Oliver, D., Balhorn, R., Granner, D. and Chalkley, R. Biochemistry 11 (1972) 3921.
17. Balhorn, R., Jackson, V., Chalkley, R. and Granner, D. (In preparation).
18. Oliver, D., Granner, D., and Chalkley, R. Biochemistry 13 (1974) 476.
19. Balhorn, R., Tanphaichitr, N., Chalkley, R. and Granner, D. Biochemistry 12 (1973) 5146.
20. Bustin, M. and Cole, R.D. J. Biol. Chem. 244 (1969) 5291.
21. Sherod, D., Johnson, G. and Chalkley, R. J. Biol. Chem. (1974) (In press).
22. Lake, R.S. and Saltzman, N.P. Biochemistry 11 (1972) 4817.
23. Lake, R.S. J. Cell Biol. 58 (1973) 317.

Note: We would like to thank the publishers of Biochem. Biophys. Res. Comm. and Biochemistry for the permission to reproduce Figures 1, 2, 3, and 4.

PROTEIN DEGRADATION IN LIVER COMPENSATORY GROWTH

Oscar A. Scornik

Introduction

Although liver cells can be successfully grown in vitro, we are still not certain that the growth control mechanisms operating in non-growing cultured cells are the same as those in the normal liver. I would like to discuss here the metabolism of proteins in the normal and regenerating livers of whole animals.

Normal liver cells synthesize proteins very actively- the equivalent of doubling their protein content in the order of one day. In the steady-state, non-growing condition all of this protein is either exported (as plasma proteins) or turned over. Net protein gain could result from an increased rate of synthesis, a decreased proportion of proteins made for export, or a lower rate of degradation. These parameters were studied in vivo, and the results seem to indicate a predominant role of protein degradation in the control of the total liver protein content.

Results and Discussion

The Course of Liver Regeneration

Fig. 1 shows the time course of the regeneration after removal of 2/3 of the tissue (2). Net protein gain (top) starts early and by the time of this study (36 hours, indicated by the arrow), it has reached a maximum, larger than 20% per day (2nd from top). At this time we are still looking at the original cells. The first wave of mitosis does not occur until the end of the second day and can be visualized here by the net increase in DNA (top). RNA and protein increase in parallel so that although the total RNA per cell becomes larger (up to the time when the enlarged cells divide) its concentration per g of tissue, or per mg of protein (3rd from top) remained the same. This applies also to the concentration of ribosomes, the RNA of which constitutes 80% of the total cell RNA in both conditions. The increase in the total number of ribosomes per organ (or per cell) is not sufficient in itself to account for the net protein gain during regeneration. Each gram of regenerating liver contains the same number of ribosomes as the normal tissue and

PROTEIN DEGRADATION

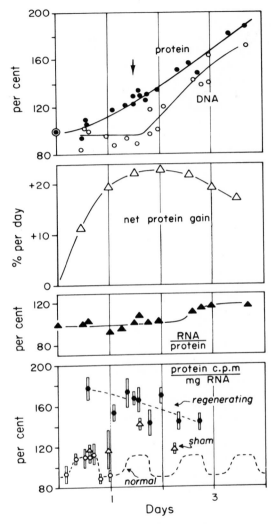

FIG. 1:

Time course of liver regeneration. The abscissa in all portions of the figure indicates the time (in days) elapsed between the partial hepatectomy and the sacrifice. Each point is the average of at least 6 animals. Top: percent increase in total protein (●) and DNA (o). Values for normal livers, expressed per 100 g body weight, are: weight, 5.5 g; protein, 1.0 g; DNA, 11.3 mg. Since 2/3 of the livers were removed at time 0, the 100% was considered 1/3 of these values. The arrow indicates the time selected for this study

FIG. 1. (Continued)

(36 hr). <u>2nd from top</u>: net protein gain, in percent of the normal value (0.055). <u>Bottom</u>: Incorporation of L-1-^{14}C leucine into the liver protein, 2 min after its intravenous injection in trace amounts. Values for hepatectomized (●), normal (o) and sham operated (Δ) animals were first calculated as protein cpm per mg of RNA and then expressed as percent of the average normal figure (3.4×10^4). Bars indicate standard deviation of the mean. For further details see (2).

yet it grows at a rate larger than 20% per day.

Efficiency of the Protein Synthesizing Machinery

The in vivo incorporation of leucine per mg of RNA is increased substantially (Fig. 1). This and a parallel increase in the in vitro incorporation in crude cell-free systems, led to the belief that the rate of protein synthesis per ribosome is substantially increased during regeneration. We have discussed before the contribution to the in vitro difference by contaminating lysosomes (1). As for this apparent increase in vivo, attempts to certify its validity through a study of the precursor leucine pool became discouraging both in technique and interpretation.

It can be explained in part by a small (ca. 10%) but consistent increase in the proportion of ribosomes in polyribosomes (2). But most of the difference reflects a more efficient labelling of the newly synthesized protein per unit ribosomes in polyribosome, such as the one illustrated by the experiment in Fig. 2. The increased labelling does not represent, however, a faster rate of translation. This I found through the use of three alternative procedures, neither of which depends on the determination of the specific activity of the pool (2,3).

In one of these procedures each animal is flooded with massive amounts of radioactive leucine in the hope of expanding the free leucine pool to the point where endogenous sources become negligible (2). For instance, in the experiment of Fig. 3A, each mouse received at time 0 the intravenous injection of 143 μmole of L- [$1-^{14}C$] leucine per 100 g of body weight. Incorporation into liver protein was linear for at least 10 min and the difference between normal and regenerating was reduced from the usual 60% (obtained with the injection of trace amounts of the precursor) to 25%. In the same experiment the polyribosome content was 13% larger in the regenerating animals. The difference between 13 and 25% can be accounted for by our incomplete success in saturating the leucine pool. As shown in Fig. 3B, if different amounts of leucine were injected and the results extrapolated to infinitely large amounts with the aid of a reciprocal plot, the extrapolated values differed by 15%. This residual difference is now fully explained by the difference in polyribosome content.

Another procedure relies on the analysis of the rate at which nascent chains are finished and released (2). A transit time can then be calculated. This was done in various ways. One of them is shown in Fig. 3C where the transit

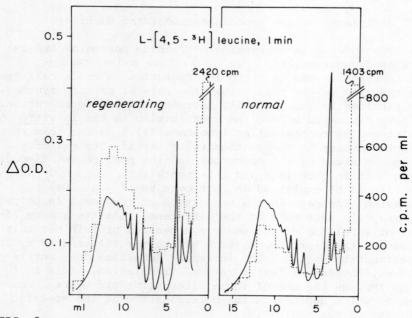

FIG. 2:

<u>Sucrose gradient analysis of normal and regenerating ribosomes</u>. A normal (<u>right</u>) and a hepatectomized (<u>left</u>) mouse were injected with a trace amount of L-[4,5-^{3}H]leucine (600 µCi per 100 g body weight); 1 min later each animal was sacrificed and the liver postmitochondrial supernate was treated with deoxycholate and analyzed in a sucrose gradient. <u>Abscissa</u>: volume collected from the top. <u>Left ordinate</u>: (ΔO.D.); difference between the absorption at 260 nm and that at 320 nm multiplied by 1.59 (8). <u>Right ordinate</u>: protein radioactivity, in cpm per ml. For further details see (2).

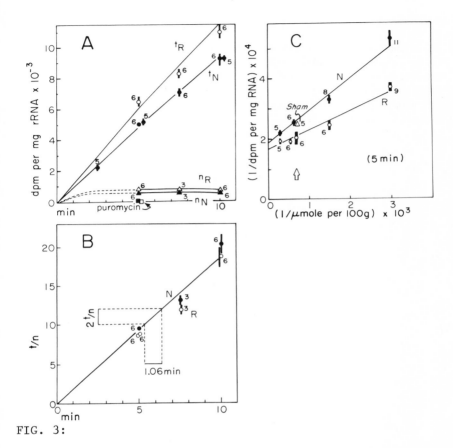

FIG. 3:

Injection of massive amounts of leucine. Bars show standard deviation of the mean. The number of animals represented by each point is also indicated. L-[1-^{14}C]leucine, 0.04 μCi/μmole/, was used in all cases. __A__. Each animal received 143 μmole of leucine per 100 g body weight. The abscissa indicates the time elapsed between the intravenous injection and the decapitation. Shown are the total liver protein radioactivity of normal (t_N) and regenerating (t_R) animals, and the respective radioactivity of the nascent chains (n_N and n_R). The squares represent the radioactivity of ribosomal samples from 3 normal and 3 hepatectomized animals injected with puromycin (6 mg per 100 g body weight) 4 min after the injection of the leucine. The animals were sacrificed 1 min later. __B__. The value of t in each animal was divided by the corresponding value of n, and the ratio t/n of each group was plotted. The transit time was measured to be 1.06 min.

FIG. 3. (Continued)

C. Animals received 33 to 286 µmole of leucine per 100 g body weight and were killed 5 min after the injection. The reciprocal of the liver protein radioactivity (ordinate) is plotted vs. the reciprocal of the amount injected (abscissa) for normal (●) and regenerating (o) animals. For further details see (2).

time is determined as the time required to incorporate in the total liver protein an amount of radioactivity equal to twice the total radioactivity of the nascent chains. The transit time thus calculated is 1.06 min for both conditions. Essentially the same result was obtained after the injection of trace doses of the precursor (2). Still a third procedure was used. The antibiotic pactamycin was injected in doses which inhibit initiation but not elongation of the nascent chains. This resulted in the rapid deaggregation of polyribosomes (Fig. 4). The initial rate of this deaggregation (Fig. 5) fits well with the transit time of 66 sec measured by the preceding procedure (3).

In summary, when the protein synthetic machinery is analyzed in normal and regenerating livers we find the concentration of ribosomes is the same, the proportion of them in polyribosomes is slightly larger, and the rate of translation remains unchanged. The lack of a substantial increase in protein synthesis per unit weight may be related to the fact that it is already very high in the normal liver. In particular, the fact that the rate of translation is unaffected is contrary to the prevalent feeling in this field. If we take a broader look, however, it is consistent with what we know in bacteria (6). In fact, the rate of translation is very similar in different mammalian cells, as indicated in the experiment of Fig. 6. In this experiment, a massive dose of leucine was injected to a mouse, and its resulting incorporation into the protein of several tissues was correlated with their respective concentration of RNA. Although both measurements were found to vary, as expected, over a wide range, the incorporation of leucine per mg of RNA was very similar in all instances.

Proportion of Proteins for Export

Next we examined what proportion of the proteins synthesized at any one time is exported and what proportion is retained as liver proteins (4). We explored this question in different ways, with the same result. One of these experiments is shown in Fig. 7. If we inject radioactive leucine into normal mice at time 0 and interrupt the pulse with pactamycin 5 min later, we find that of this pulse-labelled protein (100%) only one-half remains in the liver after 3 hrs. Of the rest, we can account for 20% by the appearance of labelled plasma proteins. The other 30% remains unaccounted for, and we suspect it represents rapidly turning-over proteins. We can now take a 3 hr time point and repeat the experiment with regenerating animals.

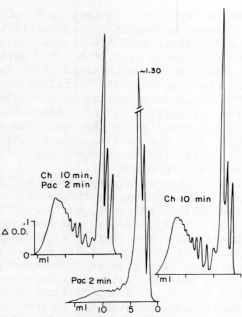

FIG. 4: <u>Deaggregation of polyribosomes by pactamycin and preventive effect of cycloheximide</u>. Each mouse received the intravenous injection of pactamycin, 400 µg (Pac); cycloheximide, 3 mg (Ch); or both, at the indicated time before sacrifice. Abscissa and left ordinate as in Fig. 2. Further details in (3).

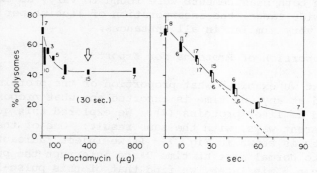

FIG. 5: <u>Dose dependence and rate of deaggregation</u>. The ordinate in both portions of the figure represents the proportion of polyribosomes relative to the total ribosomal population. The bars indicate average ± standard deviation of the mean and the number of animals represented by each bar is indicated. <u>Left portion</u>: abscissa, amount of pactamycin

FIG. 5. (Continued)

injection 30 s before sacrifice (µg per animal). Right portion: abscissa, time elapsed between the injection of pactamycin (400 µg per animal) and the decapitation. Filled bars: normal liver; open bars: regenerating livers. Further details in (3).

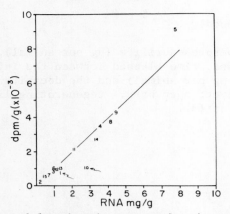

FIG. 6: <u>RNA and leucine incorporation in various tissues.</u>
A mouse was injected with a massive dose of L-[1-^{14}C]leucine
(273 µmoles per 100 g body weight, 0.1 µci per µ mole). The
concentration of RNA (ordinate) was correlated with the incorporation of the precursor (abscissa, 5 min after the...
Continues on next page.

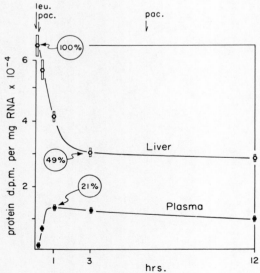

FIG. 7: <u>Redistribution of newly synthesized protein between liver and plasma.</u> Explanation in the text.

FIG. 6: (Continued)

injection) in the following organs: 1, brain; 2, diaphragm; 3, heart; 4, kidney; 5, liver; 6, lung; 7, leg muscle; 8, spleen; 9, stomach; 10, testis; 11, tongue; 12, bladder; 13, seminal vesicles; 14, small intestine; 15, abdominal wall muscle. The ratio between both variables, was found to be in the range of 8 to 12 x 10^3 dpm per mg of RNA in 13 organs. In two, brain and testis (arrows) it was approximately one-half of the others; however, the uptake of the radioactive amino acid, measured in TCA soluble extracts in those two organs was 20 to 30% of that in the others (not shown).

We find that if anything (Table 1), there is a small decrease in the proportion of long-lived liver proteins (44% of the total as opposed to 48% in the normal) and an crease (33 instead of 19%) in the proportion of plasma proteins. In other experiments (not shown) we found the same result in the absence of pactamycin and also that the increased labeling of plasma proteins is associated with proteins other than albumin.

In any case, under the conditions of these experiments decreased proportion of exportable or short-lived proteins cannot be used to account for the net protein gain. That synthesis of liver proteins should not increase at the expense of plasma proteins during regeneration is not surprising. We must remember that 1/3 of the liver has now to perform a function that the whole liver used to fulfill in the normal animal.

Protein Degradation

The above results left us with the rate of degradation of the long-lived liver proteins as the only other parameter to explore. A study of this degradation shows a very dramatic difference (5).

In the experiments shown in Fig. 8, we use precursors with a low chance of reincorporation - arginine labelled with ^{14}C in the guanido group (which in liver cells turns over at high rates in the urea cycle - circles) or leucine labelled with ^{14}C in the carboxy group (which upon the degradation of the amino acid is lost largely as CO_2-triangles). Also, we wait for 36 hrs after the injection to allow for disappearance of the exportable and short-lived proteins. The protein radioactivity with both labels is reduced to one-half in 3 days, as one would expect from the measured rates of synthesis and previous work by others. In the regenerating livers, we have to allow for dilution of the radioactivity by growth and express the results in percent of the total protein radioactivity at the time of the hepatectomy calculated in the remaining 1/3 from its measurement in the 2/3 removed in the operation (5). As shown by the open symbols and in sharp contrast with the normal livers, the regenerating tissue retains essentially all of the radioactive protein for at least a period of 5 days!

From the rates of protein synthesis and the fact that one-half of the newly synthesized protein is built into long-lived liver proteins, we expect a turnover of these proteins of at least 30% per day. Thus, a dramatic decrease of the degradation such as the one described above is sufficient to

	time	d.p.m. per mg liver RNA	
		liver	plasma
N	7 min	6.25 ± 0.34 (100%)	
	3 hr	2.99 ± 1.60 (48%)	1.19 ± 0.04 (19%)
R	7 min	15.80 ± 1.40 (100%)	
	3 hr	6.89 ± 0.62 (44%)	5.18 ± 0.44 (33%)

TABLE I. Liver protein radioactivity and plasma protein radioactivity expressed as d.p.m. per mg of liver RNA, 7 min and 3 hr after the injection of L-[1-^{14}C] leucine, in the normal (N) and regenerating (R) condition. Incorporation of the precursor was interrupted 5 min after its injection by the injection of pactamycin, as described in Fig. 7.

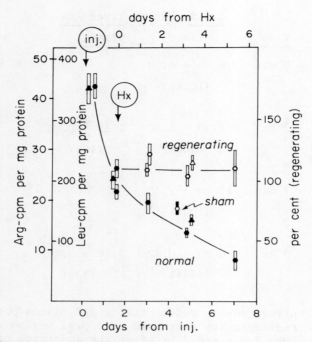

FIG. 8: Disappearance of radioactivity from the protein of normal and regenerating livers. The times of injection (inj.) of the radioactive precursor and that of hepatectomy (Hx) are indicated. Explanations in the text.

explain the net protein gain of 20% per day.

The Significance of Protein Degradation

The inverse relationship between rates of growth and protein degradation has also been found in bacteria (7) and cultured mammalian cells (15). Decreased degradation is clearly the most economical way for the liver to gain protein during compensatory growth. But why are proteins turned over in the normal liver in the first place? This question has been examined in recent reviews (9,10,11). We do not know the answer to it, but it would be interesting to consider some of the alternatives already postulated by others in relation to the findings reported in this paper.

1) It is possible that the mechanism of protein degradation has a scavenger function, that it recognizes and destroys nonfunctional proteins. Such proteins could originate by translational mistakes, or result from thermal denaturation or chemical modification of their primary structure. It could even be that proteins are programmed to inevitably "age" with time (for instance by deamidation (12)) and have then to be disposed of. This last proposition disagrees with the first order kinetics of the process (10) and fails to explain why the half-life of the same protein is so different in different tissues or in the same one under different conditions. The situation in the regenerating liver can also be taken as evidence against the inevitability of the process - the decreased degradation would be useless if it serves only to accumulate non-functional proteins.

2) Degradation of proteins could also be important as an endogenous source of amino acids (11). In the case of liver cells, we know that certain amino acids provided in high concentrations to perfused organs are able to decrease the rate of protein degradation (13). It has also been shown that amino acids injected *in vivo* (together with other substances) are able to elicit a growth response in the normal liver (14).

3) It is possible that a purpose of protein degradation is to maintain the cell mass. It has been suggested that the growth of cells is influenced by the concentration of cyclic nucleotides (15). Overall rate of protein degradation may be sensitive to the intracellular concentration of at least cAMP, as suggested by the fact that insulin decreases and glucagon increases that rate in perfused livers (13). As a mechanism for growth control, however, such a process would seem wasteful. Protein turnover is energetically expensive. With the equivalent of 3 to 4 moles of ATP hydrolyzed to ADP

for each mole of amino acid polymerized, one can estimate that normal liver cells spend in the order of 1/3 of the energy they obtain from oxidative phosphorylation on protein synthesis.

4) Finally, it has been proposed that protein degradation in non-growing cells serves to provide them with a flexibility in their composition. The liver would be able to replace its proteins by the same or different ones, depending, for example, on the composition of the diet. If this were the case, one could understand that in the growing, regenerating liver this would not be necessary, as the relative concentration of the existing proteins is diluted by growth. Variable rates of degradation of individual proteins can determine their concentration (9). It is possible that decreased rates of degradation during growth represent the sum of the effects of individual regulation of the protein components of the cell (rather than a general effect on the overall rate of protein degradation). It could be, for instance, that proteins exist in several alternative conformations and that the degradation mechanism respects those which are functionally more desirable. This possibility is particularly attractive in the context of our findings because it could provide a molecular basis for the hypothesis according to which cells grow in response to an increase in their functional load (rather than a specific growth factor). It is conceivable, for instance, that after the removal of 2/3 of the liver, the remaining 1/3 is forced to deal with higher concentrations (or more sustained ones) of a multitude of substrates; that this results in the respective enzymes to be stabilized in their functional conformation; and that this stabilization results in their decreased degradation.

The above considerations are at this time largely speculative. We simply do not yet know enough about the mechanism of protein degradation. The fact remains that net protein gain is achieved in the regenerating liver by reducing the rates of protein degradation. In this study we have focused our attention on the liver protein, but it is conceivable that degradation of other cell components is also decreased.

This work was supported by grant AM-13336 from the National Institute of Arthritis and Metabolic Diseases. The skillful assistance of Bruce Clay is gratefully acknowledged. Thanks are due to Ricardo Amils, Carlos Bedetti, Violeta Scornick and James Patrick for lively discussions on the subject of the last section of this paper.

References

1. Scornick, O.A., Hoagland, M.B., Pfefferkorn, E.C. and Bishop, E.A. J. Biol. Chem. 242 (1967) 131.
2. Scornik, O.A. J. Biol. Chem. 249 (1974) 3876.
3. Scornik, O.A. (1974) Biochem Biophys. Acta. 374 (1974) in press
4. Scornik, O.A. (1974) manuscript in preparation.
5. Scornik, O.A. Biochem. Biophys. Res. Commun. 47 (1972) 1063.
6. Coffman, R.L., Norris, T.E., Koch, A.L. J. Mol. Biol., 60 (1971) 1.
7. Mandelstam, J. Annals N.Y. Acad. Sci. 102 (1962) 621.
8. Scornik, O.A. Anal. Biochem. 52 (1973) 56.
9. Schimke, R.T. and Doyle, D. Ann. Rev. Biochem. 39 (1970) 929.
10. Siekevitz, P. J. Theor. Biol. 37 (1972) 321.
11. Goldberg, A.L., Howell, E.M., Li, J.B., Martel, S.B. and Prouty, W.B. Fed. Proc. 33 (1974) 1112.
12. Robinson, A.B. Proc. Nat. Acad. Sci. USA 71 (1974) 885.
13. Mortimore, G.E., Neely, A.N., Cox, J.R. and Guinivan, R.A. Biochem. Biophys. Res. Commun. 54 (1973) 89.
14. Short, J., Armstrong, N.B., Zemel, R. and Leiberman, I. Biochem. Biophys. Res. Commun. 50 (1973) 430.
15. Kram, R., Mamont, P. and Tomkins, G.M. Proc. Nat. Acad. Sci. USA 70 (1973) 1432.

Note: We would like to thank the publishers of the J. Biol. Chem and Bioch.Biophys. Acta. for the permission to reproduce Figures 1, 2, 3, 4, and 5.

FACTORS INFLUENCING GROWTH OF CELLS FROM REGENERATING LIVER

D.M. Hays, Y. Sera, Y. Koga, H.B. Neustein, E.F. Hays and M.O. Nicolson

The aim of the initial studies presented in this summary was to employ liver (including regenerating liver) cell cultures for the identification of serum "factors" controlling the regenerative response to hepatic excision. An attempt was made to identify the factors which influence cell survival in this system in addition to those which affect the cultural conditions of the cells following excision. Short term, primary cultures of trypsin-dispersed liver consisting of mixed cell types were used.

In monolayer culture, the ability of cells from regenerating liver tissue to survive and grow is dependent on the interaction of several factors, (in addition to those relative to the in vitro environment of the cells). These include species, animal age, size of resection, number of resections, and interval post-resection at which the cells are removed for culture. In the species employed (rat, mouse, chicken, dog, calf), regenerating liver cells have consistently shown greater potential for growth than cells from similar nonregenerating liver. The inverse relationship between animal age and the growth potential of removed liver cells, which applies to cells from both regenerating and nonregenerating liver, is consistent in all species studied and continues throughout life. Cells from livers (rat) made cirrhotic by employing either CCl_4 or CCl_4 plus choline deficiency have an increased potential for cell growth in vitro (1), although this is not as striking as the effects of liver resections, in animals of the same age and weight.

When cells are removed at intervals following a standard liver excision, quantitation of their growth in culture forms a curve, with little growth in cells removed in the initial hours, followed by a sharp peak of growth in cells after a certain interval post resection, and declining cultivatability in cells removed during the balance of the regenerative response (2,3). This is characteristic of a (>55%) hepatic resection. Following smaller resections this peak of cultivatability; is lower, extended, and occurs later in the regenerative response.

For example, in the young adult rat, during the initial 46 hours post resection (68%) the survival and growth of removed (remnant) liver cells is minimal, i.e., at the level or less than the growth of cells from nonregenerating rat liver. Studies attempting to explain the failure of growth in cells removed during this initial interval following resection have included the following. (a) Preliminary hepatic resections have been performed 24, 40, and 72 hours prior to the second hepatic resection. In this situation active cell growth does occur in cells removed during the post-resection interval after the _final_ resection, apparently under the influence of the _first_ liver resection. This suggests that this 46 hour post-resection growth failure is not due to trauma or the anesthetic agent, and that multiple resections stimulate _in vitro_ growth capacity. It was also noted in these studies that serial small hepatic resections have a stimulatory effect which is greater than a single larger resection. (b) The calf serum supplement to the media was replaced by rat serum drawn from animals during the precise post-resection period at which the tissue was removed for culture, i.e., during intervals in the immediate post-resection period, (as well as later in the course of regeneration). This serum change failed to increase the growth potential of cells removed during this initial interval, although rat serum (of all types) was as effective as calf serum in the culture of regenerating cells removed later in the regenerative response. (c) Mixtures of cells were cultured together including (#1) those from tissue removed during the early (no growth) interval (8-36 hrs.) post-hepatectomy and (#2) those removed during a period later in the response (48 hrs.) when one would anticipate active cell growth. These mixed cultures had the general growth characteristics of the cells of the #1 type, i.e., little growth _in vitro_. Thus, cells removed at the 48 hr. interval had no stimulatory effect on cells removed earlier in the response. In fact, the #1 type cell inhibited growth in the mixed cultures with the #2 type cell. It is possible that many of the cells of the #1 type are not viable for a significant length of time _in vitro_. These results may then simply be the result of reduction of the optimum cell inoculum _for growth_ of the #2 cells.

A series of partial hepatectomies in the same animal produces a sustained increase in the ability of removed cells to survive and grow _in vitro_ (4). After four large hepatic resections in the same animal, cells from the remnant liver have the cultural characteristics of weanling liver. This response is negligible in liver cells from the cirrhotic rat liver, i.e., in this system there is no response to serial resections (1).

The resection size, i.e., volume of hepatic tissue removed, influences the growth response of cells removed from the remnant liver (5). At least in the range from 25% to 68% resection, in the young adult rat, the peak of increased growth potential appears to occur earlier in the response, and reach higher levels in large as opposed to smaller resections. The duration of the increase in cultivatability on the other hand, is longer following smaller resections.

The growth characteristics of cells from regenerating liver tissue are thus dependent on the interaction of all of the factors noted previously, i.e., age, interval post-resection, size of resection, etc., each with its distinct effect on the growth pattern.

The studies described were in general short-range primary monolayer cultures employing liver tissue which was minced, the cells dispersed with trypsin and cultivated in MEM or similar media with calf (or other) serum additive. Cultures were fed at 3 day intervals and cells removed with trypsin and counted after 12 to 16 days in culture. The initial cell suspension appear to have cells of several types and the population at the time of cell counting is also mixed, including cells with the histologic characteristics of hepatocytes (among other cells). Histochemical studies of cells at the time of counting reveals some glucose-6-phosphatase activity. ^{3}H Thymidine and ^{3}H Uridine uptake can be demonstrated by radio-autographic techniques in the cells in these cultures.

In the rat and mouse, trypsin dispersed cells from regenerating liver will form colonies in soft agar. These colonies are similar to those from bone marrow cells cultivated under the same conditions, in the presence of colony stimulating factor (CSF) (6,7). In the rat, liver tissue removed 48 hours post-resection forms colonies of the greatest number and largest size. In the mouse, the interval post-resection at which colony formation is greatest appears to be later (96-120 hrs.). Three types of colonies are identified: (1) macrophage, (2) mixed granulocytic-macrophage and (3) a colony which is distinctly epithelioid in character. The first two types are similar to those formed when bone marrow cells are cultured in soft agar in the presence of a source of colony stimulating factor (CSF). However, in many instances the colony size is greater employing cultured regenerating liver cells (mouse) than in the cultures of bone marrow. When mouse liver cells are cultured in the presence of pooled mouse serum, bled 4 hours after injection of endotoxin, there is a consistent increase in the number of colonies formed. This endotoxin mouse serum is rich in CSF, the substance which has been shown to be necessary for colony formation in mouse marrow. Endotoxin rat serum did not stimulate growth of colonies from regenerating rat liver cells. This may reflect inappropriate culture conditions.

These findings suggest the presence of a cell in regenerating liver tissue which has the characteristics of the colony forming cell (CFC) of marrow. The presence of epithelioid colonies and colony growth in the absence of added CSA are features of the liver system which differ from marrow and are currently being studied.

References

1. Hays, D.M., Okumura, S. J. Surg. Res. 7 (1967) 270.
2. Hays, D.M., Matsushima, Y., Tedo, I., and Tsunoda, A. Proc. Soc. Exp. Biol. & Med. 138 (1971) 658.
3. Hays, D.M., Hirai, Y., Yokoyama, S., and Nakajima, K. J. Surg. Res. 11 (1971) 590.
4. Hays, D.M., Komi, N., Lau, R.E. Proc. Soc. Exper. Biol. and Med. 115 (1964) 106.
5. Hays, D.M., Tedo, I., Okumra, S., and Nagashima, K. Nature 220 (1968) 286.
6. Bradley, T.R. and Metcalf, D. Aust. J. Exp. Biol. Med. Sci. 44 (1966) 287.
7. Metcalf, D. Immunol. 21 (1971) 427.

STUDIES ON THE CONTROL OF GROWTH IN CULTURED PRIMARY FETAL RAT LIVER CELLS

Dieter Paul

The fact that the liver has the capacity to regenerate (1) has stimulated numerous studies on the mechanisms which lead to hepatocyte growth after partial hepatectomy in vivo. It is generally accepted that humoral factors, i.e. components in the blood of rats after partial hepatectomy are responsible for the initiation of DNA synthesis in cells in the liver remnant (2-4). Nothing is known about the nature or origin of such humoral factors. It is also unknown whether humoral factors are positive signals or whether they are inhibitors of growth suppressing agents which may be present in the blood of normal animals. Neither is it known whether such humoral factors in the blood after partial hepatectomy are involved in the control of organ size in the developing or growing animal. Many variables influence the initiation of DNA synthesis in the rat liver after partial hepatectomy (4-8), and lack of controlled experimental conditions in vivo has limited the understanding of the processes involved. Therefore it was of interest to investigate whether rat liver cells in primary cultures would be an appropriate assay system to study the regulation of DNA synthesis and cell division in hepatocytes by serum factors and to attempt to define levels of liver cell specific serum growth factors in sera of normal and partially hepatectomized rats. The possibility that serum may contain factors capable of stimulating liver growth was derived from experiments in which large doses of bovine serum Cohn fractions III and IV injected into normal rats caused a 10-fold increase of hepatic DNA synthesis (9).

Cell cultures have been useful for studies of growth control mechanisms in mammalian cells. Extensive studies with 3T3 mouse fibroblasts have shown that the cell cycle is controlled by nutrients and serum growth factors present in the culture medium (10,11). Cells stop growing in the G_1 (G_0) phase of the cell cycle when serum growth factors have been depleted by the cells (12). Therefore, the final cell density is proportional to the serum concentration in the medium (12). Also, cells stop growing in G_0 when cultured in medium containing limiting concentrations of certain nutrients (13). When a wound is scratched into a monolayer of dense, quiescent 3T3 cells and the wounded culture incubated in the depleted medium, the cells close to the edge of the wound are induced to divide (14). Analysis of the events preceding initiation of DNA synthesis in a wounded culture showed that cells which migrated from the monolayer into the denuded area (i.e., cells free of cell-cell contacts) did not synthesize DNA unless positive signals provided by serum initiated the growth cycle (15). These and other data (13), in summary, suggest that external factors (e.g. nutrients, serum growth factors) control the cell cycle in cultured mammalian cells and that cells presumably do not self-regulate their growth. These studies suggested that it should be possible to use liver cell cultures for studies of the cell cycle and as an assay system for the purification of serum factors that stimulate growth and DNA synthesis in hepatocytes.

Our studies started by culturing single fetal rat liver cell suspensions in dialyzed fetal bovine serum in arginine free Dulbecco's and Vogt's modification of Eagle's medium (16). The cells were shown to retain liver-specific urea cycle functions (i.e. cells synthesize arginine from ornithine), to synthesize and secrete albumin and to synthesize DNA in glucose deficient medium (16). The chromosome number of the cells remains diploid after a few generations (17). Although cells are plated in arginine free medium, the concentration of arginine in the medium of growing cultures (conditioned medium) is 2-4 µM. Non-hepatocytes do not grow in such a selective medium, and crossfeeding does not occur under standard culture conditions (17).

Two classes of serum factors are required for the growth of cultured fetal rat hepatocytes: one is required for survival and/or attachment, the other is required for the growth of the cells (17). The growth rate and final cell density of hepatocytes in a primary culture are serum dependent (17,18) and cells stop growing the the G_1 (G_o) phase of the cell cycle (19,20). The growth stimulating serum factor is of high molecular weight as judged by gel filtration on Sephadex G200 at neutral pH (D. Paul, unpublished) and is present in the 0-50% ammonium sulfate fraction of fetal bovine serum (17). Since liver cells can convert ornithine into arginine (16), cultures reach higher cell densities when incubated in the presence of ornithine, which suggests that the rate of arginine synthesis in fetal rat liver cells in ornithine free medium limits the growth of the cells (17). The cells do not grow when incubated in unconditioned, fresh arginine free medium in the presence of dialyzed fetal bovine serum (19).

Cells growing in conditioned arginine deficient medium are arrested and synchronized in the G_1 (G_o) phase of the cell cycle when incubated in fresh <u>arginine free medium</u> containing dialyzed serum as shown by microfluorometric analysis (20). The cells remain viable when cultured in arginine free medium because arginine is provided at levels high enough to maintain metabolism but too low to initiate DNA synthesis. This indicates that arginine and presumably other nutrients in the culture medium have regulatory functions in the control of the cell cycle defined by the intracellular levels of "regulatory" nutrients (see ref. 21,22).

In order to study initiation of DNA synthesis in resting fetal rat hepatocytes, cells were plated at low serum concentrations (2%) and incubated for 10-13 days, i.e. until cells became quiescent (19). Addition of fetal bovine serum or of high molecular weight serum fractions, or insulin to such quiescent cultures in conditioned medium results in initiation of DNA synthesis with a time course similar to that observed in the rat liver remnant after partial hepatectomy, i.e. ~ 15 hrs after start of the experiment (20). The cells have to be exposed to high serum levels for at least 12 hrs in order to become committed to enter S phase of the cell cycle (19). This is in agreement with data of <u>in vivo</u> cross-circulation experiments, which showed that the normal, unoperated partner has to be cross-circulated with blood of the partially hepatectomized partner for 12 hrs in order to initiate DNA synthesis in the normal liver (4,23).

Dialyzed serum in fresh (unconditioned) arginine free medium does not initiate DNA synthesis in resting fetal rat liver cells. However, when cells are incubated in the presence of low arginine levels (> 2 µM) <u>plus</u> dialyzed serum, DNA synthesis is initiated in an increasing proportion of cells with increasing serum levels (20). Cells become committed to synthesize DNA when incubated in the presence of arginine plus serum for > 12 hours followed by incubation in arginine free medium plus serum (Walter, S., Paul, D., unpublished). Insulin (> 0.1 µg /ml) initiates DNA synthesis in resting hepatocytes (20). Similar results were obtained by Leffert, using fetal rat liver cells in chemically defined medium (24).

The observations suggested to us that the availability of arginine inside the cell might be increased in the presence of serum. Serum or high molecular weight serum factors were shown to stimulate the uptake of (^{3}H)-arginine into TCA soluble material in resting fetal rat liver cells (20).

Arginine uptake is a saturable process at low arginine concentrations but at higher arginine levels the rate of uptake increases dramatically resulting in a deviation of linearity in the double reciprocal Lineweaver-Burk diagram

(D. Paul, unpublished). Presumably at high arginine levels the transport system begins to saturate and "non-saturable" uptake (e.g., simple diffusion) becomes increasingly significant.

The importance of intracellular arginine concentrations in the regulation of the cell cycle of fetal rat liver cells is emphasized by the fact that DNA synthesis is initiated in fetal rat liver cells by high levels of arginine (0.4 mM) in the absence of serum, i.e., in chemically defined medium, with a time course similar to that observed in serum stimulated cells (19,20,25). These data indicate that serum is not essential for the initiation of DNA synthesis as long as high enough arginine levels are present in the culture medium.

Therefore, we have concluded that in fetal rat liver cells as studied under the conditions described (1) the intracellular levels of arginine control the progression through the cell cycle, and (2) serum factors are involved in controlling the intracellular availability of arginine. Probably other nutrients have similar regulatory functions as described for arginine in fetal rat liver cells. The underlying mechanisms might be similar to those studied in bacteria during amino acid starvation (stringent control) (26).

Recently, Lieberman and coworkers demonstrated that DNA synthesis was initiated in the intact liver of normal rats by infusing a mixture of amino acids, glucagon, thyroxin and heparin (27), or by feeding a high protein diet to normal rats which were fed a protein free diet during 3 days prior to the experiment (28). Thus, amino acids and certain hormones can, under certain defined conditions, stimulate the liver to initiate the growth cycle in the intact animal. The possibility has been discussed that the intrahepatic ratio of insulin to glucagon might be a crucial initial determinant for liver regeneration (24). However, since the liver of rats eviscerated of portal splachnic organs regenerates after partial hepatectomy, it was concluded by Bucher and Swaffield that portal blood hepatotrophic factors are not the primary regulators of hepatic regeneration (29).

Most of our studies were performed by using fetal bovine serum as a source of serum factors required by fetal rat liver cells because rat serum contains substantial amounts of toxic materials whose levels vary with different serum batches (D. Paul, unpublished). Therefore, our studies in which sera obtained from different normal or partially hepatectomized rats were used for growth or DNA synthesis initiation assays as outlined in the previous sections, were difficult to evaluate. Fractionation of rat serum on Sephadex G200 yielded high molecular weight material which stimulated cell division and initated DNA synthesis in resting cells (D. Paul, unpublished). However, these growth stimulating fractions still contained variable levels of toxic materials which did not allow comparative studies using sera from different animals. Most of the material which is toxic to fetal rat liver cells can be removed from the growth stimulating activity by fractionating rat serum with chloroform/methanol: growth stimulating activity is present in the material that precipitates after adding chloroform/methanol to rat liver cells. Whether such factors are involved in regulating liver cell division in vivo remains to be determined.

I thank S. Walter, M. Henahan and J. Villela for assistance and L. White for the preparation of the manuscript. This work was supported by research grants from the American Cancer Society - California Division, The Damon Runyon Memorial Fund for Cancer Research, Inc., and by the National Institutes of Health, U.S. Public Health Service.

References

1. Prometheus, long B.C.?
2. Bucher, N.L.R., Swaffield, M.N., Moolten, F.L., and Shrock, T.R. in Baserga, R. (Ed.) Biochemistry of Cell Division, p. 139 (1969) C.C. Tomas, Springfield, Ill.
3. Lieberman, I. in Baserga, R. (Ed.) Biochemistry of Cell Division, p. 119 (1969) C.C. Tomas, Springfield, Ill.
4. Bucher, N.L.R. and Malt, R.A. in Regeneration of Liver and Kidney (1971) Little, Brown and Co., Boston.
5. Moolton, F.L., Oakman, N.J. and Bucher, N.R.L. Cancer Research 30 (1970) 2353.
6. Bucher, N.R.L., Swaffield, M.N. and DiTroia, J.F. Cancer Res. 24 (1964) 509.
7. Simek, J., Erbenova, Z., Deml, F. and Dvoracowa, I. Experientia 24 (1968) 1166.
8. Lieberman, I. and Short, J. Amer. J. Physiol. 208 (1965) 896.
9. Short, J., Zemel, R., Kanta, J. and Lieberman, I. Nature 223 (1969) 956.

10. Holley, R.W. and Kiernan, J.A. in Clarkson, B. and Baserga, R. (Ed.) Control of Proliferation in Animal Cells (1974) in press, Cold Spring Harbor Laboratory, N.Y.
11. Paul, D., Lipton, A. and Klinger, I. Proc. Nat. Acad. Sci. USA 68 (1971) 645.
12. Holley, R.W. and Kiernan, J.A. Proc. Nat. Acad. Sci. USA 60 (1968) 300.
13. Holley, R.W. and Kiernan, J.A. Proc. Nat. Acad. Sci. USA (1974) in press.
14. Todaro, G.J., Matsuya, Y., Bloom, S. Robbins, A. and Green, H. in Defendi, V. and Stoker, M. (Eds.) Growth Regulating Substances for Animal Cells in Culture, Vol. 7, p. 87 (1967), Wistar Inst. Symp. Monogr.
15. Lipton, A., Klinger, I., Paul, D. and Holley, R.W. Proc. Nat. Acad. Sci. USA 68 (1971) 2799.
16. Leffert, H.L. and Paul D. J. Cell Biol. 52 (1972) 559.
17. Leffert, H.L. and Paul D. J. Cell Physiol. 81 (1973) 113.
18. Paul, D., Leffert, H.L., Sato, G. and Holley, R.W. Proc. Nat. Acad. Sci. USA 69 (1972) 374.
19. Paul, D. and Walter, S. Proc. Soc. Expt. Biol. Med. 145 (1974) 456.
20. Paul, D. and Walter, S. J. Cell. Physiol. submitted for publication
21. Holley, R.W. Proc. Nat. Acad. Sci. USA 69 (1972) 2840.
22. Tobey, R.A. and Ley, K.D. Cancer Res. 31 (1971) 46.
23. Bucher, N.L.R., Schrock, T.R. and Moolten, F.L. Johns Hopkins Med. J. 125 (1969) 250.
24. Leffert, H.L. J. Cell Biol. (1974) in press.
25. Koch, K. and Leffert, H.L. J. Cell Biol. (1974) in press.
26. Maale, O. and Hanawalt, P.D. J. Mol. Biol. 3 (1961) 144.
27. Short, J., Brown, R.F., Husakova, A., Gilbertson, J.R., Zemel, R. and Lieberman, I. J. Biol. Chem. 247 (1972) 1757.
28. Short, J., Armstrong, N.B., Zemel, R. and Lieberman, I. Biophys. Biochem. Res. Comm. 50 (1973) 430.
29. Bucher, N.L.R. and Swaffield, M.N. Cancer Res. 33 (1973) 3189.

METABOLIC AND GROWTH-PROMOTING PROPERTIES OF SERUM TRIPEPTIDE AND ITS SYNTHETIC ANALOG

M. Michael Thaler and Loren R. Pickart

    Cell growth in most mammalian cell systems requires the presence of serum or serum proteins in the culture medium. The growth-supporting properties of serum have been variously attributed to all of the major protein fractions of serum (1-14) or, alternately, to small molecules which diffuse through dialysis membranes after partial proteolysis or heating of serum (15-18). Since purification of serum growth factors has been hampered by loss of activity during isolation from larger protein complexes, it is possible that certain serum proteins may act as stabilizers or carriers for smaller factors which are themselves responsible for the growth-promoting activities of serum. In addition, the stimulation of cellular intermediary metabolism may be another important effect of serum on cultured cells. A correlation beween cell growth, intermediary metabolism, and the albumin fraction of serum is suggested by several observations. A few examples will suffice:
    1. Cohn Fraction V albumin may be replaced with pyruvate, α-ketoglutarate and inert osmotically active agents for cultivation of mouse fibroblasts (19).
    2. Some serum functions may be replaced by insulin (20, 21). Both insulin and the non-immune suppressible insulin-like activity from the albumin fraction of serum aids cell growth (15).
    3. Those media which support cell growth in serum-free systems usually contain precursors of intermediary metabolism (e.g. pyruvate, acetate, citrate, α-ketoglutarate, succinate, fumarate, malate, oxaloacelate) which are not contained in simple media such as Eagle's Basal Medium. Examples of such media are shown in Table 1.
    Aware of the possible relationship between the growth-promoting and metabolic properties of serum, we began several years ago to look at the effects of individual serum proteins on the metabolism of acetate and palmitate in a liver slice system. The results of an early experiment are shown in Table 2. Fraction V (albumin) was the most active of the serum protein fractions in stimulating $^{14}CO_2$ production from labeled acetate. Fractions IV-4 and VI, i.e. those containing appreciable amounts of albumin, had approximately one-fourth of the effect of fraction V, while other fractions and the osmotically active polysaccharide dextran were

Media used for serum-free culture	Metabolites (mg/L) or Insulin (Units/L)
CMRL-1066 (Michl 1962, Metzgar and Moskowitz 1960, Gwatkin 1960)	Acetate (83), Pyruvate (50)
Holmes A2 (Holmes 1967)	Acetate (83), Pyruvate (100), Citrate (1)
Tozer & Pirt (Tozer and Pirt 1964)	Insulin (40)
F-12 (Ham 1965, Puck et al. 1968)	Pyruvate (110)
Neuman & Tytell (Neuman and Tytell 1960)	Pyruvate (110), Insulin (1)
MAB 87/3 (Waymouth 1965)	Insulin (8)
Modified MB 752/1 (Sinclair et al. 1963)	2-Ketoglutarate (37)

Media requiring serum for culture but containing no acetate, pyruvate, citric acid cycle intermediates, or insulin

    McCoy's 5a (Iwakata and Grace 1964)

    BME (Eagle 1955)

    BHK-21 (MacPherson and Stoker 1962)

    Gibco diploid (Gibco 1972)

    N-16 (Puck 1958)

    Swim's S-77 (Gibco 1972)

    MEM (Eagle 1959)

TABLE 1. INTERMEDIATES OF ENERGY METABOLISM IN VARIOUS MEDIA

Factor added to serum*	No of expts	CO_2 production % of control	P value
Human Fraction II	10	63	0.02
Human Fraction III-0	4	101	N.S.
Human Fraction IV-1	4	103	N.S.
Human Fraction IV-4	4	158	0.05
Human Fraction V	4	280	0.01
Human Fraction VI-1	4	150	0.05
Repurified Fraction V Human Albumin:			
20 mg/ml	4	264	
15 mg/ml	14	245	0.001
10 mg/ml	4	221	
5 mg/ml	4	170	
Human Albumin- "Vallance-Owen's"	2	251	
Human Albumin- "Vallance-Owen's"- Repurified	2	242	
Human Albumin- Crystalline	6	135	0.01
Human Albumin- Succinate Blocked	4	104	N.S.
Human Albumin- DNP Blocked	3	70	0.05
Dextran-75,000 daltons	6	96	N.S.
Egg Albumin	4	106	N.S.
$CaCl_2$-10 mg %	4	110	N.S.

*All proteins and dextran at 15 mg/ml except where noted.
N.S. = Not significant at p > 0.05.

Repurification of albumin was by the trichloroacetic acid-ethanol-diethyl ether method (Sellers et al. J. Lab. Clin. Med. 1966).

TABLE 2. EFFECT OF PROTEINS, DEXTRAN, AND CALCIUM ON THE OXIDATION OF (1-^{14}C) ACETATE BY LIVER SLICES

inactive or slightly inhibitory. It was of interest that highly purified crystalline albumin stimulated oxidation of acetate relatively weakly in comparison with less pure albumin preparations.

The stimulation of CO_2 production from acetate, palmitate, glucose and glutamic acid, was generally paralleled by the effectiveness with which their incorporation into lipids was enhanced by albumin (Table 3). A typical result with acetate is shown in Table 4. Stimulation of lipid synthesis extended to all major lipid fractions. Purified albumin was also injected into mice. These animals produced 3 to 4 times as much $^{14}CO_2$ from labeled acetate as did their saline-treated littermates (Fig. 1). DNP-blocked albumin, succinate-blocked albumin, gamma globulin had no effect, while egg albumin was quite toxic to the animals.

At this point, we began to examine factors responsible for enhancement of oxidative processes and lipid synthesis associated with albumin. The highest activity was present in the freshest albumin preparations, with gradual loss of activity during storage at $3^{\circ}C$ for several weeks. This relatively short half-life, and the observation that the purest material available, i.e. crystalline albumin had the lowest activity, suggested that the active factor could perhaps be disassociated from the larger protein.

Supernatants obtained by boiling fresh albumin or serum at pH 4.5 retained the ability to stimulate CO_2 production and lipid synthesis after removal of the coagulated proteins. The specific activity of the soluble material, expressed in terms of amino acid residue weight, was at least 55 times greater than that of the original fraction V (Table 5).

Activity was further enhanced by stepwise ultrafiltration thorugh molecular filters. Most of the activity was contained in UM-2 filtrates and UM-05 retardates, the latter being the most active. At concentrations of 1 to 2 µg/ml, UM-05 retardate stimulated CO_2 production and synthesis of all lipid classes from acetate and palmitate in liver slices and in hepatoma cells (Tables 6 and 7). The specific activity of the material in the UM-05 retardate was approximately 3×10^3 times that of fraction V. Ion exchange chromatography of UM-05 retardate from serum on Dowex 50W-X4 columns produced 7 peptide peaks, one of which contained all of the stimulatory activity (Fig. 2). A further 12-fold increase in activity was thus obtained, with the material in the active peak having a specific activity approximately 3×10^4 times that of defatted albumin (Table 5). These fractions were assayed in the more sensitive hepatoma cell system (22).

SUBSTRATE	PRODUCT			
	CO_2		Lipids	
	Control (cpm)	Increase with albumin (%)	Control (cpm)	Increase with albumin (%)
$(1-^{14}C)$ acetate	1600	242	10400	195
$D-(1-^{14}C)$ glucose	2900	132	1700	140
$(1-^{14}C)$ palmitate	470	168	31800	155
$L-(U-^{14}C)$ glutamate	9100	170	640	158

Each incubation flask contained 1 µCi substrate, 100 mg liver slices, and 2 ml serum. Incubation was for 2 hours at 37°C. Each value represents average of 2 incubations.

TABLE 3. ALBUMIN EFFECT ON OXIDATION AND LIPID SYNTHESIS FROM PRECURSORS OF HEPATIC INTERMEDIARY METABOLISM

Metabolite Measured	Control cpm	Plus Albumin cpm	P Value
CO_2	3800	11400	0.001
Phospholipids	2550	8990	0.002
Cholesterol	4370	9030	0.005
Free Fatty Acids	5620	9070	0.005
Triglycerides	3790	9190	0.002
Cholesterol Esters	2300	2770	N.S.

Each incubation flask contained 2 ml serum from fed rats, 100 mg liver slices from fed rats, and 2 microcuries ($1-^{14}C$) acetate. Incubation was for 2 hours. Added albumin (repurified human Cohn fraction V) was 15 mg/ml. Each value represents the average of 3 flasks.

N.S. = Not statistically significant at $p < 0.05$. The values in cpm are for one typical experiment. The p-values were calculated on the average of 6 such experiments.

TABLE 4. ALBUMIN STIMULATION OF ACETATE OXIDATION AND INCORPORATION INTO LIPID CLASSES BY LIVER SLICES

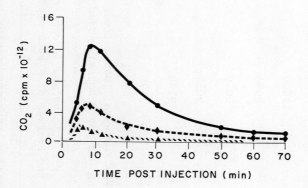

FIG. 1:

<u>Effect of human albumin and ovalbumin on production of $^{14}CO_2$ from $(1-^{14}C)$ acetate in intact mice</u>. The curves represent control animals (♦---♦), animals injected (I.P) with repurified human albumin (●——●), and animals injected with ovalbumin (▲---▲). Results of treatment with stored (time-inactivated) albumin, gamma globulin, DNP-blocked albumin and succinylated albumin did not significantly differ from control values.

Fraction	Micrograms amino acid residue per ml which doubled CO_2 production from $(1-^{14}C)$ acetate		Increase in activity over albumin
	In liver slices	In HTC cells	
Repurified human albumin	6000	--	
Supernatant from boiled albumin	50-170	--	55
Fraction passing UM-10 membrane	3-8	--	500
Fraction passing UM-2 membrane but retarded by UM-05 membrane	1-3	0.25-0.47	3000
Active fraction from Dowex 50-X4 column	--	0.02-0.04	36000

TABLE 5. INCREASE IN THE SPECIFIC ACTIVITY OF ISOLATES FROM ALBUMIN DURING FRACTIONATION PROCEDURES

Metabolite measured	Liver slices		Hepatoma cells	
	Control	Plus UM-05 fraction	Control	Plus UM-05 fraction
CO_2	2400	7400	2700	28500
Phospholipids	2350	4950	870	3370
Cholesterol	2700	4450	830	1540
Free fatty acids	1100	2900	725	1420
Triglyceride	4200	10100	2245	9220
Cholesterol Esters	740	1370	660	1540

All flasks contained 2 micrograms of the UM-05 fraction per ml and 1 microcurie sodium $(1-^{14}C)$ acetate. Liver slices (100 mg) were incubated in 2 ml serum from fed rats for 2 hours. Hepatoma cells (10^6) were incubated in Eagle's media for 1 hour. Each value represents the average of three flasks.

TABLE 6. EFFECT OF UM-05 FRACTION FROM ALBUMIN ON CO_2 PRODUCTION AND LIPID SYNTHESIS FROM $(1-^{14}C)$ ACETATE IN RAT LIVER SLICES AND HEPATOMA CELLS

Metabolite measured	Liver slices (cpm)		Hepatoma cells (cpm)	
	Control	Control plus UM-05 fraction	Control	Control plus UM-05 fraction
CO_2	3200	4700	43000	107000
Phospholipids	144900	246000	1550000	4100000
Triglycerides	195900	320000	1592000	3200000

Each flask contained 10 microcuries sodium ($1-^{14}C$) palmitate. Hepatoma cell media contained 1 mg/ml defatted human albumin. All other conditions were as in Table 6.

TABLE 7. EFFECT OF UM-05 FRACTION FROM ALBUMIN ON CO_2 PRODUCTION AND LIPID SYNTHESIS FROM ($1-^{14}C$) PALMITATE IN RAT LIVER SLICES AND HEPATOMA CELLS.

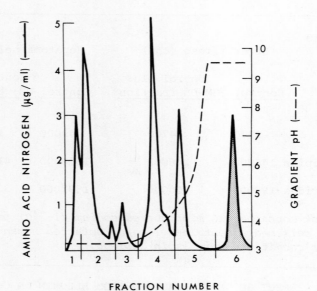

FIG. 2: Fractionation of peptides in UM-05 retardate on Dowex 50-X4 column. All the biological activity was found in column fraction 6.

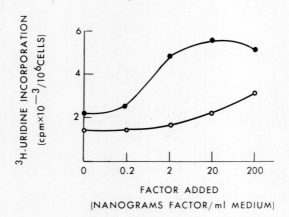

FIG. 3: Effect of serum tripeptide on RNA synthesis in hepatoma cell cultures in the presence (closed circles) and absence (open circles) of 1% serum. Experimental conditions are described in Table 10.

The active factor had a molecular weight of approximately 300 as determined with marker compounds on a G-10 column. Amino acid analysis of the active fraction showed that it was composed of glycine 31.3%, histidine 27.3%, lysine 32.8%, serine 3.9%, threonine 1.1%, aspartic acid 1.1%, alanine 1.5%, and glutamic acid 0.7%. Similar results were obtained with fractions isolated by thin-layer chromatography. The active material stained yellow with ninhydrin and isatin, suggesting the presence of N-terminal glycine, and positive Pauly reactions indicated that the imidazole ring of histidine was unsubstituted.

Two synthetic tripeptides composed of L-amino acids, glycyl-lysyl-histidine and glycyl-histidyl-lysine possessed chromatographic and staining properties which were indistinguishable from the biologic peptide in several thin-layer systems (23).

All of the metabolic effects of the tripeptide were observed in short-term experiments lasting up to 4 hours. When HTC cells were incubated for longer periods in the presence of tripeptide, marked effects on macromolecular synthesis and cell growth were observed, particularly when transferred to media containing low concentrations of serum. In medium containing 1% serum, DNA synthesis was enhanced more than 4-fold, RNA synthesis more than 12-fold and protein synthesis more than 4-fold within 24 hours (Table 8). Growth of neoplastic cells (HTC) was accelerated after 48 hours incubation with the tripeptide, nearly doubling the population in treated cultures compared with untreated (Table 9). The rise in DNA and RNA synthesis which accompanied the rise in cell number was considerably smaller than the increase in synthetic rates observed after 24 hour exposure to the factor.

In contrast with neoplastic liver cells, normal liver cells in suspension did not grow, but their survival was prolonged in the presence of tripeptide and macromolecular synthesis was increased (22). In serum-free medium the factor retained 30-40% of its activity compared with incubations containing 1% serum (Fig. 1). Optimal concentrations of tripeptide in standard 1% serum medium were below 50 nanograms/ml.

The synthetic tripeptide GHL had similar stimulatory effects on cell growth, survival and macromolecular synthesis as the native tripeptide (Table 10) (23). GLH appeared somewhat less effective, but direct comparison is difficult because of its tendency to rapid oxidation.

As pointed out previously, the tripeptide we have been studying has insulin-like effects on intermediary metabolism and cell growth. We have recently begun studies of the

Additions	DNA	RNA	Protein
		(cpm/10^6 cells)	
Basal medium (serum-free)	630	2250	4570
1% serum (control)	900	2700	5800
2% serum	1530	6200	14700
Serum tripeptide factor (200 ng/ml)	3900	32100	26100

Cells were incubated for 24 hours. Cultures were pulsed with 1 µCi of ^{3}H-thymidine (6.7 Ci/mM), or ^{3}H-uridine (5 Ci/mM), or ^{3}H-leucine (30 Ci/mM) 4 hours before the end of incubation, and macromolecular synthesis measured. Values represent averages of triplicate incubations.

TABLE 8. EFFECT OF SERUM AND OF SERUM TRIPEPTIDE FACTOR ON MACROMOLECULAR SYNTHESIS IN RAT LIVER NEOPLASTIC CELLS

Serum peptide (nanograms/ml)	Neoplastic cells (hepatoma)			Normal hepatocytes		
	Cell number* (x 10^6)	RNA (cpm/10^6 cells)	DNA (cpm/10^6 cells)	Cell number*	RNA (cpm/10^6 cells)	DNA (cpm/10^6 cells)
0	1.21	2100	570	1.34	840	110
0.2	1.32	2400	660	1.40	820	118
2	2.15	4900	1060	1.66	1430	145
20	2.38	5500	1240	1.92	1750	190
200	2.05	5350	1030	1.73	1800	175

*Initial cell concentrations were: neoplastic cells, 1×10^6 cells/flask; normal cells 2×10^6 cells/flask. Cells were incubated for 48 hours at $37^\circ C$ in 5 ml media containing 1% calf serum and added serum factor. Cultures were pulsed with 1 μc ^{3}H-thymidine (6.7 Ci/mM) or ^{3}H-uridine (5 Ci/mM) 4 hours before the end of incubation.

TABLE 9. EFFECT OF SERUM TRIPEPTIDE FACTOR ON SURVIVAL, GROWTH AND MACROMOLECULAR SYNTHESIS IN NEOPLASTIC AND NORMAL HEPATOCYTES

ADDITIONS (ng/ml)	NEOPLASTIC CELLS (HEPATOMA)			NORMAL HEPATOCYTES		
	Cell number (× 10^6)	RNA (cpm)	DNA (cpm)	Cell number (× 10^6)	RNA (cpm)	DNA (cpm)
0-(control)	1.21±0.08	2100±100	570±40	1.34±0.10	840±50	110±9
Bio-tripeptide						
0.2	1.32±0.06	2400±130	660±30	1.40±0.17	820±40	118±7
2.0	2.15±0.11	4870±270	1060±90	1.66±0.01	1430±140	145±7
20.0	2.38±0.24	5530±230	1240±30	1.92±0.02	1750±260	190±21
200.0	2.05±0.20	5350±160	1030±70	1.73±0.05	1800±80	175±13
Synthetic tripeptide (Gly-His-Lys)						
0.2	1.29±0.02	2570±180	550±70	1.45±0.05	910±30	107±2
2.0	2.40±0.13	7500±160	1250±90	1.60±0.08	1100±110	140±11
20.0	4.67±0.61	10940±380	2140±220	1.88±0.06	1560±80	212±4
200.0	1.91±0.04	8380±390	1490±40	1.45±0.06	1370±80	135±10

Initial cell concentrations were: neoplastic cells, 1.00× 10^6 cells per flask; normal cells 2.04 × 10^6 cells per flask. Cells were incubated for 48 hours at 37°C in 5 ml media containing 1% calf serum and the added peptide factor. Cultures were pulsed with 1 microcurie of (^3H) uridine or (^3H) thymidine four hours before the end of the incubation. All samples were in triplicate.

TABLE 10. 48 HOUR EFFECT OF BIOLOGICAL AND SYNTHETIC TRIPEPTIDES ON SURVIVAL, GROWTH, AND MACROMOLECULAR SYNTHESIS IN NEOPLASTIC AND NORMAL HEPATOCYTES

Factor	Concentration (µg/ml)	Increase in insulin binding (% of control)
Serum tripeptide	1.0	22
Glycyl-histidyl-lysine	1.0	24
Putrescine	500	13
Spermidine	500	28
Spermine	500	65

TABLE 11. EFFECT OF TRIPEPTIDE FACTORS AND POLYAMINES ON INSULIN BINDING TO RECEPTORS IN HUMAN PLACENTAL PLASMA MEMBRANES

Additions	Insulin molecules per cell
None	60
Serum tripeptide (1µg/ml)	145

TABLE 12. EFFECT OF TRIPEPTIDE ON INSULIN BINDING TO HEPATOMA CELLS (HTC) IN MEDIUM CONTAINING 0.5% SERUM

interactions between tripeptide and insulin at the cell surface, using an insulin-somatomedin receptor assay devised by Underwood and co-workers (24). The tripeptide appears to increase the binding of insulin to receptors in human placental membranes. We have also found that polyamines such as putrescine, spermine and spermidine which have been reported to stimulate growth, (7,25,26) also increase insulin binding in this system (Table 11).

The effect of tripeptide on insulin binding by HTC cells was tested in monolayer cultures maintained for 24 hours in serum-free medium, followed by 24 hours in medium containing 0.5% bovine serum and tripeptide at a concentration of 1 μg/ml. Radioiodinated insulin (0.05 ng = 20,000 cpm) was added during the last hour. The cells were then washed in cold PBS, and bound radioactivity determined. Non-specific binding of insulin was determined from cultures to which an excess of unlabeled insulin was added together with labeled hormone. The results, shown in Table 12, indicate that the tripeptide increases the binding of insulin to HTC cells (27).

The enhancement of insulin binding to its receptors by serum tripeptide and the polyamines suggests that these factors may act as cofactors in stimulation of growth by insulin and insulin-like peptides such as somatomedin or NSILA. The somatomedins compete with insulin for cellular receptors (28) and may therefore respond to similar co-factors. Whether these small, highly basic molecules function in the regulation of growth _in vivo_ is still undetermined, but the characterization of the serum tripeptide and its synthetic analogs should encourage the search for additional simple molecules which may exert pharmacological effects on cell growth and survival _in vivo_.

Supported by NIH grant #HD-03148. M.M. Thaler is the recipient of a NIH Research Career Development Award.

References

1. Clark, G.D. and Stoker, M.G.P., In "Growth Control in Cell Cultures,"(G.E.W. Wolstenholme and J. Knight, Eds.) Churchill Livingstone, London (1971) 17.
2. Jainchill, L. and Todaro, G. Expt. Cell Res. 59 (1970) 137.

3. Holley, R.W. and Kiernan, J.A. In "Growth Control in Cell Cultures," (G.E.W. Wolstenholme and J. Knight, Eds) Churchill Livingstone, London (1971) 3.
4. Holmes, R. J. Cell Biol. 32, (1967) 297.
5. Michl, J. and Svobodova, J. Expt. Cell Res. 58 (1969) 174.
6. Puck, T., Waldren, C. and Jones, C. Proc. U.S. Nat. Acad. Sci. 59 (1968) 192.
7. Waymouth, C. Nat. Cancer Inst. Monogr. 26 (1967) 1.
8. Birch, J.R. and Pirt, S.J. J. Cell Sci. 5 (1969) 135.
9. Ellem, K.A. and Mironescu, S. J. Cell Physiol. 79 (1972) 389.
10. Holmes, R. and Wolfe, S.W. J. Biophys. Biochem. Cyto. 10 (1961) 389.
11. Matsuya, Y. and Yamane, I. Expt. Cell Res. 50 (1968) 652.
12. Matsuya, Y. and Yamane, I. Proc. Soc. Expt. Biol. Med. 135 (1970) 893.
13. Todaro, G. and Green, H. Proc. Soc. Expt. Biol. Med. 116 (1964) 688.
14. Healy, G. and Parker, R. J. Cell Biol. 30 (1966) 539.
15. Pierson, R.W. and Temin, H.M. J. Cell Physiol. 79 (1972) 319.
16. De Luca C., Habeeb A.F.S. and Tritsch, G.L., Expt. Cell Res. 43 (1966) 98.
17. Eagle, H. Proc. U.S. Nat. Acad. Sci. 46 (1960) 427.
18. Gwatkin, R.B. Nature 186 (1960) 984.
19. Metzgar, D.P. and Moskowitz, M. Proc. Soc. Expt. Biol. Med. 104 (1960) 363.
20. Temin, H. In "Growth Regulatory Substances for Animal Cells in Culture," (V. Defendi and M. Stoker, Eds) Wistar Inst. Press, Philadelphia, 7 (1967) 103.
21. Hershko, A., Mamont, A.P., Shields, R. and Tomkins, G.M. Nature New Biol. 232 (1971) 206.
22. Pickart, L.R. and Thaler, M.M. Nature New Biol. 243 (1973) 85.
23. Pickart, L.R., Thayer, L. and Thaler, M.M. Biochem. Biophys. Res. Comm. 54 (1973) 562.
24. Underwood, L. E., Marshall, R.N., Voina, S.J. and Judson, J.V.W. Ped. Res. 8 (1974) 102.
25. Pegg, A.E. Biochim. Biophys. Acta. 232 (1971) 630.
26. Pohjanpelto, P. and Raina, A. Nature New Biol. 235 (1972) 247.
27. Pickart, L., Youlton, R., Byers, L. and Thaler, M.M. (In Press).

28. Marshall, R.N, Underwood, L.E., Voina, S.J. and Van-Wyk, J.J. Laurentian Hormone Conference, August 1973 (In Press).

Note: We would like to thank the publishers of Biochem. Biophys. Res. Comm. and Nature for the permission to reproduce Table 3 and Figures 2 and 3.

THE USE OF SELECTIVE MARKERS IN THE STUDY OF DIFFERENTIATED
GENE FUNCTION IN NORMAL AND MALIGNANT CELLS OF HEPATIC ORIGIN

David Rintoul and John Morrow

Introduction

It is commonly accepted that the somatic cells of an adult organism possess all the genetic information present in the fertilized egg (1). If this statement is correct, then the problem of differentiation to a large extent reduces to a question of understanding the nature of the regulatory forces responsible for the selective activation of different structural genes in different tissues. One model of such a control system would suggest the existence of regulatory loci producing substances which either repress or activate the various structural genes. The observation in several systems that the union of an undifferentiated cell with a differentiated one through cell hybridization results in suppression of differentiated gene function, suggests that negative control is the appropriate explanation. This is true of the properties of melanin production (2), glycerol-3-phosphate dehydrogenase inducibility (3), tryosine aminotransferase induction (4) growth hormone production (5), and aldolase isozymes (6), among others. However, Minna et al. (7) have shown that some neuronal properties are retained in Neuroblastoma x L cell hybrids.

Furthermore, Davidson (8) has shown that while hybrids between pigmented and nonpigmented cells are nonpigmented, a hybrid consisting of two pigmented and one fibroblast genome will be nonpigmented. This, together with the fact that differentiated functions return to somatic cell hybrids after chromosome loss, suggests that the control of differentiation resides in the existence of diffusable repressors, whose constant production is required, and which can be titrated out by an excess of structural genes.

A corollary of this model would be that reduced hybrids which reexpress the differentitated gene in question would sometimes offer a new form of that gene product, namely a previously unexpressed form from the fibroblast parent (9,10).

A complete understanding of the nature of the repressor substance awaits its chemical elucidation through the

Note: This work was supported by grants from the Public
       Health Service (1-R01-Ca12310-02).

techniques of molecular biology. The cell hybridization experiments do, however, offer the possibility of understanding the manner in which control systems interrelate within the cell genome. Thus, different patterns of segregation in somatic cell hybrids might indicate whether a number of differentiated functions are controlled by the same regulatory factor, or whether each are under separate and independent control.

We have investigated a number of differentiated functions with this eventual goal in mind. To this end we have hybridized liver cells of malignant and nonmalignant origin with mouse fibroblasts and followed the expression of liver specific functions in the hybrids.

Methods and Materials

These are reported in detail in previous publications (11,12). Briefly, the cell lines were FLC, a line of mouse fetal liver cells isolated by Waymouth in defined medium supplemented with dexamethasone, the BrdU resistant mouse L cell line, LMTK$^-$ (13), and the rat hepatoma HTC. Culture methods included the use of autoclavable Eagle's medium, HAT medium (14) and Viokase (GIBCO). Tryptophan pyrrolase was assayed by the method of Knox and Piras (15) and aldehyde dehydrogenase by the method of Deitrich (16). Histochemical staining was our own procedure, adopted from Wajntal and DeMars (17).

Results

The FLC line was hybridized with LMTK$^-$ through a selective technique which made use of the slow growth properties and differential adhesiveness of the fetal liver cells. Because the FLC are difficult to dislodge from the glass substrate, a mixture of the two cell lines was made and HAT added to the medium. The LMTK$^-$ cells soon died in this selective medium, and the only cells remaining were the FLC and any hybrids that might have arisen. When the bottle was lightly viokased, the parent FLC remained attached to the glass while the hybrids were dislodged. After several such treatments, a line of hybrid cells free of the parent FLC was obtained, and their hybrid character was confirmed by karyotypic analysis. (Figure 1)

Our approach for selection of hybrids between HTC and LMTK$^-$ was just the reverse of the previous technique. Large (2×10^6) numbers of the two parental cell lines were mixed together in HAT medium. Since the HTC cells attach lightly

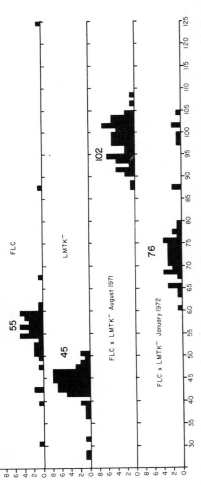

FIG. 1: Abscissa: no. of chromosomes; ordinate: no. of cells. Chromosome counts of the FLC and LMTK⁻ cell lines, with counts of the hybrid cell line when first isolated and after 5 months in culture. Numbers above the histograms are modal chromosome numbers. (Data from ref. 11).

FIG. 2: Zymograms of alcohol dehydrogenase isozymes of HTC, LMTK⁻, and HL-1, a cloned somatic cell hybrid.

to culture vessels, several treatments with Viokase were sufficient to remove them, leaving the dead LMTK⁻ cells and the hybrids. Subsequent cloning insured the genetic purity of the hybrids. (Figure 2)

The use of differential adhesiveness as a parameter in hybridization enables us to bypass the use of inactivated Sendai virus and the possible membrane disruption that might result from its use. Thus, any morphological differences between the hybrid and its parents are probably due to the genome of the parents and not due to the hybridization technique itself. In the absence of viral effects on the cells, it seems that the semi-differentiated morphological appearance of both the FLC and the HTC lines is recessive genetically. In both instances hybrids possessed the fibroblastic morphology characteristic of the LMTK⁻ parent. (Figures 3, 4 and 5)

When cultures of FLC were stained with Periodic acid-Schiff reagent, all clones examined were observed to contain glycogen granules. The LMTK⁻ parent cells were not observed to synthesize such granules and, likewise, no granules were observed in the hybrids. Therefore, this liver specific characteristic is not expressed in the hybrids. Similarly, the results of the tryptophan pyrrolase assay indicate that both the presence of the enzyme and its inducibility are recessive in hybrids. When assayed at various stages of growth, the FLC showed a 3-5 fold increase in the enzyme while the LMTK⁻ and the hybrids showed no activity and no induction when assayed at 8 and 14 weeks after hybridization. Table 1 shows the typical results obtained from induction of the three cell types during the exponential phase of growth assayed at $25°C$. It can be seen that the inducibility and, in fact, the presence of the enzyme, appear to be recessive in the hybrids.

A similar pattern was encountered when the HTC-LMTK⁻ were tested for the presence of aldehyde dehydrogenase. Although initially activity was observed in 6 week old hybrid lines, when the hybrid clones were stained immediately after hybridization, no activity was detected. An explanation of the subsequent appearance of the enzyme can possibly be found in the loss of chromosomes, which occurs in these hybrids at a prodigious rate. The modal number, which should be near 100, the sum of the modes of parent lines, was 93 at the time of the first karyotyping at six weeks. Two weeks later, the mode had dropped to 81. However, at this time the deviation from the mode had also increased, so it seems possible that not all the cells were losing chromosomes. More likely the population was being overgrown by cells with lower chromosome

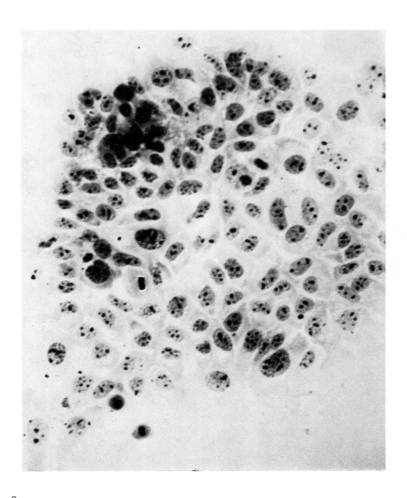

FIG. 3:

FLC, haematoxylin X800

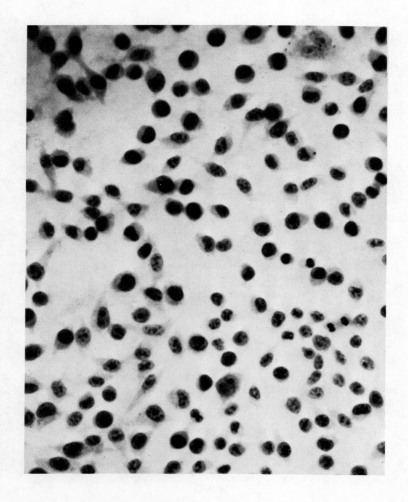

FIG. 4:

LMTK⁻, haematoxylin X800

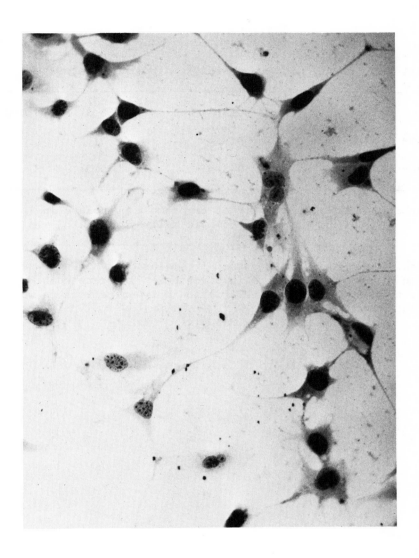

FIG. 5:

Hybrids (FLC x LMTK⁻) haematoxylin X800

Cell Line	Tryptophan pyrrolase activity (plus steroid)	Tryptophan pyrrolase activity (minus steroid)
FLC	5.90(3)	1.53(3)
LMTK$^-$	0(3)	0(3)
Hybrids	0(3)	0(3)

TABLE 1. Specific activity of tryptophan pyrrolase in FLC, LMTK$^-$, and their somatic cell hybrid.

Specific activity is expressed in moles of kynurenine per h/mg of protein. Numbers in parentheses indicate number of experiments performed. Variation between replicates was about 15%.

numbers. These cells were still hybrid in character since at no time did the modal number drop into the range of the parent cells, and subsequent counts at 11 weeks indicated that none of the cells had chromosome complements in the range of the parent cells. (Figure 6)

Investigation of the electrophoretic properties of the aldehyde dehydrogenase enzyme in the parent line and in the hybrids using both starch and acrylamide gels indicate that a new form of the enzyme appears in the hybrids, which migrates more slowly that that of the HTC. Three equally intense bands of aldehyde dehydrogenase activity can be seen in the hybrids at six weeks, and a double band was found in the HTC cell line. The fastest band in the six week old hybrids migrated at the same rate as the slow HTC band, while the other two bands migrated about 1 and 2 centimeters behind, respectively. The patterns once established in the hybrids seemed quite stable. As long as 18 months after the original cloning the hybrids still retained the original three band pattern.

Although as expected the LMTK$^-$ line showed no evidence of the enzyme, we have been able to demonstrate the enzyme in preparations extracted from mouse liver with EDTA and buffer. The mouse liver form migrated behind the HTC type in acrylamide gel preparations.

The possibility that the additional bands observed in the hybrids were merely subunits of the HTC enzyme was explored. Figure 7 shows the results of a heat inactivation study, where it can be noted that the enzyme in the hybrids is considerably more heat labile than the HTC enzyme. These results indicate that the enzyme in the hybrids is significantly different from that in the HTC parent and is probably not entirely of HTC origin.

Discussion

We have observed the repression of tryptophan pyrrolase and its inducibility, glycogen accumulation, "differentiated" morphology and aldehyde dehydrogenase in two types of liver x fibroblast hybrids following the general pattern that differentiated characters are not expressed in hybrids between differentiated and nondifferentiated cells. Our result indicates that the malignant potential of the cells is not a factor affecting gene expression in hybrids since both sets of hybrids showed extinction for the characters in question. In the case of the HTC-LM/TK$^-$ hybrids the results indicate that chromosome loss is necessary before aldehyde dehydrogenase can be expressed, as evidenced from the karyo-

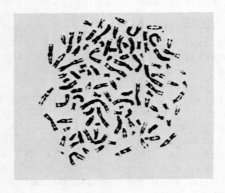

FIG. 6:

Giemsa staining pattern of a typical metaphase from an uncloned hybrid line after six weeks of growth in culture. (Data from ref. 12).

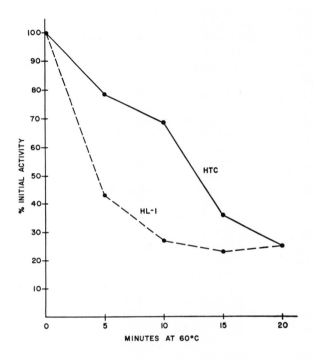

FIG. 7:

Heat inactivation curves of HTC and hybrid (6-week) aldehyde dehydrogenase activity. Extracts of both cells were incubated at 60° for various lengths of time and assayed for aldehyde dehydrogenase as explained in the text. Each point is an average of two separate experiments.

typic analysis. Precedents for this hypothesis were found by Bregula et al. (18) and Wiener et al. (19) who reported that the ability for form tumors in malignant by non-malignant hybrids was correlated with the loss of chromosomes. Weiss and Chaplain (9) found that tyrosine aminotransferase inducibility reappeared in hepatoma-fibroblast hybrids after loss of chromosomes. In addition, Bertolotti and Weiss (10) reported that rat liver alcohol dehydrogenase was expressed in only one of five hepatoma-fibroblast hybrids, and this hybrid clone had undergone extensive chromosome loss compared to the other clones. Klebe et al. (20) reported the reversible expression of mouse esterase in RAG-W138 hybrids correlated with the presence of a human chromosome, C_{10}.

Although negative control of the characters studied in the FLC-LM/TK$^-$ hybrids appears to be the most likely hypothesis, our data do not rule out the possibility that the characters are expressed in hybrids, but that the structural genes for the enzymes in question had been eliminated by chromosome loss prior to the assay. Since the tryptophan pyrrolase assay required large cell numbers and many generations, this problem does not appear to be amenable to resolution.

The appearance of different bands of aldehyde dehydrogenase in the hybrids may be due to the reactivation of repressed structural genes in the mouse parent. Peterson and Weiss (21) have shown the existence of mouse albumin in a hepatoma-fibroblast hybrid, where the fibroblast, the mouse parent, did not produce albumin. This apparent activation of repressed genes in the hybrids might also be paralleled by the work of Carlsson et al. (22) who detected chick myosin in rat embryonic myotubes containing reactivated chick erythrocyte nuclei. Our results suggest that the reactivated gene hypothesis is the most likely. Presently we are reanalyzing our hybrids and parental line using the acrylamide techniques in an effort to demonstrate conclusively the existence of the hybrid bands and the two parental types.

In the FLC-LM/TK$^-$ hybrid the chromosome number declined much more rapidly than has been observed in other mouse-mouse hybrids. Thus, after five months of continuous culture the modal chromosome number had declined from 102 to 76 chromosomes. This chromosome loss may be related to internal timing of chromosome replication; since the FLC parent has a much longer generation time than the LM/TK$^-$ parent, it may be that the FLC chromosomes failed to complete their replication in time for mitosis and were preferentially lost.

Recently we have investigated the possibility that arginine independence might be imployed as a selective marker in hybridization studies between liver and fibroblastic cells. Since the liver possesses the urea cycle, such cells should be arginine independent if they retain this differentiated gene function. Furthermore, hybrids could be selected for the retention of liver specific chromosomes and the loss of those that suppress this function. However, when both the FLC line and RLC, another hepatic cell line derived from an adult liver by Gerschenson et al. (23) were tested, the results were ambivalent. When these cells were cultured in medium containing Sephadex-treated serum from which arginine had been omitted, the cells initially proliferated for a generation or two and then declined, and were unable to sustain continued growth in the absence of this amino acid. However, their decline was not as precipitous as in the LM/-TK$^-$ line, suggesting that the cell may possess some minimal arginine synthesis.

Conclusions and Summary

Several differentiated characters which we have investigated behave as genetic recessives when cells of hepatic origin are fused with fibroblastic cells. These and other liver specific characters should be of value in investigating models of gene regulation based on the existence of specific repressor substances and also for determining linkage relationships of liver specific genes and their regulatory elements.
The characters studied included the characteristic morphology of both cell lines, tryptophan pyrrolase and its inducibility, and accumulation of glycogen granules in the fetal liver x fibroblast hybrids. The enzyme aldehyde dehydrogenase was studied in the HTC x fibroblast hybrids. In the latter case the enzyme reappeared after extensive chromosome loss, suggesting that the negative control model is valid here. Furthermore, a new form of enzyme appeared which could be the results of the derepression of genes in the mouse fibroblast parent.
Finally, the results of experiments measuring the ability of cells of hepatic origin to proliferate in the absence of arginine, possibly due to the presence of the urea cycle enzymes, suggest that the cells may have a slight but insufficient capacity to manufacture this amino acid.

---

1. Gurdon, J. B. Scientific American (December) 24 (1968).
2. Silagi, S. Cancer Res. 27 (1967) 1953.
3. Davidson, R. L. and Benda, P. Proc. Natl. Acad. Sci. 67

(1970) 1870.
4. Schneider, J. A. and Weiss, M. C. Proc. Natl. Acad. Sci. 68 (1971) 127.
5. Sonnenscheim, C., Tashjian, A. H., Jr. and Richardson, U. F. Genetics 60 (1968) 227.
6. Bertolotti, R. and Weiss, M. C. J. Cell. Physiol. 79 (1972a) 211.
7. Minna, J., Gazier, D. and Nirenberg, M. Nature New Biol. 235 (1972) 225.
8. Davidson, R. L. Proc. Natl. Acad. Sci. 69 (1972) 951.
9. Weiss, M. C. and Chaplain, M. Proc. Natl. Acad. Sci. 68 (1971) 3026.
10. Bertolotti, R. and Weiss, M. C. Biochimie 54 (2) (1972b) 195.
11. Rintoul, D., Colofiore., J. and Morrow, J. Exptl. Cell Res. 78 (1973a) 414.
12. Rintoul, D., Lewis, R. F. and Morrow, J. Biochem Genet. 9 (1973b) 375.
13. Kit, S., Dubbs, D., Pierkarski, L. and Hsu, T. Exptl. Cell Res. 31 (1963) 297.
14. Littlefield, J. Science 145 (1964) 709.
15. Knox, W. E. and Piras, M. M. J. Biol. Chem. 242 (1967) 2952.
16. Deitrich, R. A. Science 173 (1971) 334.
17. Wajntal, A. and DeMars, R. Biochem Genet. 1 (1967) 61.
18. Bregula, U., Klein, G. and Harris, H. J. Cell Sci. 8 (1971) 673.
19. Wiener, F., Klein, G and Harris H. J. Cell Sci. 8 (1971) 680.
20. Klebe, R. J., Chen, T. and Ruddle, R. Proc. Natl. Acad. Sci. 66 (1970) 1220.
21 Peterson, J. A. and Weiss, M. C. Proc. Natl. Acad. Sci. 68 (1971) 127.
22. Carlsson, S. A., Luger, O., Ringertz, R. R. and Savage, R. E. Exptl. Cell Res. 84 (1974) 47.
23. Gerschenson, L. E., Andersson, M., Molson, J. and Okigaki, T. Science 170 (1970) 859.

Note: We would like to thank the publishers of Exp. Cell Research for the permission to reproduce Figures 1, 2, 3, and 4.

REGULATION OF THE CORTICOSTEROID INDUCIBILITY OF TYROSINE
AMINOTRANSFERASE IN SOMATIC CELL HYBRIDS

Carlo M. Croce, Gerald Litwack and Hilary Koprowski

In a number of rat hepatoma cell lines, the enzyme
tyrosine aminotransferase (TAT) is inducible by corticosteroid hormones (1). Recently it has been shown that the
fusion and the hybridization of rat hepatoma cells with
mouse fibroblasts result in the suppression of the corticosteroid inducibility of TAT in the heterokaryocytes (2) and
in the hybrid cells (3). This type of somatic cell hybrid
loses, preferentially, rat chromosomes, while retaining all
the chromosomes of the parental mouse fibroblasts, chromosomes in which TAT cannot be induced by corticosteroid
hormones (4). We have recently shown that somatic cell hybrids between rat hepatoma cells (Fu5AH) and normal human
diploid fibroblasts lose, unidirectionally, human chromosomes
(5), while retaining all the chromosomes of the rat hepatoma
parental cells. It therefore became possible to study the
effect of the presence of specific human chromosomes derived
from an undifferentiated cell parent (human diploid fibroblast) on the corticosteroid inducibility of TAT.
Investigating the corticosteroid inducibility of TAT in
a series of rat-human hybrids, we found that the presence of
the human X-chromosome in the hybrid clones resulted in the
modulation of the inducibility of TAT (6). Those hybrid
clones which had lost the human X chromosome displayed inducible TAT activity (6). In addition, experiments to determine if the suppression of the TAT inducibility in the rat-human hybrid clones was due to a decrease in dexamethasone
receptor activity indicated that the rat-human hybrids, in
which TAT was not inducible, displayed the same dexamethasone receptor activity of the rat hepatoma parental cells
(6).
Therefore, to understand the mechanism of the suppression of the TAT inducibility in the hybrid cells, we tested
a series of rat hepatoma-mouse fibroblast and rat hepatoma-human fibroblast somatic cell hybrids for TAT inducibility
by dexamethasone, presence of dexamethasone-receptor activity and transfer of the dexamethasone-receptor complex to the
nucleus. In addition, to localize the X linked gene(s) in-

volved in the suppression of the TAT inducibility on the X chromosome, we also hybridized rat hepatoma cells deficient in hypoxanthine guanine phosphoribosyltransferase (HGPRT) with human fibroblasts derived from a subject with a translocation of the long arm of the X chromosome to the chromosome D-14 (KOP cells) (7). Since the human gene for HGPRT is located on the long arm of the X chromosome and since this portion of the X chromosome is translocated to the human chromosome D-14, it is expected that all the hybrids selected in hypoxanthine-aminopterin-thymidine (HAT) medium (8) will contain the human t (14q, Xq) chromosome.

The rat hepatoma cells (Fu5AH) used in these studies were resistant to 15 µg/ml of 8-azaguanine. They were fused with either Cl 1D mouse fibroblasts, deficient in thymidine kinase and resistant to BrdU (9), or human KOP fibroblasts in the presence of βpropiolactone inactivated Sendai virus at pH 8.0 (10). Immediately after fusion, HAT medium was added to the cultures. Two weeks after fusion, the rat-mouse hybrid colonies were picked up and the cells were grown and subsequently cloned. Each clone was derived from a single hybrid colony. The rat hepatoma-human fibroblast cultures were subcultured several times in HAT medium to get rid of the human fibroblasts, and after 4-6 weeks the hybrid cell colonies were picked up and grown; the cells were later cloned.

As shown in Table 1, none of the rat hepatoma-human fibroblast cell hybrids displayed any inducible TAT activity. In addition, the TAT baseline activity was absent in all the rat-mouse hybrid cells. Karyological analysis of these rat-mouse hybrid clones following Giemsa banding staining indicated that those hybrids contained the full complement of mouse chromosomes and 60-80% of the rat chromosome present in the parental rat hepatoma cells.

The rat-human hybrid clone which was found to contain the t (14q, Xq) chromosome and no other human chromosome (Fig. 1), displayed a TAT baseline activity higher than that of the parental rat hepatoma cells, but a very low inducible TAT activity.

As shown in Table 2, the hybrid clones in which TAT was not inducible by dexamethasone displayed dexamethasone receptor activity higher than in the rat hepatoma parental cells. It is likely that the dexamethasone receptors, coded by both parental genomes, are expressed in the hybrids.

Since all the hybrid clones in which TAT was not inducible had dexamethasone receptor activity, we wanted to determine if the transfer of the dexamethasone-receptor complex to the nucleus proceeded at the same extent in inducible and

Cell Line	Dexamethasone	TAT specific activity*	Average Chromosome no.
Parental			
Fu5AH	−	0.5	52
	+	2.2	
Cl-1D	−	0	52
	+	0	
KOP	−	0	46
Hybrids			
Rat-mouse Cl 5	−	0	99
	+	0	
Rat-mouse Cl 7	−	0	95
	+	0	
Rat-mouse Cl 8	−	0	101
	+	0	
Rat-mouse Cl 10	−	0	89
	+	0	
Rat-mouse Cl 16	−	0	91
	+	0	
Rat-mouse Cl 17	−	0	98
	+	0	
Rat-mouse Cl 18	−	0	103
	+	0	
Rat-mouse Cl 20	−	0	83
Rat-human Cl 17	−	0.7	79
Rat-human Cl 7	+	1.1	

TABLE 1. Induction of TAT in Rat-Mouse and Rat-Human Hybrid Cells

Almost confluent monolayers of parental and hybrid cells were kept in serum-free medium with or without 10 μM dexamethasone for 18 hr. The cells were collected after centrifugation at 800g for 10 min at $4°C$, washed and resuspended in extraction medium (0.15 M KCl, 1 mM EDTA, 0.1 mM potassium phosphate buffer at pH 7.6). This suspension was sonicated by two 10 sec bursts with a chilled probe of a Branson sonifier at a setting of 5A. The sonicated suspension was immediately centrifuged in a RC-2 Sorvall at 20,000g for 15 min at $4°C$. The supernatant solutions were tested for TAT by the method of Canellakis and Cohen (11) and for proteins by the method of Lowry et al (12).

*TAT specific activity = μmol p-hydroxyphenyl pyruvate formed per 10 min ($37°C$) per mg protein.

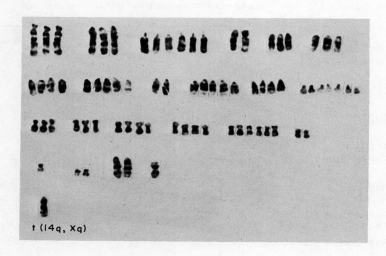

FIG. 1:

Rat-human hybrid Cl 7. Seventy-five rat chromosomes are present in this hybrid. Only the human t (14q, Xq) chromosome is present in this clone.

Cell Line	Receptor Activity (% hormone binding)	TAT Inducibility
Fu5AH	21	++++
Cl-1D	45	−
KOP	4	−
Rat-mouse Cl 5	71	−
Rat-mouse Cl 18	67	−
Rat-human Cl 7	32	+

TABLE 2. Corticosteroid Receptor Activity of Parental and Hybrid Cell Lines

Cytosols of the cell lines were incubated for 90 min at $0°C$ with 2 nM 1,2,4-^{3}H-dexamethasone (New England Nuclear Corp.). A load sample of 0.5 ml was added on a 10 ml syringe column of Sephadex G-25 and eluted with 1 mM phosphate buffer (pH 7.6). Pre-bound, bound and unbound pools were collected and radioactivity of the pools was determined in a scintillation mixture containing Triton-X, toluene and methanol fluor (13). Percentage binding was calculated by the following formula:

$$\frac{\text{c.p.m. bound pool}}{\text{c.p.m. load sample}} \times 100 = \% \text{ binding.}$$

Using radioactive dexamethasone at this level in glucocorticoid responsive cells in culture we observed that only one major dexamethasone-bound macromolecule appears on ion-exchange chromatography which accounts for about 95% of the binding (14). This fraction is identical to the binder II dexamethasone binding protein of rat liver cytosol which has been shown to be the major hormone receptor protein (15). In addition, when this level of radioactive dexamethasone is used in vitro binding experiments with adrenalectomised liver cytosol, there is only one major peak on ion-exchange chromatography which is characterized as binder II, the hormone receptor (unpublished experiments). Thus, the values reported for binding of ^{3}H-dexamethasone in cytosol represent about 95% hormone receptor.

non-inducible hybrid cells. As shown in Table 3, the nuclear transfer of ^{3}H-dexamethasone in the non-inducible rat-mouse hybrid cells proceeded at a higher rate than in the rat hepatoma parental cells.

In the rat-human hybrid clone 7, the bound dexamethasone nuclear transfer proceeded to the same extent as in the rat hepatoma parental cells (unpublished results). The fact that in the rat-human clone 7, which contained only the t (14q, Xq) chromosome, TAT was very slightly inducible, suggests that gene(s) on the long arm of the human X chromosome have a role in the modulation of the TAT inducibility by corticosteroid hormones, since this role has been excluded for the human chromosome D-14 (6).

Our results indicate that the lack of TAT inducibility in interspecific hybrid cells is not due to a decrease in dexamethasone receptor activity. On the contrary, the hybrid clones displayed more dexamethasone receptor activity than the parental rat and mouse cells. In addition the lack of TAT inducibility cannot be attributed to a decrease in the transfer of the bound dexamethasone to the nucleus of the hybrid cells, since hybrid cells exhibit more nuclear ^{3}H-dexamethasone than do the parental cells. Therefore, it is evident that the suppression of TAT inducibility by corticosteroid hormones must take place either at the level of the binding of the dexamethasone-receptor complex to the chromatin or at the transcriptional or post-transcriptional level. These problems are now under investigation.

We thank Irene Kieba, Nancy Shapiro and C. A. Wishman for technical assistance. This work was supported in part by grants from the US Public Health Service Division of Research Resources, the National Cancer Institute and the National Institute and the National Institute of Arthritis, Metabolic and Digestive Diseases.

---

1. Thompson, E. B., Tomkins, G. M. and Curran, G. F. Proc. Nat. Acad. Sci. U.S.A. 56 (1966) 296.
2. Thompson, E. B. and Gelehrter, T. D. Proc. Nat. Acad. Sci. U.S.A. 68 (1971) 2589.
3. Schneider, J. A. and Weiss, M. C. Proc. Nat. Acad. Sci. U.S.A. 68 (1971) 127.
4. Croce, C. M., Koprowski, H. and Litwack, G. Nature 249 (1974) 839.
5. Croce, C. M. Kieba, I. and Koprowski, H. Exp. Cell Res. 79 (1973) 461.

6. Croce, C. M., Litwack, G. and Koprowski, H. Proc. Nat. Acad.Sci. U.S.A. 70 (1973) 1268.
7. Ricciuti, F. and Ruddle, F. H. Nature New Biol. 241 (1973) 180.
8. Littlefield, J. W. Science 145 (1964) 709.
9. Dubbs, D. R. and Kit, S. Exp. Cell Res. 33 (1964) 19.
10. Croce, C. M., Koprowski, H. and Eagle, H. Proc. Nat. Acad. Sci. U.S.A. 69 (1972) 1953.
11. Canellakis, Z. N. and Cohen, P. P. J. Biol. Chem. 222 (1956) 53.
12. Lowry, O. H., Rosebrough, N. J., Farr, A. L. and Randall, R. J. J. Biol Chem. 143 (1951) 265.
13. Litwack, G. and Singer, S. in Biochemical Actions of Hormones (edit. by Litwack, G., Academic Press, New York) 2 (1972) 113.
14. Singer, S., Becker, J. E. and Litwack, G. Biochem. Biophys. Res. Commun. 52 (1973) 943.
15. Litwack, G., Filler, R., Rosenfield, S. A., Lichtash, N., Wishman, C. A. and Singer, S. J. Biol. Chem. 248 (1973) 7481.

Note: We would like to thank the publishers of Nature for the permission to reproduce Table 1 and Figures 1, 2, and 3.

Cell Line	2 h	% Nuclear transfer 4 h	6 h	18 h
Fu5AH	17.8	25.9	34.1	20.5
Cl-1D	26.9	26.5	23.4	28.1
Rat-mouse Cl 5	33.6	28.1	30.6	41.4
Rat-mouse Cl 8	ND**	56.3	42.1	48.6
Rat-mouse Cl 10	ND**	43.1	54.5	49.3

TABLE 3. Nuclear Transfer of ^{3}H-Dexamethasone in Rat-Mouse Hybrids

Almost confluent monolayers of parental and hybrid cells were exposed to 3.3 X 10^{-8} M 1, 2, 4-^{3}H-(N)-dexamethasone (specific activity 21.2 Ci $mmol^{-1}$) in serum-free medium. At different intervals the cells were trypsinised, centrifuged and washed three times in ice cold phosphate-buffered saline (PBS). The cells were resuspended in hypotonic reticulocyte standard buffer (RSB), composed of 10 mM NaCl, 10 mM Tris-HCl, pH 7.4, and 1.5 mM $MgCl_2$, and homogenised. The cytoplasmic fraction was centrifuged in a RC-2 Sorvall centrifuge at 20,000g for 15 minutes at 4°C. The nuclei were stripped of the outer nuclear membrane by double detergent procedure according to the method described by Penman. The percentage of nuclear transfer of the dexamethasone-receptor complex was determined by the following formula:

$$\% \text{Nuclear transfer} = \frac{\frac{c.p.m.}{mg \text{ of nuclear proteins}}}{\frac{c.p.m.}{mgm \text{ of nuclear proteins}} + \frac{c.p.m.}{mgm \text{ of cytosol proteins}}} \times 100$$

We conclude that the activity of the hormone receptor has indeed been measured by our techniques for the reasons set forth in Table 2. Furthermore, nuclear translocation experiments ascertain the presence and biological activity of the receptor as shown by the results described in this table.

EXPRESSION OF HEPATIC FUNCTIONS IN SOMATIC CELL HYBRIDS

Gretchen J. Darlington, Hans Peter Bernhard and Frank J. Ruddle

The expression of differentiated functions in somatic cell hybrids has been examined in a variety of crosses where one parent was derived from a rodent hepatoma. We isolated, from a murine hepatoma, a strain of cells which continues to synthesize several serum proteins. It was our intent to examine the interaction of the hepatoma genome in combination with that of a non-hepatic human cell. In mouse X human hybrids, the human chromosomes are segregated, while those of the mouse are retained. Each hybrid would contain a partial set of the human chromosomal complement. Therefore, the effect of particular human chromosomes on the liver properties of the mouse hepatoma could be observed in this system.

Hepatoma cells from the BW 7756 tumor carried in C57 Leaden mice at Jackson Laboratories were adapted to growth outside the body by alternate passage from in vivo to in vitro conditions according to the protocol of Buonassisi et al. (1). The cells had a generation time of about 24 hours in Waymouth's MAB 87/3 medium (2). The mass population was cloned and subsequently treated with 6-thioguanine. A drug resistant colony, deficient in hypoxanthine: guanine pyrophosphorylase activity was isolated. This population was designated Hepa la and was used as the parental line.

The modal number of chromosomes in Hepa la is 58. Figure 1 shows a typical karyotype. In each case, the chromosome of the left has been stained with quinacrine mustard, then subsequently stained for centric heterochromatin (right, white background). There are 4 to 6 metacentric chromosomes in the complement, none of which can be confused with any of the human biarmed chromosomes. One long acrocentric (first chromosome, second row) with interdigitated regions of heterochromatin was present in over 96% of the cells and served as a marker.

Hepa la cells were characterized with regard to the liver properties they possessed in vitro. Table I enumerates the traits which have been examined. Aldolase B, alcohol dehydrogenase and xanthine oxidase were assayed by starch gel electrophoresis and found to be absent. Phenylalanine hydroxylase activity was measured by Dr. Donald Haggerty at UCLA and was not present. Tyrosine aminotransferase (TAT) was present in Hepa la cells in culture for many months following

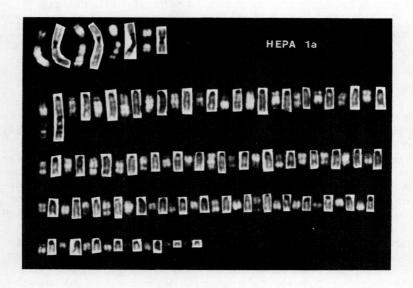

FIG. 1:

Hepa 1a karyotype. Metaphase chromosomes of Hepa 1a were stained with quinacrine mustard (left) then subsequently stained for centric heterochromatin (right).

EXPRESSION OF HEPATIC FUNCTIONS

aldolase B	−
alcohol dehydrogenase	−
xanthine-oxidase	−
phenylalanine hydroxylase	−
carbamyl phosphate synthetase	−
ornithine transcarbamylase	−
tyrosine aminotransferase (TAT)	−
inducibility of TAT	−
esterase-1	−
esterase-2	+
pseudo cholinesterase	+
aryl hydrocarbon hydroxylase (AHH)	+
inducibility of AHH by phenobarbital	+
α-fetoprotein	+
transferrin	+
ceruloplasmin	+
albumin	+

TABLE 1. Hepa 1a Phenotypes

isolation of the cells from the tumor. The activity at that time was 2 to 3 fold the level observed in fibroblasts. Inducibility by 10^{-5}M dexamethasone was also observed in Hepa 1a. Subsequently, however, the line was no longer sensitive to the inducing agent. The basal level of enzyme level remained slightly higher than that of non-hepatic cells. Because of the instability of this trait in Hepa 1a, we have not pursued the study of TAT activity as an hepatic marker. Aryl hydrocarbon hydroxylase activity was determined by Dr. Daniel Nebert of the NIH. Induction of this enzyme by phenobarbital is liver specific.

Esterase-2 activity is present in Hepa 1a. This mouse enzyme is found primarily in tissues of the liver and kidney. Esterase-1, also found in liver could not be observed in the hepatoma cells in culture.

Several serum proteins were produced by the Hepa 1a line. Serum albumin, α-fetoprotein, transferrin, and ceruloplasmin are all synthesized and secreted into the culture medium. The amount of serum albumin secreted is approximately 230 ng/mg cell protein/hour.

Figure 2 shows the electrophoretic mobility of the various serum proteins in a polyacrylamide gel. The serum proteins are collected in serum-free supernatant medium after 24 to 48 hours incubation of the Hepa 1a cells. The medium is then concentrated as much as 40 fold in a Schleicher-Schuell membrane or on a PM30 Amicon filter. Following electrophoresis, the gel was stained with coomassie blue, a dye which identifies proteins. The first channel contains adult mouse serum. Transferrin is the most slowly migrating set of bands. Transferrin in Hepa 1a (Channel 3) moves consistantly slower than does adult transferrin (Channel 1) but is identical in mobility to transferrin in amniotic fluid (Channel 5).

Ceruloplasmin was specifically stained with o-dianisidine on a replicate gel and the band indicated in Fig. 2 corresponds in mobility to the ceruloplasmin activity as specifically stained with the substrate.

A series of bands identified as α-fetoprotein are seen in Hepa 1a and in mouse amniotic fluid. Adult mouse serum lacks this protein as expected. Both transferrin and α-fetoprotein are reported to contain sialic acid residues (3). Treatment with sialidase removes these compounds from the serum protein thus altering the electrophoretic mobility as can be seen in Channels 2, 4, and 6 (4). In addition to electrophoretic mobilities, both of these proteins were identified by immunologic means.

Mouse serum albumin (MSA) was present in the supernatant

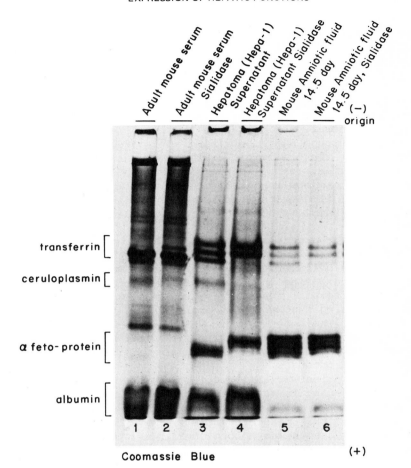

FIG. 2:

<u>Acrylamide gel showing serum proteins</u>. Electrophoresis of adult mouse serum, Hepa-1 supernatant (concentrated approximately 40 X) and mouse amniotic fluid in polyacrylamide (7.5%) according to the protocol of Tischfield <u>et al</u>. (6). Following electrophoresis, the gel was stained with Coomassie blue.

medium of Hepa 1a cells. In figure 2, the very heavy band, closest to the anode is albumin. Because the procedure for preparing the serum proteins for analysis requires concentration of the medium, albumin is in excess when the other proteins are represented in the appropriate amounts. Consequently, most quantitative analysis for albumin has been done using the Laurell (5) or rocket immunodiffusion technique. In general, the procedure involves electrophoresis of a sample through agar containing anti mouse serum albumin. The area under the precipitin arc is a measurement of the amount of albumin present in the sample. The kind of antiserum used determines the qualitative specificity of the assay. Figure 3 illustrates the Laurell procedure for detecting albumin. Sample 1 contains 500 ng of MSA; sample 2 is 250 ng.

Following characterization of the hepatoma line, we examined a series of somatic cell hybrids with regard to the expression of liver functions. Hepa 1a was fused with human leucocytes using inactivated Sendai virus. Six to eight weeks after the fusion, five colonies were isolated in medium containing hypoxanthine, aminopterin, and thymidine (HAT medium). The colonies were designated Hal 3,5,6,7a and 7b. The latter two were taken from the same flask. They have some similar properties and are most probably not independent fusion products.

The Hal series was analyzed for the 30 or so enzymes which are different in their electrophoretic mobilities between mouse and man. The hybrid nature of the colonies was demonstrated in this manner. Table 2 summarizes those enzymes for which the human homopolymeric or a hybrid heteropolymeric band was seen in the Hal cell extracts. All hybrids had the human X-linked enzymes, glucose-6-phosphate dehydrogenase, (G-6PD) phosphoglycerate kinase (PGK) and hypoxanthine: guanine phosphoribosyl transferase (HGPRT) with the exception of Hal 7a. This colony had only the human form of HGPRT. The presence of this enzyme is selected for by growth in HAT medium. Only a few other human isozymes were observed in the Hal hybrids, suggesting that the input from the human genome was greatly reduced.

Chromosomal analysis of the hybrids showed that Hal 3 and 6 had modal numbers very similar to Hepa 1a (Table 2). Hal 7a and 7b, however had modal chromosome numbers of 106 and 110 respectively. The karyotypes contained two Hepa 1a acrocentric marker chromosomes in the majority of cells examined. Therefore, these colonies contain a double dose of the Hepa genome.

From other studies it has been established that the

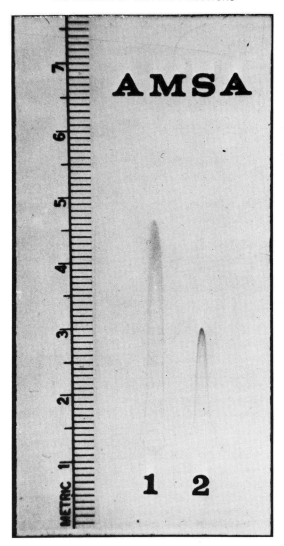

FIG. 3:

<u>Laurell immunoelectrophoretic procedure for identification of mouse serum albumin.</u> Electroimmunodiffusion was performed in 1% sodium barbital buffered agarose at pH 8.6 (5). Antimouse serum albumin (AMSA) was added to the agarose at a final concentration of 0.5% AMSA. Separation was carried out for 6 h at 10 V/cm. The dried slides were stained with amido black and photographed at 10-fold magnification for the measurement of the peak heights. AMSA = Anti-mouse serum albumin(rabbit) MSA = Mouse serum albumin, ethanol extract of Cohn fraction V. 1 = 500 ng MSA, 2 = 250 ng MSA.

	Human Isozymes Present	Modal Chromosome Number	Human Chromosomes Expected from Isozyme Data	Human Chromosomes Present from Cytological Examination*
Hepa 1a	–	58	–	–
Hal 3	GPI HGPRT,GPD PGK	59	19,X	X(13/36)
Hal 5	HGPRT,GPD,PGK	–	X	not done
Hal 6	SOD-1,ME-1, SOD-2 HGPRT, GPD, PGK	60	6,X,21	6(17/47), X (29/47), 8 (23/47)
Hal 7a	NP, ME-1 HGPRT	106	rearranged 6 rearranged X 14	none (0/39)
Hal 7b	NP HGPRT,GPD PGK	110	14,X	X(2/29)

TABLE 2. Human Enzyme Phenotypes and Chromosomes Retained by Hal Hybrids

The enzymes examined electrophoretically in hybrid cell extracts were adenylate kinase (AK), adenosone deaminase (ADA), alcohol dehydrogenase (ADH), esterase-2 (Es-2), glutamic oxaloacetic transaminase (m-GOT), glucose-6-phosphate dehydrogenase (GPD), glucose phosphate isomerase (GPI), hypoxanthine guanine phosphoribosyltransferase (HGPRT), mitochondrial superoxide dismutase (SOD-1), cytosol superoxide dismutase (SOD-2), lactate dehydrogenase A (LDH A), lactate dehydrogenase B (LDH B), malic enzyme (ME-1), malate dehydrogenase (MDH), mannose phosphate isomerase (MPI), nucleoside phosphorylase (NP), peptidases (PEP) A, B, C, and D, phosphoglucomutase-1 (PGM-1), phosphoglycerate kinase (PGK), and xanthine oxidase (XOX).

**The numerator of the ratio in parentheses represents the number of cells in which the chromosome was observed. The denominator is the total number of cells examined.

structural locus for glucose phosphate isomerase is carried on chromosome 19. When Hal 3 was examined cytologically following quinicrine mustard staining, no chromosome 19 was observed. In other cases, a discrepancy between the kinds of human chromosomes expected based on the human isozyme retention and the kinds of chromosomes identified from direct karyotypic analysis was observed. Superoxide dismutase-2, located on chromosome 21 was present in Hal 6, although the chromosome was not found. The most striking deviation of the observed chromosomal complement from the expected one was seen in Hal 7a. In this clone no human chromosome could be identified; nor could any rearranged figure be found which carried a portion of a human chromosome. From the disruption of the linkage groups of the loci on chromosome 6 (superoxide dismutase-1 and malic enzyme), as well as the X chromosome, it was apparent that much rearrangement had taken place.

Liver phenotypes expressed by the Hal hybrids were examined in most of the colonies. Table 3 lists some of these characteristics. The presence of Es-2 was established using starch gel electrophoresis. Mouse serum albumin was detected by immunoelectrophoresis. Murine α-fetoprotein and transferrin were identified on polyacrylamide gels. All five hybrids continued to express Es-2, mouse serum albumin, mouse α-fetoprotein, and mouse transferrin. In this series of hybrids, no extinction of the murine hepatic phenotypes was observed. It should be noted that the great reduction in the number of human chromosomes retained by the hybrids leaves open the possibility that extinction could occur in ones which retained a larger human complement.

The Hal hybrids were also tested for the presence of some traits for which Hepa 1a was negative. These included alcohol dehydrogenase, xanthine oxidase, and aldolase B. The hybrid populations were also negative.

In addition to the presence of mouse albumin, we also observed the expression of human serum albumin (HSA) in Hal 7a and 7b. Figure 4a shows the identification of human serum albumin by identity of the Hal 7a supernatant product (well 1) upon double diffusion with human serum albumin (well 6) against anti-human serum albumin. Figure 4b demonstrates the presence of human serum albumin by immunoelectrophoresis. Thus not only did the murine hepatic characteristics continue to be expressed, but the homologous trait was activated in the human genome. The production of human serum albumin was approximately 1% the amount of MSA secreted by Hal 7a. This low level of production was possibly due to the fact that only a portion of the cells in the Hal 7a population were expressing HSA.

	ES-2	MSA	HSA	Mα-FP	MTRF
Hal 3	M	+	−	+	+
Hal 5	M	+	−	n.d.	n.d.
Hal 6	M	+	−	+	+
Hal 7a	M	+	+	+	+
Hal 7b	M	+	+	n.d.	n.d.

TABLE 3. Hal Phenotypes

Abbreviations used are: Es-2, esterase 2; M, murine; MSA, murine serum albumin; HSA, human serum albumin; Mα-FP, murine α-fetoprotein; MTRF, murine transferrin; n.d., not determined.

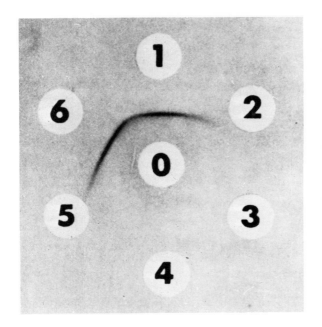

FIG. 4a:

Identification of human serum albumin (HSA) by double immunodiffusion. Exponentially growing cultures were incubated for twenty-four hours in serum free medium. The growth medium was harvested, dialyzed, concentrated approximately 200 fold and analyzed by double immunodiffusion. Identity of the protein secreted by the hybrid lines Hepa 1a and purified HSA is demonstrated by the complete fusion of the precipitin bands formed between anti HSA (0) and supernatant medium from Hal 7a (1) and HSA (6). Cell free medium (4) and growth medium from the parental line Hepa 1a (5) are shown as controls to demonstrate the absence of cross reacting materials. The absence of cross reaction between anti HSA and MSA was tested for MSA concentrations ranging from 10 to 10^3 µg/ml by immunoprecipitation. (0) Monospecific crystallin goat anti HSA (Schwarz and Mann, New York, N.Y.) 10 µg (1) Hal 7a. (2) Hepa 1a. (3) MSA, Cohn Fraction V. 5 µg. (4) Cell free growth medium (5) Fetal calf serum (Flow Laboratories, Rockville, Md.) 10 µg protein. (6) HSA, Cohn Fraction V, 5 µg.

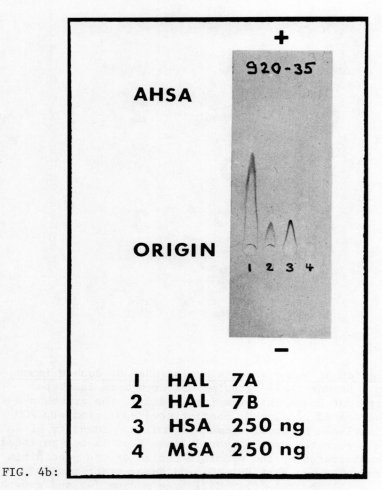

FIG. 4b: Electroimmunodiffusion of hybrid supernatant. Supernatant medium from Hal 7a and 7b were concentrated approximately 100 fold and wells 1 and 2 each contained 5 λ of sample. Well 3 contained 250 ng of human serum albumin; well 4 contained 250 ng of mouse serum albumin. Conditions of electrophoresis and diffusion were as described in Fig. 3.

The activation of the human genome suggested that the human structural locus for albumin could be mapped. Because only two hybrid clones synthesized HSA and no human chromosomes were identified in these colonies, it was impossible to make an assignment of the HSA gene to a particular chromosome. Some information concerning linkage can be derived from the isozyme phenotypes. Both Hal 7a and 7b expressed human nucleoside phosphorylase (NP), while none of the remaining populations were positive for this enzyme. In subsequent generations of Hal 7a, both human NP and human serum albumin were lost. The concurrent segregation of these two phenotypes gives the barest suggestion of linkage. Clearly a larger series of hybrids expressing HSA will need to be examined before this association is established.

Activation of human albumin presents a model for the ability to map the structural locus for albumin. Potentially, any specialized human function which is subject to activation can be assigned to a linkage group using somatic cell hybrids.

---

1. Buonassisi, V., Sato, G. and Cohen, A. I. PNAS **48** (1968) 1184.
2. Waymouth, C. in Tissue Culture (Ed.) C. V. Romakrishnan; Dr. W. Junk, Publishers, The Hague (1965) 168.
3. Gustine, D. L. and Zimmerman, E. F. Amer. J. of Obst. & Gyn. **114** (1972) 553.
4. Zimmerman, E. F. and Madappally, M. M. Biochem. J. **134** (1973) 807.
5. Laurell, C. B. Anal. Biochem. **15** (1966) 45
6. Tischfield, J. A., Beynhard, H. P. and Ruddle, F. H. Analyt. Biochem **53** (1973) 545.

Note: We would like to thank the publishers of Science and Developmental Biology for the permission to reproduce Figure 1 and 2 and Table 1.

EXTINCTION, RE-EXPRESSION AND INDUCTION OF LIVER SPECIFIC FUNCTIONS IN HEPATOMA CELL HYBRIDS

Mary C. Weiss

My purpose in this communication is to give a brief review of the observations which have been made on the expression of differentiated functions in somatic hybrid cells, with emphasis on those which concern tissue specific proteins of the liver. There are several reviews which cover the experiments using various other types of differentiated cells, including pigment, glial and neuroblastoma cells (1,2,3).

We have heard a lot at this meeting about both normal liver cells in culture, and abnormal ones, derived from hepatomas. Workers from a number of different laboratories have chosen to use the abnormal ones, because of the ease with which they can be cultured and cloned, and because of their stability in the expression of a wide range of liver specific proteins. Hepatoma cells, thus, have been and are being used as a model system for the analysis of the control of differentiation.

This analysis has followed so far two different paths: one is through the study of differences between hepatoma lines (4) or of clonal variability within hepatoma lines (5); the other is a genetic approach, which uses cell hybridization. This technique, a tool for the genetic analysis of somatic cells, is available to us today thanks to the pioneering work of Barski, Ephrussi, Littlefield and Harris (1).

Workers from several different laboratories have used cell hybridization to study a variety of types of differentiated neoplastic permanent lines. The results of these experiments have demonstrated that several different types of interactions can be observed when parental cells, which differ in their functional states as a consequence of the path of differentiation which they have followed, are fused to form proliferating mononucleate hybrid cells. We will consider these various types of interactions in order to see what they can tell us about the nature of the differentiated state.

To begin with, I would like to describe how the genetic analysis of differentiation by cell hybridization proceeds. A cloned line of a differentiated cell type, which produces in a stable fashion one or more well defined tissue specific proteins, is hybridized with a cell type (such as a fibroblast) which does not produce the proteins in question, and which may be derived from an animal of the same or a different species. Since the various lines of differentiated cells are permanent and heteroploid (usually hyperdiploid), their karyotype is defined as the stem line number of 1s; if this number of chromosomes is doubled, the line is 2s (hyper-tetraploid). The analysis can proceed as follows:

1) Establishing the properties of hybrid cells formed by crossing 1s cells of the two parental lines, and to determine whether or not the tissue specific functions are expressed in hybrids containing the complete chromosome sets of both parents.

2) Determining the effect of chromosome loss from the hybrid cells upon the expression of the functions.

3) Determining the effect of modification of the ploidy of the parental cells (dosage effect) upon phenotypic expression of the hybrid cells.

Extinction

In hybrid cells which contain one complete chromosome set from each parent, the most frequently observed result is the extinction (or supression) of the tissue specific protein which was produced by only one of the parental cells. This was first described by Davidson, Ephrussi and Yamamoto for pigment production by melanoma-fibroblast hybrids (6). Similar findings were subsequently described for other types of systems, and it has been possible to show that extinction is not the consequence of the presence in the hybrid cells of an inhibitor of enzyme activity, total inactivation of the genome of the differentiated parent, loss of chromosomes specifying the protein in question, or cell fusion and the consequent increase in ploidy (1).

In Dr. Ephrussi's lab in Gif-sur-Yvette, my co-workers and I have undertaken the analysis of clones of the H4IIEC3 line derived by Pitot et al (7) from the Reuber H35 rat hepatoma (8). These rat hepatoma cells have been crossed with mouse fibroblasts (9), diploid rat epithelial cells (10) and Chinese hamster fibroblasts (11), and the hybrids tested for activity of four liver specific enzymes produced by the hepatoma cells: the liver specific isozyme of alcohol dehydrogenase (1-ADH) (12), aldolase B (13), and two amino-transferases which are inducible by glucocorticosteroid hormones, tyrosine aminotransferase (TAT) (k,10) and alanine aminotransferase (AAT)(14). None of the hybrid clones tested showed significant activity or inducibility of any one of the four enzymes: all of these functions were extinguished, whether the non-expressing parent was an aneuploid line of a different species or a diploid line of the same species.

Thompson and Gelehrter (15) have shown that the extinction of hepatoma TAT can be demonstrated in heterokaryons: the enzyme disappears very soon after fusion, before any chromosomes have been lost, and this event does not require nuclear fusion.

Although the mechanism of extinction is unknown, the original interpretation of Davidson, Ephrussi and Yamamoto (6), that it reflects the action of a diffusible intracellular regulatory substance, remains the most tenable working hypothesis.

While in nearly all cases extinction appears to be total, occasionally it is only partial: a reduced amount of the tissue specific protein may be produced by the hybrid cells, as was demonstrated for rat albumin production by rat hepatoma-mouse fibroblast hybrids (16) (see below)

Re-Expression

Extinction is not due to an irreversible change: chromosome loss from hybrid cells can lead to the re-expression of a previously extinguished enzyme, as was first demonstrated by Klebe, Chen and Ruddle for the kidney specific esterase ES-2 (17).

The four liver specific enzymes mentioned above, all of which are subject to extinction in hybrid cells containing the complete chromosome sets of both parents, can be re-expressed. Moreover, the different enzymes are re-expressed independently.

The first hybrids for which we observed re-expression of the liver specific enzymes were derived from a cross of rat hepatoma cells with diploid epithelial rat cells (10). The latter, designated BRL, were isolated by Dr. Hayden Coon by cloning of a suspension of liver cells from a young Buffalo rat (18). These epithelial rat liver cells do not show activity of any of the liver specific enzymes which we have tested, nor do they produce albumin (19). As is the case for numerous similar lines described at this meeting, it is not known whether BRL cells are "dedifferentiated" hepatocytes, whether they were derived from some epithelial elements of the liver which are distinct from the hepatocytes, or whether they arose from some as yet poorly defined stem cell or immature hepatocyte. The hybrids derived from the cross of rat hepatoma cells with BRL cells produced none of the liver specific enzymes, but like the hybrids of rat hepatoma-mouse fibroblasts, they did secrete albumin (19). These hybrid cells contained very close to the sum of the parental number of chromosomes, and the karyotype was very stable, undergoing little change over several months of continuous cultivation. However, these hybrid cells presented a property which permitted us to select for those cells which had lost chromosomes: like the BRL parental cells, and unlike the hepatoma ones, the hybrids were contact inhibited. By permitting well nourished hybrid cultures to remain at confluence for several months, it was possible to isolate (and re-clone) foci of cells which grew in multiple layers: these "variant" hybrids had all lost many chromosomes (20-40%), and a few of them re-expressed the previously extinguished liver specific enzymes (10,11,12,20).

I will describe now the properties of some of these "variant" segregant subclones because they permit us to draw some conclusions concerning the regulation of some of the liver-specific enzymes. Table 1 shows the chromosome numbers and enzyme activities of the parental cells (Fu5-5 rat hepatoma cells and BRL-1), of the hybrid clone BF5 which contains the complete sets of chromosomes of both parents, and of three clones of variant segregants which re-express one or more of the liver specific enzymes. Cells of clone BFR-1-1 show no basal activity of TAT, but the enzyme is produced when the cultures are induced with dexamethasone. The specific activity of TAT in the induced cells is very low compared to that of the parental hepatoma cells, but nevertheless an examination of the parameters of induction of TAT in BF5-1-1 and Fu5-5 cells showed numerous similarities between the two celltypes (10); moreover, the enzyme of the hepatoma cells and that of induced BF5-1-1 cells is precipatated by a specific antiserum (unpublished experiments) which was given to us by Dr. Gordon Tomkins. These cells therefore re-express TAT inducibility in the absence of a high baseline activity, an observation compatible with the suggestion of Tomkins et al (21) that a specific gene mediates the inducibility of this enzyme. BF5-1-1 cells also re-express an intermediate basal level of AAT (14), but this enzyme is not inducible. The fact that TAT but not AAT is inducible in these cells demonstrates that independent mechanisms are responsible for the induction of the two aminotransferases. Finally, cells of this clone show no activity of either L-ADH or aldolase B.

Among numerous subclones isolated from BF5-1-1, one designated BF5-1-1α, shows a further reduction in chromosome number, and re-expresses nearly all of the liver specific enzymes tested: like the BF5-1-1 cells, from which they were derived, they show TAT inducibility, but in additon, a high basal activity and inducibility of AAT (14), some L-ADH activity, and high aldolase B activity (20). These observations strengthen the interpretation that the various liver specific enzymes are independently re-expressed, for they demonstrate that BF5-1-1 cells retained the potential to re-express high levels of AAT and adolase B in the absence of overt expression. Therefore, although BF5-1-1 cells re-expressed only one enzyme, they retained the genetic information necessary to re-express others, as shown by the properties of the subclone derived from it.

The last hybrid variant segregant clone listed in Table 1, BF5- 4, resembles BF5-1-1-α in the re-expression of AAT (14), L-ADH (12) and aldolase B (20), but differs in that these cells do not show TAT inducibility (20).

All of these cases of re-expression of hepatoma enzymes are correlated with re-expression of hepatoma-like morphology: Fig.1 shows phase contrast photomicrographs of cultures of each of the clones listed in Table 1. It will be noticed that phase dense, granular cytoplasm and a large central nucleolus, characteristic of hepatocytes in culture, is seen only in cells of Fu5-5, BF5-1-1-α and BF5- 4.

Similar observations of the independence of re-expression of liver specific enzymes have more recently been obtained for hybrids of rat hepatoma cells X Chinese hamster fibroblasts. These hybrids, unlike those described above, do show re-expression of high basal activity of TAT in some clones, but as in the BF5 hybrids, inducibility of TAT can be re-expressed independently of a high basal activity of the enzyme (11). Moreover, in these hybrids, aldolase B is re-expressed independently of TAT, and L-ADH is re-expressed in each subclone which produces one of the other enzymes (11).

Croce et al (22) have also studied the re-expression of TAT inducibility, using hybrids of rat hepatoma cells X human fibroblasts, which undergo rapid and preferential loss of human chromosomes. They have described a negative correlation between the presence of the human X chromosome and the re-expression of TAT inducibility, concluding that the human X chromosome controls enzyme inducibility (22).

Dosage Effects

We have already seen that hybrids of Syrian hamster melanoma X mouse fibroblasts are unpigmented (6). However, if the ploidy of the melanoma parental cells is increased, relative to that of the fibroblast parent, and 2$\underline{s}$ melanoma cells are crossed with the same mouse fibroblasts, pigmented hybrids are obtained, as shown by Fougere, Ruiz and Ephrussi (23), and by Davidson (24). These authors have concluded that the relative ploidy (gene dosage) fo the expressing and non-expressing parental cells determines the phenotype

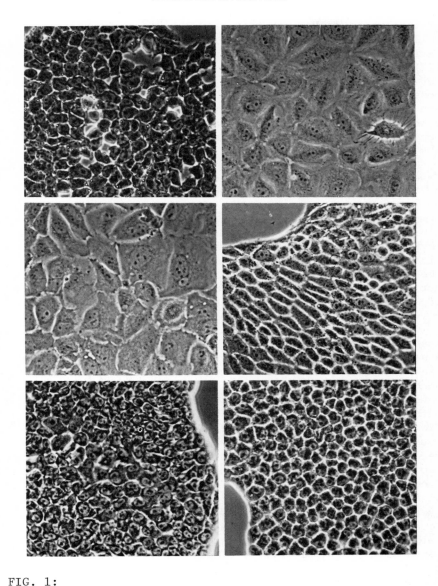

FIG. 1:

Phase contrast photomicrographs of living cultures, all at the same magnification, of Fu5-5 (upper left), BRL-1 (upper right), BF5 (middle left), BF5-1-1 (middle right), BF5-γ4 (lower left) and BF5-1-1-α (lower right).

of the hybrid cells. In this case the use of 2s melanoma
cells as the expressing parent resulted in the absence of
extinction, and in some cases, probably also in re-espression
(23). The pigmented hybrids were not always stable, and some
gave rise to unpigmented progeny (24). Since it has not been
possible to demonstrate a karyological difference between
the pigmented and unpigmented hybrids, Fougere et al have
concluded that the fibroblast chromosomes which are responsible for extinction of melanogenesis probably are not
numerous (23).

Increase of the ploidy of the expressing parent can
have a second consequence: it can result in the induction
of "silent" genes of the non-expressing parent to produce
a protein foreign to its tissue of origin. This was first
demonstrated for albumin production by rat hepatoma-mouse
fibroblast hybrids (16). Hybrids were formed by crossing 1s
(hyperdiploid) and 2s (hypertetraploid) rat hepatoma cells
with 3T3 mouse fibroblasts (nearly tetraploid). All of
the hybrid clones from the 1s cross produced some rat albumin,
but no mouse albumin; among five clones isolated from the 2s
cross, one produced both rat and mouse albumins, two produced
only mouse, and two produced neither type of albumin (16).

In an effort to clarify the dosage requirements for
the induction of mouse albumin, we have recently undertaken
a new cross (25), using as the non-expressing parent a permanent line of mouse leukemic cells, characterized by a modal
number of only 39 chromosomes (the diploid number for the
mouse is 40). These mouse leukemic cells were crossed with
both 1s and 2s rat hepatoma cells. Twelve hybrid clones
from the 2s cross were studied: each of them produced both
mouse and rat albumin. Nine hybrid clones from the 1s cross
were studied: six of them produced both rat and mouse albumin,
two only mouse, and one only rat. In these hybrids, the induction of mouse albumin is almost systematic and is observed
in the 1s as well as the 2s hybrids (25). To return to the
question of the dosage requirements for induction, it will be
recalled that the 3T3 cells used in the first experiments (16)
were nearly tetraploid, i.e. had almost twice as many chromosomes as the leukemic cells. Therefore the ration of hepatoma to mouse chromosomes was nearly the same in the 2s
hepatoma-3T3 and the 1s hepatoma-leukemia hybrids, and in
both series, the induction of mouse albumin was observed.

The quantity of albumin produced by the 2s hepatomaleukemia hybrids was very different from one clone to another.
Although the method of analysis used (double diffusion in
agar) is not quantitative, a rough idea of the amount of
albumin produced in the culture medium can be estimated from
the dilution of concentrated medium required to approach

antigen-antibody equivalence. Thus, it was found that a few of the hybrids produce at least as much albumin as the 2s hepatoma cells, and others much less (25). Therefore, those that produce much less would be classified as "partially extinguised", and nevertheless induction of mouse albumin was observed. This observation suggests that in the same cell there may be operating at the same time a mechanism of "extinction" and one of "activation".

Discussion

Now that we have seen the various phenomena and interactions which have been observed in hybrid cells, let us go back over them to see what they can tell us about the nature of the differentiated state. It will be recalled that all of these observations concern hybrids made by crossing cells of different histogenetic origin (with the possible exception of the hepatoma-BRL hybrids described above), and that the parental cells differed qualitatively in that only one of them produced the tissue specific protein(s) analyzed (1).

Extinction of differentiated functions is the most frequently observed result in such hybrid cells, and occurs whether the parental cells are derived from the same or different species, and whether the non-expressing parent is heteroploid or diploid. Extinction occurs rapidly and does not require nuclear fusion (15). Since extinction is dosage dependent (23,24), and since its maintainence requires the continuous presence of chromosomes of the non-expressing parent (17), there is little doubt that it represents a regulatory mechanism (1). The level in the cell at which this mechanism operates is still unknown, as has been discussed by Ephrussi (1).

Re-expression provides a formal proof that extinction is not the consequence of loss of chromosomes specifying the tissue specific protein, nor of the fusion of rare de-differentiated cells which may have existed in the parental cell population. The re-expression of a tissue specific protein, the potential for whose expression is inherited through tens or even hundreds of generations in the hybrid cells, demonstrates the extraordinary stability of the epigenotype of the differentiated cells: hybridization with a cell which does not express the protein results in extinction of the expression of the protein, but does not alter the determined potential of the differentiated epigenotype (1). The independence of re-expression of the diverse enzymes which characterize hepatic differentiation demonstrates that in the hybrid cells there are multiple regulatory mechanisms, probably at least one for each enzyme, and suggests that the mechanism of extinction may be specific (20). This conclusion is reinforced by the report that the extinction of TAT inducibility is mediated by a single specific chromosome (22).

Dosage effects demonstrate that there are quantitative aspects of the regulation of differentiated functions in hybrid cells (24): a given function, such as melanogenesis, is systematically extinguished in 1s-1s melanoma-fibroblast hybrids, but not if the melanoma parent is 2s. In the latter case, there may be absence of extinction, or re-expression (23).

Dosage effects leading to the induction of a "silent" gene to produce a protein (albumin) foreign to its tissue of origin demonstrates: a) the absence of species specificity in the response to an inducing signal, and b) that genes which are not active as a consequence of determination and differentiation can be "activated" (16). Moreover, the fact that hybrids which show partial extinction of albumin production may synthesize nevertheless the proteins specified by both parental genomes suggests that mechanisms for extinction and for induction may be present and active simultaneously in a single cell (25).

In summary, all of these results demonstrate a surprising stability of the determined state: there are no indications of modification by cell hybridization of the determined potential of any cell type. A possible exception to this conclusion may be found in the cases where a genome has been induced to produce a protein it shouldn't. However, it remains to be determined whether or not induction reflects a stable change in the epigenotype of the non-expressing parent, or whether the maintainence of the production of the induced protein requires the continuous presence of some "activating" element from the expressing parent (26) just as the maintainence of extinction requires the continuous presence of some "repressing" element of the non-expressing parent.

It is a pleasure to acknowledge the participation, in all of the phases of the work performed in this laboratory, of my coworkers Roger Bertolotti, Michèle Chaplain, Steve Malawista, Jerry Peterson, Jerry Schneider and Robert Sparkes. Numerous stimulating discussions with them and with Dr. Ephrussi have helped to clarify interpretations. Dr. Ephrussi and Jenny E. Brown kindly read the manuscript. Work from this laboratory has been supported by grants from the Delegation Generale a la Recherche Scientifique et Technique and the Action Thematique Programmee du C.N.R.S.

	Total no. of	TAT		AAT			
Cell line	chromosomes	Basal	Induced	Basal	Induced	ADH	Aldolase B
Fu5-5	52(51-53)	38.5	224.0	143	324	102	1.2
BRL-1	42	0.7	0.6	0	0	0	<0.1
BF5	92(91-93)	0.5	0.6	5	7	6	<0.1
BF5-1-	63(60-66)	0.4	5.7	44	55	6	<0.1
BF5-1-1-	60(57-62)	0.6	5.8	123	520	N.D.**	2.1
BF5- 4	55(54-55)	0.7	1.2	71	216	22	3.6

TABLE I. Chromosome numbers and enzyme activities of parental and hybrid cells.*

Footnotes to Table 1

*Data taken from references 10,12,14,20 and unpublished data of Bertolotti and Weiss. Activities for all enzymes (except aldolase B) are expressed as milliunits of specific activity, as defined in the original publications. TAT was induced by exposure of cultures to 10^{-6} M Dexamethasone for 24 hours (except BF5-1-1: 7 hours); AAT was induced by a 48 hour incubation in 10^{-7} M Dexamethasone. The aldolase B activity given is the calculated specific activity due to B protomers, as described in ref. 13.

**N.D.: not done. Electrophoresis of total ADH of extracts of this line and of Fu5-5 and BF5-γ 4 reveals a band of activity corresponding in its mobility to the enzyme from rat liver; these extracts show in addition a band which corresponds in its mobility to the enzyme from rat stomach, a band which is also found in extracts of BF5 and BF5-1-1. The assay used to determine the values given in this column measures total ADH activity.

1. Ephrussi, B. Hybridization of Somatic Cells. Princeton Univ. Press, Princeton, N. J. (1973).
2. Davidson, R.L. In vitro 6 (1971) 411; in Genetic Mechanisms of Development. Ed. Ruddle, F. H. (Academic Press, N.Y.) (1973) 295.
3. Davis, F.M. and E.A. Adelberg Bact. Rev. 37 (1973) 197.
4. van Rijn, H., M.M. Bevers, R. Van Wijk and W.D. Wicks J. Cell Biol. 60 (1974) 181.
5. Aviv, D. and E.B. Thompson Science 177 (1972) 1201.
6. Davidson, R.L., B. Ephrussi and K. Yamamoto Proc. Nat. Acad. Sci. U.S.A. 56 (1966) 1437; J. Cell. Physiol. 72 (1968) 115.
7. Pitot, H.C., C. Peraino, P.A. Morse and V.A. Potter Nat. Cancer Inst. Monogr. 13, (1964) 229.
8. Reuber, M.D. J. Nat. Cancer Inst. 26 (1961) 861.
9. Schneider, J.A. and M.C. Weiss Proc. Nat. Acad. Sci. U.S.A. 68 (1971) 127.

10. Weiss, M.C. and M. Chaplain Proc. Nat. Acad. Sci. U.S.A. 68 (1971) 3026.
11. Weiss, M.C., R.S. Sparkes and R. Bertolotti Somatic Cell Genetics (in press).
12. Bertolotti, R. and M.C. Weiss Biochimie 54 (1972) 195.
13. Bertolotti, R. and M.C. Weiss J. Cell Physiol. 79 (1972) 211.
14. Sparkes, R.S. and M.C. Weiss Proc. Nat. Acad. Sci. U.S.A. 70 (1973) 377.
15. Thompson, E.B. and T.D. Gelehrter Proc. Nat. Acad. Sci. U.S.A. 68 (1971) 2589.
16. Peterson, J.A. and M.C. Weiss Proc. Nat. Acad. Sci. U.S.A. 69 (1972) 571.
17. Klebe, R. J., T. Chen and F. H. Ruddle Proc. Nat Acad. Sci. U.S.A. 66 (1970) 1220.
18. Coon, H.G. Carnegie Inst. Wash. Yearbook 67 (1969) 419.
19. Weiss, M.C., R. Bertolotti and J.A. Peterson in Molecular Genetics and Developmental Biology Ed. Sussman, M. Prentice-Hall, Inc., Englewood Cliffs, N.J. (1972) 425.
20. Bertolotti, R. and M.C. Weiss Differentiation (in press).
21. Tomkins, G. M., T.D. Gelehrter, D. Granner, D. Martin,Jr., H.H. Samuels and E.B. Thompson Science 166 (1969) 1474.
22. Croce, C.M., G. Litwack and H. Koprowski Proc. Nat. Acad. Sci. U.S.A. 70 (1973) 1268.
23. Fougere, C., F. Ruiz and B. Ephrussi Proc. Nat. Acad. Sci. U.S.A. 69 (1972) 330.
24. Davidson, R.L. Proc. Nat. Acad. Sci. U.S.A. 69 (1972) 951.
25. Malawista, S.E. and M.C. Weiss Proc. Nat. Acad. Sci. U.S.A. 71 (1974) 927.
26. Kao, F.T. and T.T. Puck Proc. Nat. Acad. Sci. U.S.A. 69 (1972) 3273.

EXPRESSION OF LIVER-SPECIFIC AND OTHER DIFFERENTIATED FUNCTIONS IN HYBRIDS BETWEEN CULTURED HEPATOMA CELLS AND L CELLS

E. Brad Thompson and Marc E. Lippman

INTRODUCTION

Somatic cell hybrids have become a useful system in which to study cellular control mechanisms. By hybridizing a pair of cell lines with contrasting functions, one can study whether each function is dominant or recessive. In addition, if appropriate mutants can be obtained, complementation analysis can be carried out. As segregation takes place in the hybrids, gene localization on a specific chromosome can sometimes be determined by observing coincidence of altered function with loss of that chromosome.

In the study of differentiated cellular processes, cell hybridization has been used to compile a sizeable number of observations (1). We believe the next level of effort must be to use such hybrids in biochemical studies to unravel the detailed mechanisms behind the observed control. One major hindrance in doing so has been the paucity of information about what, in fact, are the control elements in animal cells. Ideally, one might hope to hybridize pairs of cells isogenic but for known mutations in specific loci involving known or postulated control elements. Somatic cell hybridization is in a more rudimentary state. One is still searching to define the control elements of the cell, and fine mapping is still impossible. It does seem that proteins are exchanged between nuclei in heterokaryons (2,3). Perhaps these exchanges include specific and/or general control factors which account for the early shutdown or activation of genes in the heterologous nucleus. But little more can be said at present.

With a few exceptions, it has been expedient to utilize completely different cell lines - often from differing species - to obtain hybrids. Interspecific hybrids have some special advantages for giving insights into cellular control of differentiation. In addition to allowing the investigator to search for positive and negative control elements, they permit inquiry as to whether these factors exist and operate across species barriers. The concomitant concern such hybrids raise is whether they are truly representative of natural controls. Time, and future experiments, will be necessary to answer this question. In a few instances, hybrids between differentiated and undifferentiated cell lines of the same inbred species have been prepared. In this type of hybrid, one can be somewhat more secure that the factors influencing gene expression are natural ones. At least the cells initially contained the same genetic information. Complicating factors arise even here, however, because of the non-euploidy of many established lines and because of the unexplained alterations in gene expression sometimes seen in established cell lines. The fact that the phenotype of any cell line never precisely parallels that of its in vivo tissue counterpart must caution the investigator about assuming that all the control functions seen in vitro are the "natural" ones.

As more is learned of the specific steps in gene expression, hybrids become increasingly valuable. Mutants at known points can be sought; new control points can be defined by complementation analysis. The mechanism of action of steroid hormones is providing one important prototype for clarifying gene expression in animal cells. Although the general problem of gene expression in eukaryotes remains a mystery (4), enzyme and RNA induction by steroids seems to be one system in which progress is being made (5,6). A general model for steroid action has evolved in which a specific soluble cytoplasmic protein receptor for the hormone plays an essential role. In this report, we will describe our recent results with hybrids between cells capable of responding to glucocorticoids in different ways. Analysis of the receptor content and glucocorticoid responses of these hybrids gives rise to interesting conclusions about steroid specificity, and the role of steroid receptors in mediating cellular responses to these hormones.

Before relating our experiments, some details of the general steroid-action model should be considered. Current evidence indicates that the target tissues for each class of steroid--estrogenic, androgenic, glucocorticoid, or mineralocorticoid--contain cytoplasmic receptors that bind specifically the active forms of steroids of that class. Thus, uterus receptor binds estradiol, other functionally active estrogens, and certain compounds capable of blocking estrogen action. Liver specifically binds hydrocortisone and its analogs, and so forth. Saturation curves of steroids binding to their receptors in responsive tissues have repeatedly shown that most tissues contain a single class of receptors with respect to affinity for their specific steroid. The affinity constants of binding are appropriate for the circulating level of each type of steroid to be affecting cell physiology by varying the degree of saturation of receptors and thus varying cell response. Figure 1 is a general diagram of the receptor model of steroid action. The diagram has been simplified by omitting many details that have been described in specific systems. Many of these may be quite important, but it is not yet known whether they apply to all steroids.

The main steps, shown in the figure, are as follows: Steroid freely enters the cell from the interstitial space and is at once bound by the soluble protein receptor (R) in the cytoplasm. The steroid-receptor complex (SR) then undergoes one or more temperature-dependent steps, to become the active complex (SR*). Only the active complex can enter and bind in the nucleus. The steps leading to formation of this complex are not completely known, but they appear to involve a change in size of the protein (5-7). The active complex enters the nucleus where it binds to a nuclear receptor site (NR) on chromatin. This site is still not clearly defined, but both nuclear protein and DNA appear to be involved (8-10). Subsequently, various specific responses to the steroid appear, such as increased synthesis of rRNA and tRNA, enzymes, etc. Exactly how the receptor complex effectuates the induction of all these substances is still not known. The time from entry of steroid to nuclear binding is no more than 2-10 min. Specific cytoplasmic products have been detected as early as 20-30 min. after addition of steroid, and nuclear events, even earlier (6,11,12).

The central role of the cytoplasmic receptor has been shown in many ways. Besides those mentioned above (steroid and tissue specificity, appropriate affinities), in several systems loss or absence of receptor has been shown to be coincident with loss of response to steroid. When lymphoid cells in tissue culture, normally killed by glucocorticoids, were selected for resistance, they were found to contain altered and/or diminished receptors (13). Similar results have been reported for L cells (14). Human leukemic blasts, found to be glucocorticoid resistant in vitro and in vivo, had little or no glucocorticoid receptor, while glucocorticoid-sensitive blasts contained receptors (15). Estrogen-sensitive mammary tumors also contain estrogen receptors; estrogen-autonomous tumors do not (16). HTC cells found to have spontaneously lost inducibility of tyrosine aminotransferase by steroid had about half the usual amount of receptor (17). From this list, it can be seen that in cells and tissues in which glucocorticoids exert inhibitory effects (lymphoid cells, fibroblasts), just as with the trophic responses, receptors are involved. This leads to the possibility that steroids induce some lethal changes in these tissues, a notion supported by experiments with compounds used to

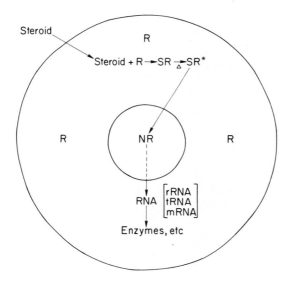

FIG. 1:

Simplified diagram of steroid hormone acting in a cell by complexing with soluble cytoplasmic receptor. Outer circle represents the plasma membrane of the cell; inner circle, the nuclear membrane. Free receptor in the cytoplasm is shown by R, inactive steroid-receptor complex by SR, activated complex by SR*, and nuclear steroid-receptor complex by NR.

block various aspects of macromolecular synthesis (18-20). Furthermore, in these kinds of tissues, steroidal inhibitory effects take an hour or more to appear, timing consistent with induction of a lethal element. This brief review of steroid action serves as introduction to the cell hybrids to be described next.

STEROID-MEDIATED FUNCTIONS IN HTC X L CELL HYBRIDS

We have been studying effects of steroids in hybrids between cells with sharply contrasting responses to these hormones. One parent has been the HTC line, a rat hepatoma; the other, mouse L cells. HTC cells retain at least one typical differentiated hepatic response to glucocorticoids, viz. the induction of tyrosine aminotransferase (21). Growth, overall RNA and protein synthesis, and glucose utilization are unaffected in HTC cells by hydrocortisone and related steroids. These same hormones in L cells, however, markedly reduce cloning efficiency, reduce DNA synthesis by about 50%, and reduce glucose uptake by 10-20% (22,23, and A. Venetianer and E. B. Thompson, unpublished results). Table 1 compares several characteristics of HTC and L cells. Both, it should be noted again, contain specific cytoplasmic glucocoritcoid receptors.

Workers in several laboratories have found that crosses between HTC cells (or another similar, inducible rat hepatoma line) and mouse, rat, or human non-inducible cell lines result in hybrids in which tyrosine aminotransferase cannot be induced (24-28). Therefore, the proposal has been made that the differentiated responses induced by steroid will always prove to be recessive in hybrids. In the cross between HTC and L cells, we have examined for the first time contrasting responses to a single hormone in hybrid cells. These studies raised two questions: Would the L cell-specific inhibitory responses prove recessive along with the HTC-specific aminotransferase induction? Furthermore, if one or both responses were lost, would the dominant effect be due to loss of the specific cytosol steroid receptor, as often has been the case in non-hybrid cells?

Hybrids were obtained by standard techniques. Equal numbers of HTC and L cells were mixed either with or without inactivated Sendai virus. The HTC subline used lacked hypoxanthine-guanosine phosphoribosyl transferase, having been selected for growth in 10^{-4} $\underline{M}$ 6-mercaptopurine. The LB82 line, which lacks thymidine kinase, was used as the other parent. Hybrids were selected by growth in selective medium containing thymidine, hypoxanthine, and methotrexate. The

HTC-specific	L	HTC
Thymidine Kinase	Absent	Present
Hepatic tyrosine aminotransferase	Absent	Present
Induction of tyrosine aminotransferase by glucocorticoids	Absent	Present
Production of C'2	Absent	Present
L-cell-specific		
Hypoxanthine-guanine phosphoribosyl transferase	Present	Absent
Cyclic 3'5' AMP		
Content	Higher	Lower
Increase after Prostaglandin E	Yes	No
Inhibitory responses to glucocorticoids	Present	Absent
Common to HTC and L cells		
Glucocorticoid receptor in cytoplasm	Present	Present
Contact inhibited growth	No	No

TABLE 1. Some Differentiated Characteristics of L and HTC Cells

Legend: Conclusions based on data in references (5,14,23,&27).

parental lines, plated alone in the numbers used in their fusion, with or without Sendai virus treatment, gave rise to no colonies in the selective medium. When putative hybrid colonies arose, they were verified by both chromosomal and biochemical markers (27). The receptor studies to be described were carried out on one such clone, HL5, at a time when its chromosomal complement had fallen from the initial 1S + 1S number of 106 to about 80.

The results obtained with the HTC X L cell hybrid clone, HL5, were both surprising and instructive (23). As expected and already documented, the HTC-specific aminotransferase induction was lost (27). No tyrosine aminotransferase activity which could be blocked by monospecific antiserum to the true hepatic enzyme could be found in the hybrid, nor could enzyme activity be induced by incubation with the potent glucocorticoid, dexamethasone (Table 2). The L cell responses in the HL5 hybrid, however, were found to be intact (Table 2). Dexamethasone treatment lowered thymidine and uridine incorporation and glucose uptake. The presence of these responses argued that steroid receptors had to exist in these cells, to mediate the L cell-specific functions.

Assay for glucocorticoid receptors in the cytosol fraction of the HL5 hybrid clone showed a saturation level of steroid specifically bound equal to or greater than the amount found in either parent (Figure 2). The affinity for dexamethasone appeared also to be quite similar. Thus, it appeared that the hybrids possessed adequate cytoplasmic receptor that was functional at least for certain steroid responses.

There seemed to be three possibilities: one, a single, fully functional steroid receptor was present and the loss of tyrosine aminotransferase and its induction were due to some post-receptor effect in the sequences shown in Figure 1. Two, there were specific receptors for L cell that differ from those for HTC cell events. If this were so, the L cell receptor might still be present, but the HTC cell receptor lost. Or three, both receptors might be present with the aminotransferase responses lost at a post-receptor step. No assay existed, however, capable of distinguishing between the two putative types of receptor. We decided to explore the nuclear binding capabilities of the receptor in the cytosol fraction of the hybrids. As was discussed in the Introduction, the binding of steroid-receptor complex to nuclei requires activation of the complex, a temperature-sensitive step. To carry out this binding reaction in vitro, one must mix steroid with cytosol and hold at $20°C$ to obtain activated complex capable of nuclear binding. At higher temperatures

	Parental Cells		Hybrid
Induction of tyrosine aminotransferase	HTC	L	HL5
Basal	6.7[a]	1.3	1.6
+10^{-6} M dexamethasone PO_4, 18 hr	16.3	1.6	1.4
% Immunoreactive (hepatic)[b]	90%	0-10%	0-10%
Inhibitory responses			
Glucose utilization, % inhibition at 2 hr	1.3	14	22
Maximum thymidine incorporation			
Control	11200cpm	14400cpm[c]	11000cpm
+ dexamethasone PO_4	9500cpm	7600cpm	3900cpm
% inhibition	15%	47%	65%

TABLE 2. Responses of HTC Cells, L Cells, and Hybrid Clone HL5 to a Glucocorticoid Hormone[d]

[a] Enzyme activity expressed as nmol product formed per min per mg protein.
[b] Represents inactivation of the enzyme by antiserum monospecific to authentic hepatic enzyme. 0-10% indicates virtually no reaction.
[c] Determined in LA9 cells, since LB82 cells lack thymidine kinase.
[d] Table based on data in references (23, 27).

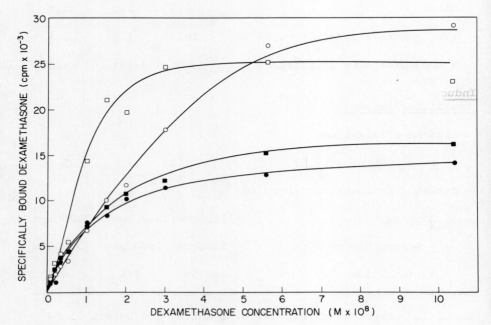

FIG 2:

Binding curves of cytoplasmic glucocorticoid receptors, showing saturation values in HTC cells (•—•), L cells (■—■), HL5 hybrid cells (□—□), and a mixture of HTC and L cell cytosols (○—○). From Lippman and Thompson (23); reprinted with the permission of the publisher.

active complex also is formed but is rapidly degraded, while at 20°C it is stable for the duration of the experiment. A model of the events measured in the nuclear binding assay is shown in Figure 3. In practice, one measures nuclear uptake of steroid-receptor complex by a competitive binding assay, diagrammed in Figure 4. Nuclei and cytosol are prepared as previously described (29). Each is then divided in half. Both halves of each are incubated at 0-2° for 2 hours with sufficient ^{3}H-dexamethasone to saturate any specific receptor sites. In addition, to one half of each is added $\geqslant$ 100-fold excess of non-radioactive dexamethasone. These are called the "competed" samples since the unlabelled steroid competes for the specific binding sites. After 2 hours, nuclei and cytosol are mixed, "uncompeted" nuclei with "uncompeted" cytosol and "competed" nuclei with "competed" cytosol. The combined fractions are then incubated an additional 75 min. at 20°C, after which nuclei are collected by centrifugation, washed thoroughly and sonicated. Aliquots are then assayed for protein and trichloroacetic acid-precipitable radioactivity. Specifically bound counts represent the difference between the "uncompeted" and "competed" samples, that is, between total nuclear bound counts and non-specifically bound counts (23,29).

In this assay, one can demonstrate apparent saturation of nuclear binding sites with steroid-receptor complex, as is shown in Figure 5. We see no evidence for inhibition of binding at super-saturating levels of cytosol. Our results indicate that once nuclear saturation is reached, addition of more cytosol of that type neither increases nor decreases the radioactive steroid bound. We found that sites for L cell steroid-receptor complex existed in HTC nuclei and sites for HTC complex in L cell nuclei, as well as for the homologous combinations. Whether the sites being measured represent in any part the physiologically significant ones utilized by whole cells we do not know. The point is, however, that the in vitro nuclear binding assay provides a means for assaying effective quantity and quality of the cytosol receptors.

We therefore attempted to see whether nuclear binding could be used to distinguish L cell and HTC cell receptors. Nuclei were prepared and their saturation level for either L cell or HTC cell receptor complex determined by a saturation curve as in Figure 5. An identical aliquot of nuclei was then incubated with a saturating quantity of cytosol plus steroid from the other cell source. In this circumstance, additional steroid was found bound to the nuclei. The incremental amount of new cytosol was incubated alone with nuclei. An example of such results is shown in Figure 6. It was found that either L cell or HTC nuclei could bind either

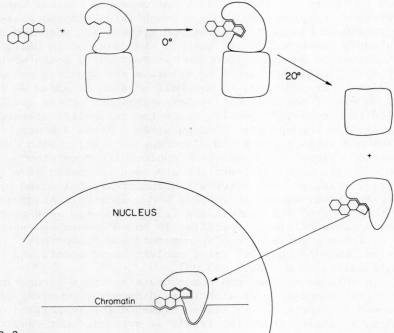

FIG 3:

<u>Diagram of supposed events in nuclear binding assay.</u> Steroid plus receptor are allowed to interact at 0°. At 20°, possibly with a reduction in size due to loss of a subunit, the steroid-receptor complex is activated. The activated complex enters the nucleus, where it binds to as yet undefined sites on the chromatin.

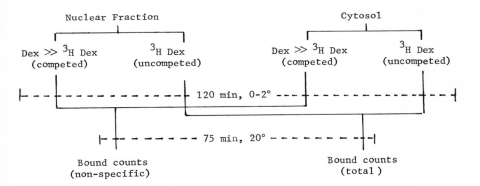

FIG. 4:

Flow diagram of nuclear binding assay.

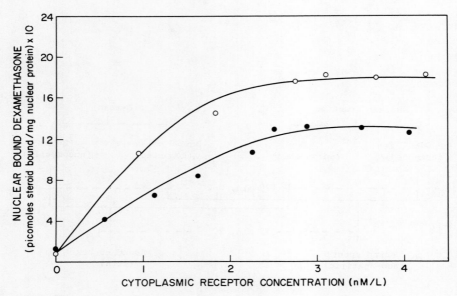

FIG. 5:

<u>Saturation of specific nuclear sites in a cell-free system by activated cytosol receptor-steroid complex</u>. HTC cell cytosol plus dexamethasone in HTC nuclei (o). L cell cytosol plus dexamethasone in L cell nuclei (●). From Lippman and Thompson (29), reprinted with the permission of the publisher.

cytosol. When saturated with one cytosol, however, additional binding from the other cytosol could be demonstrated. These results suggest that the receptors in L and HTC cell cytosols differ from one another with respect to their in vitro nuclear binding. Therefore, we used nuclear binding as a means to distinguish whether HTC X L cell hybrids contain L cell receptor, HTC cell receptor, or receptor of both types.

We carried out such studies on the HL5 hybrid clone. We found that if nuclei were saturated with HL5 cytosol as a source of steroid-receptor complex, no additional binding from HTC or L cell cytosols could be demonstrated. On the other hand, if one saturated first with either HTC or L cell cytosol, additional binding from the hybrid cytosol could be demonstrated (Figures 6 and 7). We concluded that the hybrid appeared to contain both types of receptor.

Thus, we have excluded the possibility that the loss of inducibility of tyrosine aminotransferase in one HTC X L cell hybrid is due to loss of HTC-specific glucocorticoid receptors. Negative control must be exerted at some subsequent step (Figure 8). Since we have never been able to recover TAT induction in these hybrids, we cannot exclude the trivial possibility that the structural gene for TAT is simply lost. It should be noted, though, that in other hybrids loss of inducibility occurred very soon after hybridization, even before the first DNA replication (25). Also, in two other studies, inducibility of the aminotransferase returned after partial chromosomal segregation (28,30). We therefore consider loss of the TAT structural gene to be unlikely.

Looking beyond the question of control in hybrids, these results suggest that different cell lines from mouse and rat contain qualitatively different glucocorticoid receptors. This is the first evidence supporting the idea that receptors for a given class of steroid might be varied in function. It could be that we are measuring tissue differences (fibroblast vs hepatocyte), species differences (mouse vs rat), or functional differences (inhibitory vs stimulatory). Which of these alternatives is correct remains to be seen. It is intriguing, however, that there might be different types of receptor for different glucocorticoid responses. If this were so, it would suggest participation of the receptor itself in whatever nuclear events mediate differing tissue-specific responses. Until now, the type of induction (of one or another enzyme, of cell killing, of tRNA or rRNA, etc.) has been thought to be determined by interaction of a single cytosol receptor with different control elements residing in the nucleus. Control supposedly was exerted within the nucleus by exposing or masking the necessary nuclear binding

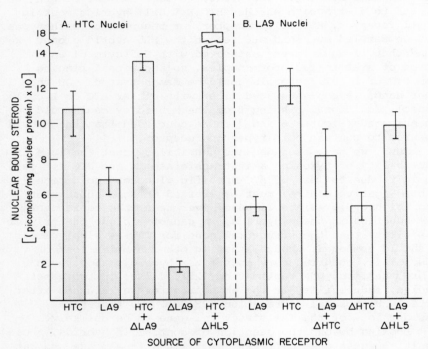

FIG. 6:

<u>Binding of steroid-receptor complexes to HTC or L cell nuclei</u>. Bars represent level of binding from cytosol source shown on abscissa. All cytosols were at saturation levels except for those preceded by a "Δ". Vertical lines represent range of 4-8 determinations. From Lippman and Thompson (23), reprinted with permission.

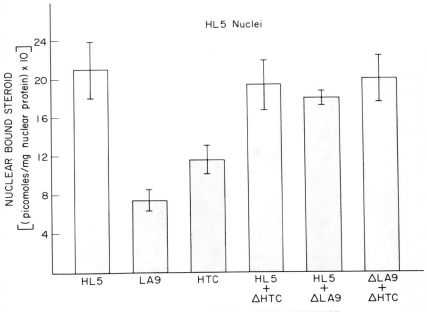

FIG. 7:

Binding of steroid-receptor complexes to HL5 hybrid nuclei.
Details as in Fig. 6. From Lippman and Thompson (23),
reprinted with permission.

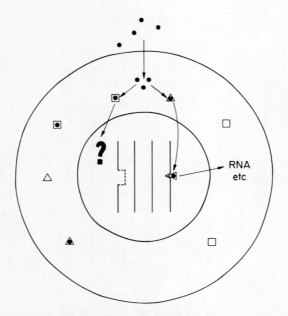

FIG. 8:

Model interpreting results in HL5 hybrids. A hybrid cell is shown containing two classes of cytoplasmic receptors, each of which recognizes a specific set of receptor sites. Steroid (●) entering the cell can bind with either class of receptor to form activated complex. L cell type complex (△) can enter the nucleus, bind, and evoke its known responses. Activated HTC-type complex (▣), however, does not evoke its known response due to a block at some unknown subsequent step (?). Although the two classes of nuclear sites are shown as mutually exclusive, this is merely a convenience. There might be some overlapping sites in either cell which could recognize either active complex.

sites. Our results suggest that selective control may also involve a cytoplasmic element entering the nucleus complexed to steroid and participating in the choice of nuclear sites and, hence, cellular responses.

One functional test of this hypothesis would be to develop either an HTC or an L cell line unable to respond to steroids because of a cytosol receptor defect, and then to fuse this cell with a receptor-containing cell. If one receptor serves as well as another, the source of the receptor-positive line should make no difference. But if receptors are response-specific, only L cell receptors should restore function to receptor-negative L cells, and similarly for HTC cells. We have begun to carry out such experiments. A clone of LB82 cells was selected for its ability to grow in the presence of 10^{-6} M dexamethasone. Assay for cytosol receptor in this clone showed only very little to be present. Then, by standard methods, hybrids were prepared between these receptor-defective L cells, and normal, receptor-positive, steroid-responsive HTC cells. The resulting hybrids were of considerable interest (23). Induction of tyrosine aminotransferase, as well as basal enzyme, was lost. Cytosol receptor could be measured in the hybrid at normal levels. The hybrid remained insensitive, however, with respect to the L cell-specific responses to steroid. That is, growth, DNA synthesis, RNA synthesis, and glucose uptake were not inhibited by dexamethasone. The steroid receptor provided by the HTC cells could not substitute functionally in the receptor-defective L cell.

It might be, of course, that the defective L cell had two or more lesions, including one that would prevent all its responses, even in the presence of effective receptor. It might even possess a dominant factor, inhibiting all steroid responsiveness. Our recent results do not favor this idea. The same steroid-resistant L cell clone has been fused with receptor-positive, steroid-sensitive L cells. All of the hybrid clones isolated from this cross have been found to be steroid-sensitive. (A. Venetianer and E. B. Thompson, MS in preparation.) Clearly, the resistant L cell does not possess a dominant inhibitory factor, and apparently L cell receptor functions in L X L hybrids. Experiments to determine whether there might be a cis-acting lesion in the steroid-resistant L cell remain to be done.

In sum, these experiments have been instructive in several ways. They have shown that the loss of tyrosine aminotransferase induction in HTC X L cell hybrids is not due to loss of glucocorticoid receptors. Furthermore, the inhibitory responses of L cells to steroids are not necessary for the extinction of the aminotransferase, since extinction occurs in the hybrid between HTC cells and the non-responsive L cells. These same hybrids suggest that the presence of the L cell receptor is unnecessary for extinction of the aminotransferase. Besides eliminating these various possibilities for the control seen, our results with the HTC X L hybrids have raised some interesting new possibilities. HTC and L cell receptors seem to be physically and functionally different. It this is confirmed to be the case, one must consider the control of differentiated responses to steroids at a new level. Heretofore, theories have concentrated on the assumption that tissue responses would be controlled by: 1) having functional cytosol receptor or not, or 2) controlling access of activated steroid-receptor complex to the genome. Our results with the HTC X L hybrids suggest that within the general class of receptors for gluccorticoids, there may be subclasses, and these participate in selecting the cellular responses displayed. These results ought to be extendable to all types of steroids to which differing tissue responses have been observed. Finally, our results indicate that the cytosol receptor itself must be interacting with the nuclear control mechanisms, and not just bearing steroid to the nucleus.

---

1. Davis, F. M. and Adelberg, E. A. Bacteriol. Rev. 37 (1973) 197.
2. Carlsson, S. A., Ringertz, N. R. and Savage, R. E. Exp. Cell Res. 84 (1974) 255.
3. Chatterjee, S. and Narashima Rao, M. V. ibid 235.
4. Thompson, E. B. In Current Topics in Biochemistry 1973. Ed. by C. B. Anfinsen and A. N. Schechter. Academic Press Inc., New York (1974) 187.
5. Thompson, E. B. and Lippman, M. E. Metabolism 23 (1974) 159.
6. O'Malley, B. W. and Means, A. R. Science 183 (1974) 610.
7. Jensen, E. V. and DeSombre, E. R. Science 182 (1973) 126.
8. Baxter, J. D., Rousseau, G. G., Benson, M. C., Garcea, R. L. Ito, J. and Tomkins, G. M. Proc. Nat. Acad. Sci. (U.S.A.) 69 (1972) 1892.
9. Spelsberg, T. C., Steggles, A. W., Chytil, F. and O'Malley, B. W. J. Biol. Chem. 247 (1972) 1368.
10. Puca, G. A., Sica, V. and Nola, E. Proc. Nat. Acad. Sci. (U.S.A.) 71 (1974) 979.
11. Granner, D. K., Thompson, E. B. and Tomkins, G. M. J. Biol. Chem. 242 (1970) 1472.

12. DeAngelo, A. B. and Gorski, J. Proc. Nat. Acad. Sci. (U.S.A.) 66 (1970) 693.
13. Rosenau, W., Baxter, J. D., Rousseau, G. G. and Tomkins, G. M. Nat. New Biol. 237 (1972) 20.
14. Pratt, W. B. and Ishii, D. N. Biochem. 11 (1972) 1401.
15. Lippman, M. E., Halterman, R. H., Leventhal, B. G., Perry, S. and Thompson, E. B. J. Clin. Invest. 52 (1973) 1715.
16. McGuire, W. L., Chamness, G. C., Costlow, M. E. and Shepherd, R. E. Metabolism 23 (1974) 75.
17. Levisohn, S. R. and Thompson, E. B. Nat. New Biol. 235 (1972) 102.
18. Makman, M. H., Nakagawa, S., Dvorkin, B. and White, A. J. Biol. Chem 245 (1970) 2556.
19. Makman, M. H., Dvorkin, B. and White, A. Proc. Nat. Acad. Sci. (U.S.A.) 68 (1971) 1269.
20. Mosher, K. M., Young, D. A. and Munck, A. J. Biol. Chem. 246 (1971) 654.
21. Thompson, E. B., Tomkins, G. M. and Curran, J. Proc. Nat. Acad. Sci. (U.S.A.) 56 (1966) 296.
22. Pratt, W. B. and Aronow, L. J. Biol. Chem. 241 (1966) 5244.
23. Lippman, M. E. and Thompson, E. B.J.Biol. Chem. 249 (1974) 2483.
24. Schneider, J. A. and Weiss, M. D. Proc. Nat. Acad. Sci. (U.S.A.) 68 (1971) 127.
25. Thompson, E. B. and Gelehrter, T. D. Proc. Nat. Acad. Sci. (U.S.A.) 68 (1971) 2589.
26. Benedict, W. F., Nebert, D. W. and Thompson, E. B. Proc. Nat. Acad. Sci. (U.S.A.) 69 (1972) 2179.
27. Levisohn, S. R. and Thompson, E. B. J. Cell Physiol. 81 (1973) 25.
28. Croce, C. C., Litwack, G. and Koprowski, H. Proc. Nat. Acad. Sci. (U.S.A.) 70 (1973) 1268.
29. Lippman, M. E. and Thompson, E. B. Nature 246 (1973) 352.
30. Weiss, M. C. and Chaplain, M. Proc. Nat. Acad. Sci. (U.S.A.) 68 (1971) 3026.

Note: We would like to thank the publishers of Nature and the J. Biol. Chem. for the permission to reproduce Figures 3, 5, 6 and 7.

EXPRESSION OF ARYL HYDROCARBON HYDROXYLASE INDUCTION IN LIVER- AND HEPATOMA-DERIVED CELL CULTURES

Ida S. Owens, Akira Niwa and Daniel W. Nebert

As our contribution to this symposium, we wish first to review the various properties of the monoxygenase enzyme systems that are currently accepted by pharmacologists. Secondly, we demonstrate the presumed importance of one monooxygenase--aryl hydrocarbon (benzo [a]pyrene) hydroxylase-- in the possible metabolic activation of chemical carcinogens. Thirdly, our previous studies are described in which we use cultures derived from fetal rat primary liver cells and from rat or mouse hepatoma. Lastly, we present data indicating that certain differences do exist between cultured liver or hepatoma cells and liver from the intact adult animal.

When one speaks of monooxygenase "activity," this represents the integrity of an electron pathway between several membrane-bound components. Reducing equivalents are supplied by NADPH and probably NADH, and one atom of molecular oxygen is incorporated into the hydrophobic substrate to render an intermediate or product which is more polar and hence more readily conjugable and excreted from the body (1,2). Any compound possessing a sufficiently high chloroform-to-water partition ratio will be attracted to the lipoidal membrane and then will be metabolized; these enzyme systems therefore metabolize most xenobiotics, as well as hemin, bilirubin, indoles, steroids (including cholesterol), thyroxine, sympathomimetic amines, and fatty acids (1,2). The function of the lipoidal microsomal membrane presumably is to attract the hydrophobic substrate to the site within the cell at which metabolic conversion to a more polar intermediate or product occurs. These monooxygenase "activities" are extremely sensitive to differences in sex, age, strain, and species, and to diffferences in the hormonal or nutritional state of the animal (1). Certain enzyme "activities" are subject to circadian rhythmicity. Many environmental factors such as cigarette smoke, insecticides, or even cedar or pine wood bedding (3) used in animal cages may change what one might consider the "control enzyme activity."

It used to be said (1) that there were two well defined classes of microsomal enzyme "inducers." There are now known to exist at least four classes of compounds which distinctly differ in their mechanisms by which they stimulate monooxygenase "activities." Once class, exemplified by the polycyclic hydrocarbons, "induces," or stimulates, a rise in the "activity" of several monooxygenases. Virtually all tissues, with the possible exception of nervous tissue and adrenal cortex, appear to possess these inducible enzyme "activities." Maximally induced activity is attained one to two days after a single dose. There is little, if any, proliferation of the endoplasmic reticulum. The second class, typified by phenobarbital and more than 200 other compounds, affects not only membrane-bound monooxygenases, but mitochondrial-bound and cytosol enzymes as well. The effect is primarily in the liver. Hepatic biliary excretion is also increased. Maximally induced activity usually requires three to seven days of continuous doses of phenobarbital. There is a marked proliferation of the smooth endoplasmic reticulum as determined by electron microscopy (1). Another class of monooxygenase "inducers" includes a group of steroids, the most potent of which is pregnenolone-16 -carbonitrile (4). A fourth class of inducers includes the biogenic amines, such as tryptamine, norepinephrine, or isoproterenol; these compounds induce aryl hydrocarbon hydroxylase activity in liver cell cultures, but the inductive effects in the intact animal have not been observed (5). The physiological significance of this class of inducers therefore remains unclear. If liver cells in culture (5) or intact animals (1) are given several "types" of inducers, one will find that the monooxygenase "activity" reach is often the sum of each of the levels maximally induced by a single inducer. Thus, each of these classes of inducers appears to be acting through some distinctly different mechanism.

Fig. 1 illustrates the hydroxylase assay used in an _in vitro_ system. The membrane-bound-enzyme -- as the cellular or tissue homogenate or the microsomal fraction -- in the presence of NADPH, NADH, molecular oxygen, and certain divalent cations, hydroxylates the substrate benzo[a]pyrene to more than a dozen alkali-extractable products (6). The rate of formation of the 3-hydroxy derivative is determined spectrophotofluorometrically and equated with hydroxylase activity. The hydroxylase system is inducible in cell culture by many polycyclic hydrocarbons dissolved in the growth medium, and these same inducers are also substrates for the enzyme system, resulting in numerous oxygen-containing intermediates and products (7). It is convenient that commonly used inducers such as MC^2 or benz[a]anthracene and their metabolites do not interfere with the spectrophotofluorometric determination of 3-hydroxybenzo[a]pyrene production _in vitro_. With benzo[a]pyrene as the substrate _in vitro_, the "aryl hydrocarbon hydroxylase" activity is equated with the rate of formation of 3-hydroxybenzo[a]pyrene; this phenolic product

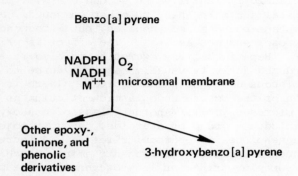

FIG. 1:

Enzymatic conversion of benzo[a]pyrene to phenolic products in vitro.

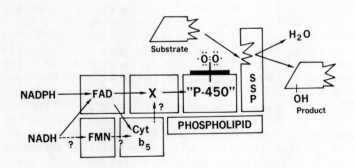

FIG. 2:

Theoretical model depicting the microsomal electron pathway in substrate hyrdoxylation, such as aryl hydrocarbon hydroxylase "activity". Components that may be essential to the

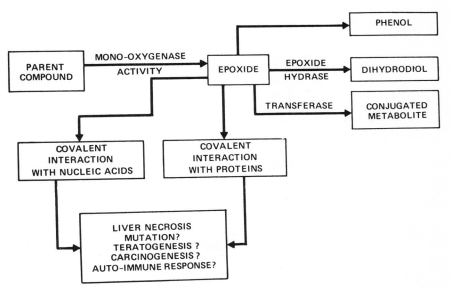

FIG. 3.

Possible metabolic pathways for formation and disappearance of arene oxides of aromatic hydrocarbons.

FIG. 2. (legend continued)

integrity of mixed-function oxygenase activity include FAD- and FMN-containing flavoproteins, cytochrome b_5, unknown factor(s) (X), phospholipid, and the terminal cytochrome P450, which may be composed of a hemoprotein and a "substrate-specific" protein (SSP) component. This model was developed several years ago as the result of valuable discussions with Drs. Howard S. Mason, Tsuneo Omure, and Volker Ullrich.

may be formed either by a direct hydroxylation or in a two-step process via an arene oxide. "Induction" simply denotes a relative accumulation of hydroxylase activity. This may represent an increase in the rate of de novo synthesis of a protein, in the rate of activation of enzyme activity from pre-existing components, or both, compared with the rate of degradation. Until technical difficulties regarding the solubilization of the enzyme system are resolved further, one cannot at present distinguish between increased enzyme synthesis de novo or activation and decreased degradation.

The number of necessary components in the aryl hydrocarbon hydroxylase system is not certain, but it is important to keep in mind that the enzyme activity" we measure (Fig. 2) is the result of a multicomponent electron chain--similar in some ways to the oxidative phosphorylation chain involving mitochondrial cytochromes. Thus, a FAD-containing flavoprotein is reduced by NADPH and NADH; electrons are then passed, perhaps via cytochrome b_5 and other unknown structural or functional factors designated by X, ultimately to the terminal oxidase cytochrome P450, where the substrate is hydroxylated at the active heme site in the presence of activated oxygen and a theoretical "substrate-specific" protein (SSP). In a liver cell, the hemoprotein P450 comprises 5 to 20% of the total microsomal protein (8,9), and this concentration is 10 to 20 times greater than the sum of all the mitochondrial cytochromes. Cytochrome P450 is so named because its reduced complex with CO has a Soret maximum at about 450 nm.

The oxidative metabolism of polycyclic aromatic hydrocarbons (Fig. 3) proceeds via the reactive arene oxide (epoxide) intermediates, which can isomerize to phenols, be converted enzymatically to trans-dihydrodiols or glutathione conjugates, or become covalently bound to cellular nucleic acids and proteins. Both the hepatotoxicity (10-12) and carcinogenicity (13-17) of such xenobiotic compounds are dependent on metabolic activation and appear to be associated with the degree of reaction of the arene oxides with cellular macromolecules. Thus the cytotoxic or carcinogenic effects of a xenobiotic would presumably be decreased or prevented by selective increases in epoxide hydrase (18) and glutathione-epoxide transferase (19) activities and/or selective decreases in aryl hydrocarbon hydroxylase (20) activites.

Previously it has been shown in this laboratory that aryl hydrocarbon hydroxylase activity in fetal rat hepatocytes in culture is stimulated by PB, the insecticide 2,2-bis(p-chlorophenyl)-1,1,1-trichloroethane (p,p'-DDT), benz-[a]anthracene, MC, or biogenic amines such as norepinephrine or isoproterenol (5,21-23). PB, MC, and biogenic amines all produce rises in the hydroxylase activity that are additive or synergistic when two or three of these types of inducers are combined in the culture medium (5,21,22). The action of PB, polycyclic hydrocarbons, and biogenic amines on the hydroxylase induction appears to be transcriptional,(3) because each of the induction processes is inhibited when actinomycin D is added simultaneously with the inducer initially (5,23). With PB, MC, or norepinephrine, there is also a posttranslational effect in which the regular rate of decay of the induced hydroxylase activity is impeded (5,23). Either PB or MC or norepinephrine as a second inducer can also direct at some posttranscriptional level a further rise in hydroxylase activity after treatement of the hepatocytes with the first inducer (5).

Fig 4 illustrates the kinetics of induction of hydroxylase activity by either PB or benz[a]anthracene as a function of age of the hepatocyte cultures. Stimulation of the enzyme by either type of inducer was optimal between the 2nd and 4th days after plating the cells. Also, the basal levels of the enzyme in cells grown in control medium alone were maximal during this time. When either PB or benz[a]-anthracene was present in the growth medium at optimal concentration, the rates of microsomal enzyme induction were not statistically different; thus, the time required for a doubling of oxygenase activity in cells treated with either inducer was approximately 3.0 hours.

By the 4th day of culture this enzyme activity in the cells decreased and the response to either PB or benz[a]-anthracene was clearly less. The fall in the constitutive and inducible hydroxylase activity must be due to some nonspecific eventrelated to the problem of growing viable, normal liver cells in culture for prolonged period of time.

The hydroxylase induction by PB reached higher levels of specific activity, in proportion to how early in the age of the cultures the PB was added. Therefore, the enzyme in hepatocytes exposed to PB from the time if initial plating reached maximal levels, even though the rate of oxygenase induction during the first 24 hours was slow. This observation was not seen with benz[a]anthracene. Hence, whereas this phenomenon probably reflects some difference in the mechanism of microsomal enzyme induction by PB compared with that by benz[a]anthracene, the reason for this finding remains unexplained.

One advantage of fetal rat liver primary cultures is that the basal and inducible hydroxylase activities are many

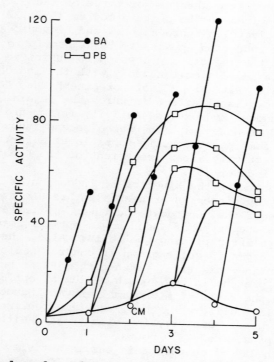

FIG. 4: <u>Hydroxylase</u> <u>induction</u> <u>in</u> <u>primary</u> <u>cultures</u> <u>of</u> <u>fetal</u> <u>rat</u> <u>hepatocytes</u> <u>by</u> <u>2.0</u> <u>mM</u> <u>PB</u> <u>or</u> <u>13</u> <u>μM</u> <u>benz(a)anthracene</u>, <u>as a function of age of the cultures</u> (22). The logarithmic phase of liver cell growth occurred between 1 and 4 days after plating the cells. The basal levels of hydroxylase activity in hepatocytes grown in control medium alone (CM) are also shown. In <u>this figure</u> and in <u>subsequent figures</u>, both the hydroxylase activity and protein concentration were determined in duplicate for the homogenate from cells scraped from one cell culture dish 60 mm in diameter as previously described (20-22). Usually two values of hydroxylase specific activity were obtained from each of two dishes harvested at each time point. <u>One unit</u> of aryl hydrocarbon hydroxylase activity is defined (20) as that amount of enzyme catalyzing per min at 37° the formation of hydroxylated product causing fluorescence equivalent to that of 1 pmole of 3-hydroxybenzo(a) pyrene. The specific activity of duplicate samples varied less than 10%.

times greater than those found in previous cultures that have been studied (24). Therefore, with the use of various inhibitors of cell metabolism such as actinomycin D and cycloheximide (5,23), subtle changes in the oxygenase activity can be measured with confidence. The disadvantages with fetal rat hepatocytes, however, include: (i) the sacrifice of 12-24 pregnant rats in our laboratory each week; (ii) the weekly expense for this number of animals; (iii) our dependence upon animal suppliers for healthy animals of the correct gestational age; and (iv) differences in the basal and inducible hydroxylase activities from week to week--presumably due to dissimilarities in the mean gestational age of the fetuses used, thereby causing different growth rates of the cultures and differences in yields of viable hepatocytes. Thus, if the hydroxylase were inducible equally well in hepatic established cell lines and if the kinetics of induction and the responses to various inducers in such established cell lines are shown to be similar to those parameters in fetal rat liver primary cultures, clearly such an experimental system would alleviate these disadvantages.

H-4 II-E, a rat cell line derived (25) from Reuber hepatoma H-35, was generously provided by Dr. E. Brad Thompson, National Cancer Institute, Bethesda, Md. MH_1C_1, a clone of epithelial cells from the transplantable Morris hepatoma 7795 originally produced in the Buffalo strain of rat (26) was purchased from the American Type Culture Collection Cell Repository, Rockville, Md. Hepa-1, a mouse cell line derived from the transplantable hepatoma BW 7756 originally produced in the C57/LJ mouse (27), was kindly given to us in 1971 by Dr. Gretchen Darlington, Department of Biology, Yale University, New Haven, Conn. TRL-2-Cl-2 and ERL-2-Cl-3 cell lines (28) derived, respectively, from normal liver of 10-day-old and 8-week-old rats of the DB-6 strain, were generously provided by Dr. Yoji Ikawa, National Cancer Institute, Bethesda, Md.; these two lines had been growing in McCoy's 5a medium with 10% fetal calf serum and were adapted in our laboratory to Eagle's minimal essential medium with 10% fetal calf serum. The BD-6 strain of rats was brought from Europe to this country by Dr. W.J. Burdette, M.D. Anderson Hospital and Tumor Institute, Houston, Texas, and is now commercially available from Flow Laboratories Inc., Rockville, Maryland. Waymouth MAB medium (specially ordered from Grand Island Biological Company, Grand Island, N.Y.) with 10% fetal calf serum was used during the induction studies with Hepa-1 cells, whereas H-4-II-E, MH_1, C_1, TRL-2-Cl-2, and ERL-2-Cl-3 cultures were grown in Eagle's minimal essential medium containing 10% fetal calf serum.

The rate at which the induced hydroxylase activity accumulates in the various established hepatoma- and liver-derived cell lines treated with aromatic hydrocarbons, PB, or biogenic amines is illustrated in Fig. 5. Note that the maximally induced enzyme levels in Fig. 5B are at least 20 times less than those in Fig. 5A. However, Dr. Akira Niwa in our laboratory has recently developed clones of the TRL-2-Cl-2 line having an inducible hydroxylase specific activity of more than 80. The general trend observed was that the polycyclic hydrocarbons were better inducers than the biogenic amines and that PB was the least effective type of inducer. PB is also less effective than the aromatic hydrocarbons or biogenic amines in the fetal rat hepatocyte primary cultures (5). That PB did cause detectable increases in the hydroxylase activity is presumptive evidence that each of these five lines is of hepatic origin (22). In each these lines, the optimal inducer concentrations in the growth medium and the kinetics of enzyme induction by each type of inducer were strikingly similar to what has already been characterized in fetal rat liver cultures (5,22). In addition we have found (29,30) that the responses of the hydroxylase induction process to inhibitors of RNA and protein synthesis and the apparent independence of transcription and translation appear the same as these parameters described in fetal rat hepatocytes (23). All five cell lines appear as homogeneous epithelial cells resembling normal hepatic cells. H-4-II-E cultures grow as individual cells and are the most rapidly proliferating of these five cell lines, with a generation time (t_G) of 18-22 hr. Hepa-1 cultures grow in colonies and the TRL-2-Cl-2 line appears as individual cells, and both grow less rapidly than H-4-II-E cells (t_G between 24 and 30 hr.). MH_1C_1 cells grow in colonies and have a t_G of 35-45 hr. ERL-2-Cl-3 cultures are individual cells and at the present time grow in Eagle's medium very slowly (t_G greater than 40 hr.) (24).

There appears to be an association between aromatic hydrocarbon-inducible hydroxylase activity and cytochrome P_1 450 formation[4] in hepatic and in several nonhepatic tissues of the mouse and rat. This change in the CO-binding hemoprotein and presumably in the enzyme active-site(s) can be detected spectrally (20,33-36) by changes in the EPR spectra at temperatures below $10°$ K (36,37), by preferential inhibition of benzo[a]pyrene hydroxylation in vitro (38,39), and by studies in which several microsomal solubilized subfractions are combined to reconstitute aryl hydrocarbon hydroxylase activity (40). In PB-treated animals, the

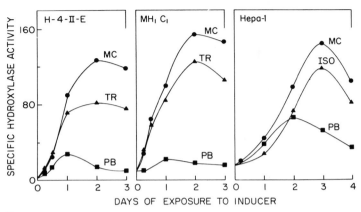

FIG. 5a:

<u>Kinetics of the hydroxylase induction in three established cell lines derived from rat Reuber hepatoma (H-4-II-E), rat Morris hepatoma ($MH_1 C_1$) and mouse hepatoma (Hepa-1) by 1.0 µM MC, 1.0 mM tryptamine (TR), 2.0 mM PB or 1.0 mM isoproterenol (ISO) in the growth medium (24).</u> The basal enzyme specific activities in cells grown in control medium alone ranged in various experiments from 0.4 to 3.5 in H-4-II-E, from 3.2 to 8.8 in $MH_1 C_1$ and from 5.0 to 24 in the Hepa-1 line.

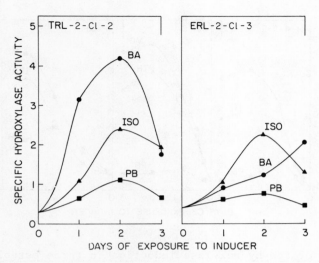

FIG. 5b:

Kinetics of the hydroxylase induction in two established cell lines derived from normal liver of 10-day-old (TRL-2-CL-2) and 8-week-old (ERL-2-CL-3) rats by 13 µM benz[a]-anthracene (BA), 0.50 mM isoproterenol (ISO) and 2.0 mM PB in the growth medium (24). The basal enzyme specific activities in cells grown in control medium alone ranged in various experiments from >0.02 to 1.0 in TRL-2-Cl-2, and from >0.02 to 0.64 in the ERL-2-Cl-3 cell line. The highest specific activities we have observed for the hydroxylase are: 35 with 13 µM benzo[a]pyrene as the inducer and 7.2 with 13 µM benz[a]anthracene as the inducer in the TRL-2-Cl-2 line; and 3.1 with 13 µM benzo[a]pyrene as the inducer and 7.2 with 13 µM benz[a]anthracene as inducer in the ERL-2-Cl-3 line.

hydroxylase activity more closely approximates that of the control animal (20,34-40)--although differences between the PB-treated and control animal are detectable. In the remainder of this presentation, we wish to show that, in cell cultures derived from fetal rat liver or from rat or mouse hepatoma, the hydroxylase activity is induced by any of a wide variety of compounds. Moreover, the induced enzyme activity--even in cultures treated with PB--appears to be always associated with cytochrome $P_1 450$.

We examined many compounds that are known (38) to inhibit or otherwise to interact with cytochrome P450-mediated monooxygenase activity. Of the more than one dozen compounds studied in cell culture, all of them caused an accumulation of hydroxylase activity in fetal rat liver primary cultures and in the H-4-II-E or Hepa-1 established cell lines. Representative examples are illustrated in Fig. 6. When either cycloheximide (0.7 or 3.5 μm) or actinomycin D (0.4 or 4.0 nM) was added simultaneously with the various compounds, the process of hydroxylase induction always appeared to be sensitive to the inhibitors of protein and RNA synthesis, respectively. There were minor variations in the "sensitivity" of the induction process to the different concentrations of cycloheximide (Fig. 6) or actinomycin D (data not illustrated). These variations might be related to differences in affinity of the compound for cytochrome P450, since 1-(2-isopropylphenyl)-imidazole (41) and metyrapone (42) are known to have higher affinities than MC, for example, and especially to have higher affinities than β-naphthoflavone.

Other lipophilic compounds which induced in culture the hydroxylase activity are illustrated in Fig. 7. Whereas only the optimal inducing compounds are shown in Figs. 6 and 7, between three and five different concentrations of each chemical were studied during a 48-hour period. Generally, the dose just below that which caused morphological changes reflecting cytotoxicity was the optimal inducing dose. Palmitoleic acid ethyl ester, palmitoyl chloride, palmitoleyl acetate, or palmityl acetate--alone or in combination with PB and over a concentration range of 0.01 to 1.0 mg per ml of growth medium--did not significantly alter the induction of hydroxylase activity, when compared with cells exposed to control medium or PB alone. Spermine and crude extracts of bovine brain--both over a range of 0.01 to 1.0 mg per ml of culture medium--depressed rather than enhanced the induction of hydroxylase activity by PB. Similar results were found in the fetal rat liver primary cultures and in H-4-II-E and Hepa-1 cultures (29,30).

FIG. 6:

Effect of 0.70 µM (●) or 3.5 µM (○) cycloheximide on aryl hydrocarbon hydroxylase (AHH) induction and gross protein synthesis in fetal rat liver primary cultures (29). Cycloheximide was added simultaneously with the inducers 1.0 µM MC, 50 µM β-naphthoflavone (BNF), 500 µM metyrapone (MTP), 100 µM 1-(2-isopropylphenyl)imidazole (IPI), or 200 µM 2,5-diphenyloxazole (PPO). No cytotoxicity was observed during the 24-hour experiment. Gross protein synthesis was determined as previously described (5,22); 30-min pulses at 4, 12, and 24 hours were performed. None of the compounds had any significant effect on the degree of inhibition of protein synthesis caused by cycloheximide. For the illustration at far right, each symbol and brackets represent the mean ± standard deviation for 10 tissue culture dishes (2 each per inducing compound).

FIG. 7:

Induction of aryl hydrocarbon hyrdoxylase (AHH) activity in fetal rat liver primary cultures by various hydrophobic compounds (29). The optimal inducing concentrations in this experiment are indicated for MC, piperonyl butoxide (PBO), sodium laurate (LAUR), allylisopropylacetamide (AIA), aniline (ANIL), aminopyrine (AP), and diethylstilbestrol (DES). In each case except for sodium laurate, the hydroxylase activity after 24 hours of exposure to the compound was much greater than that after 48 hours of exposure. Slight cytotoxicity was detectable with diethylstilbestrol or aniline present; otherwise, no cytotoxicity was found at these optimal inducing concentrations.

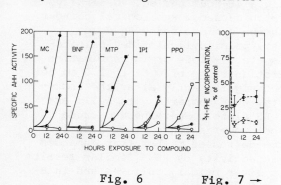

Fig. 6

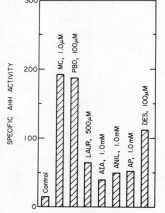

Fig. 7 →

In the intact mouse or rat, compounds such as ANF, β-naphthoflavone, 2,5-diphenyloxazole, and lindane inhibit in vitro the MC-inducible hydroxylase activity (38,39) and are presumably interacting with the cytochrome $P_1 450$ enzyme active-site. The compounds metyrapone and SKF 525-A inhibit in vitro the basal hydroxylase activity (38,39) and are presumably interacting with some other cytochrome P450 binding site. The PB-inducible hydroxylase activity behaves more like the basal hydroxylase activity than the MC-induced enzyme (38), since metyrapone and SKF 525-A are much better inhibitors than ANF or 2,5-diphenyloxazole. We thus posed the question: in cell culture, is the hydroxylase activity which is induced by these various compounds of the "a*" or "b*" type?[5]

Fig. 8 shows that, in cells treated with PB, MC, 1-(2-isopropylphenyl) imidazole, or metyrapone, the induced enzyme activity in each case was more sensitive to inhibition by the "a* compounds" ANF and 2,5-diphenyloxazole than by the "b* compounds." Even the enzyme in cells grown in control medium alone was more sensitive to inhibition by "a* compounds" than "b* compounds" in vitro (data not shown). This is in direct contrast to results found in the intact animal (38,39) and suggests that the hydroxylase activity which we regard as the basal enzyme activity in control cultures may in fact already be induced by something in the growth medium: a factor in the serum or the phenol red indicator dye might be suspect as the "inducer." An alternative possibility is that a certain amount of cytochrome $P_1 450$ and "a* hydroxylase activity" exists normally in fetal rat liver or hepatoma tissue but is usually obscured by much larger amounts of cytochrome P450 and "b* hydroxylase activity." The induced enzyme activity was also more sensitive to inhibition by two other "a* compounds"-- β-naphthoflavone and lindane--than by the "b* compounds (data not illustrated).

The data in Fig. 8 would be supported by the finding that all inducing compounds, even PB, cause the formation of cytochrome $P_1 450$ in cell culture. This is precisely what was found (Fig. 9). The degree to which the Soret peak was shifted to the blue was highly correlated with the content of CO-binding cytochrome in the 450 nm region and also with the hydroxylase activity induced in H-4-II-E cultures by PB, metyrapone, and MC. Similar data were observed with 1-(2-isopropylphenyl) imidazole or norepinephrine as the inducing compounds in Hepa-1 cultures (data not illustrated.

Fig. 10A demonstrates the appearance of the basal, PB-inducible, or MC-inducible hydroxylase activity in the rat as a function of age, and Fig. 10B illustrates the

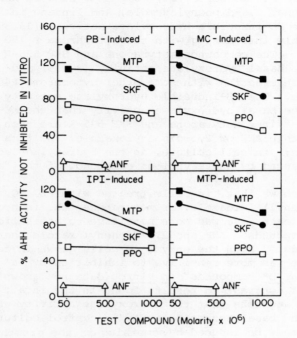

FIG. 8:

Effects of various compounds as in vitro inhibitors of aryl hydrocarbon hydroxylase (AHH) activity induced by PB, MC, 1-(2-isopropylphenyl)imidazole (IPI), or metyrapone (MTP)(29). The fetal rat liver primary cultures were first treated for 48 hours with 2.0 mM PB, 1.0 μM MC, 50 μM 1-(2-isopropylphenyl)imidazole, or 500 μM metyrapone. The specific hydroxylase activity, which represents "100%," was 76, 136, 51, and 160 units, respectively, per mg of total cellular protein. To cell homogenates in the 1.0 ml reaction mixture in vitro, an "a* compound" ANF or 2,5-diphenyloxazole (PPO) or a "b* compound" metyrapone or SKF-525A (SKF) was added prior to addition of the substrate benzo[a]pyrene, as described previously (38).

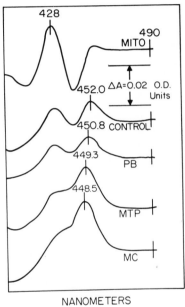

FIG. 9:

CO-difference spectra of fractions from rat Reuber hepatoma H-4-II-E cultures treated with control medium alone (CONTROL) 2.0 mM PB, 500 μM metyrapone (MTP), or 3.3 μM MC for 48 hours (29). Specific hydroxylase activities were 1.4, 36, 88, and 122 units, respectively, per total cellular protein. The fraction containing nuclei and mitochondria (MITO) and sedimenting at a centrifugation of 1,075 x g for 10 min is shown at top; this fraction is from control cells. Protein concentrations of the suspensions in the cuvettes (from top to bottom) were 6.2, 9.2, 4.4, 4.2, and 4.8 mg per ml, respectively. CO-binding cytochrome concentrations in the 450 nm region, as compared with the baseline at 490 nm, were estimated as (second from top to bottom) 12, 23, 54, and 70 pmoles, respectively, per mg of "microsomal" protein. Wave length measurements were standardized with the use of a holmium oxide crystal.

sensitivities of these three enzyme activities to ANF or metyrapone. The "b* hydroxylase activity" in the control rat and the "a* enzyme activity" in the MC-treated rat are very distinct from the time of birth. However, the hydroxylase activity in PB-treated rats appears to be a combination of the two, until about 2 weeks post partum -- at which time the adult-type "b* hydroxylase activity" becomes apparent.

From the data shown in this report, we conclude that the PB-inducible hydroxylase activity in the liver of the intact animal is not the same as the PB-inducible enzyme activity in liver cell culture. This conclusion supports the finding that no proliferative changes in the endoplasmic reticulum were detectable in fetal rat liver primary cultures which had been treated with PB. The mechanisms of induction of hydroxylase activity by PB in vivo and in cell culture might be different, and therefore, one should be cautious about extrapolating findings from one experimental system to the other. For example, it is likely that different forms of monooxygenase activities and "cytochromes P450" exist. Comparing MC versus phenobarbital as an inducer (Fig. 11), various laboratories have demonstrated that hydroxylations may occur in different chemical positions on the molecule for such substrates as biphenyl (43), testosterone (44), bromobenzene (45), and n-hexane (46). Such differences in the metabolite profile of a polycyclic hydrocarbon might result in marked differences in the reactivity of intermediates and therefore might result in marked dissimilarities in the carcinogenicity of a given compound. It is quite likely that different reactive intermediates intercalate with different DNA base-pairs, thereby damaging or activating different genomes. The carcinogenicity in vivo (47) and the mutagenicity in vitro (48) of MC--but not benzo[a]pyrene or 7,12-dimethylbenz[a]-anthracene--appears to be closely associated with "a* hydroxylase activity" and new cytochrom $P_1 450$ formation in mice. It is therefore possible that a metabolite of MC generated by cytochrome $P_1 450$ is more carcinogenic than a metabolite of MC formed by cytochrome P450. Because the hepatic cytochrome in the adult animal is predominantly P450, experiments with polycyclic hydrocarbon treatment of cultured liver or hepatoma cells may not reflect the actual situation in the intact animal, with respect to carcinogenesis or hepatotoxicity studies.

The data in this report indicate that numerous xenobiotics may stimulate the "a* hydroxylase activity" and cytochrome $P_1 450$ formation. These observations are in agreement with the suggestion (49) that this monooxygenase

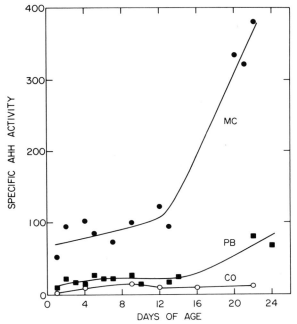

FIG. 10a:

The ontogenetic expression of hepatic aryl hydrocarbon hydroxylase (AHH) activity in vivo in rats treated with MC, PB, or corn oil (CO) alone. MC treatment consisted of a single dose 48 hours prior to assay; PB treatment consisted of daily doses for 5 days. Thus, values on 1-day-old rats indicate that the mother received MC or corn oil intraperitoneally for the final 5 days of gestation.

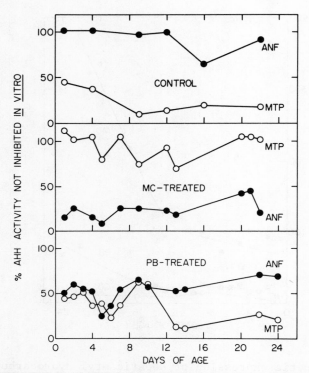

FIG. 10b: <u>The appearance of "ANF-sensitive" or "metyrapone-sensitive" hepatic hydroxylase activity as a function of age in these young rats</u>. The inhibition of benzo(a)pyrene hydroxylation <u>in vitro</u> by 50 μM ANF or 100 μM metyrapone (MTP) was studied with whole liver homogenates in the reaction mixture in the same manner as described in Fig. 8 & ref. (38). The same results were also found with liver microsomal fractions from postnatal rats. Differences greater than 30% were regarded as statistically significant ($\bar{P}$>0.05). For the data illustrated (Figs. 10a & 10b), a total of 18 PB-treated, 15 MC-treated, & 6 control litters were studied. The livers from all fetuses in one litter were combined to make a single homogenate. "One hundred per cent" of the specific hydroxylase activity from control, MC-treated, and PB-treated rats aged 9 days, for example, represents 15, 100, & 27.6 units, respectively, per mg of liver homogenate protein (29).

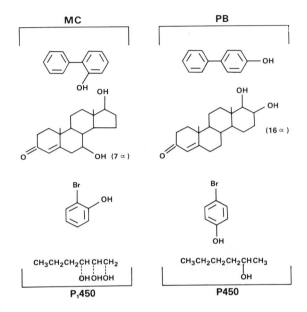

FIG. 11:

Chemical structures of differences in metabolite formation when these four substrates are incubated in vitro with liver microsomes from MC- or PB-treated rats.

system may function primarily in the organism's portals of entry (i.e. skin, lung, and intestine) and in liver where the system mediates the biotransformation of numerous noxious exogenous compounds. Since the "a* hydroxylase activity" occurs in hepatoma-derived cultures as well as in fetal rat liver cell cultures, it would be interesting if one can ever find in PB-treated cell cultures a predominance of "b* hydroxylase activity," or the induction of any other P450-mediated (rather than $P_1$450-mediated) drug-metabolizing enzyme activity. Perhaps the properties which are found in fetal tissues and in tumors that allow such cells to grow well in culture are not compatible with the conditions under which PB causes its effects in the intact animal: for example, PB-induced proliferation of the endoplasmic reticulum and the increases in certain cytochrome P450-mediated monooxygenase activities (1).

Footnotes:

1. The term in vitro denotes a cell-free reaction observed within a flask or test tube and is distinct from the description of events occurring in cell culture.
2. The abbreviations used are: MC, 3-methylcholanthrene; PB, sodium phenobarbital; the hydroxylase, aryl hydrocarbon hydroxylase; ANF, β-naphthoflavone; metyrapone, 2-methyl-1,2,3-pyridyl-1-propanone; and SKF 525-A, 2-diethylaminoethyl-2,2-diphenylvalerate HCl.
3. Whereas the inducing compounds may be stimulating the rate of synthesis of specific mRNA species, these studies (5,23) do not in fact distinguish between a mechanism whereby the compounds act to amplify specific genes or to allow transport or stabilization of (otherwise) rapidly degraded induction-specific RNA which is continuously synthesized yet rapidly degraded within the nucleus.
4. We refer to "cytochrome $P_1$450" (31)--also called "P-448" (32) or "P-446" (33)-- as that species of CO-binding hemoprotein which increases in concentration in response to aromatic hydrocarbon treatment either in vivo or in cell culture.
5. For purposes of conserving space, we wish to refer to "cytochrome $P_1$450-associated" aryl hydrocarbon hydroxylase activity as "a* hydroxylase activity" and the control enzyme activity as "b* hydroxylase activity." Thus "a* hydroxylase activity" is more sensitive to inhibition by ANF or 2,5-diphenyloxazole in vitro, whereas

"b*" hydroxylase activity" is more sensitive to inhibition by metyrapone or SKF 525-A in vitro (38).
6. Harold L. Moses, Jacques E. Gielen, and Daniel W. Nebert, unpublished data.

We acknowledge the valuable contributions of Drs. Jacques E. Gielen and William F. Benedict to this work.

References

1. Conney, A.H. Pharmacol. Rev. 19 (1967) 317.
2. Daly, J.W., Jerina, D.M. and Witkop, B. Experientia 28 (1972) 1129.
3. Vesell, E.S. Science 157 (1967) 1057.
4. Lu, A.Y.H., Somgyi, A., West, S., Kuntzman, R. and Conney, A.H. Arch. Biochem. Biophys. 152 (1970) 457.
5. Gielen, J.E. and Nebert, D.W. J. Biol. Chem. 247 (1972) 7591.
6. Sims, P. Biochem. Pharmacol. 16 (1967) 613.
7. Nebert, D.W. and Gelboin, H.V. J. Biol. Chem. 243 (1968) 6242.
8. Omura, T. and Sato, R. J. Biol. Chem. 239 (1964) 2370.
9. Omura, T., Sato, R., Cooper, D.Y., Rosenthal, O. and Estabrook, R.W. Fed. Proc. 24 (1965) 1181.
10. Brodie, B.B., Reid, W.D., Cho, A.K., Sipes, G., Krishna, G. and Gillette, R. Proc. Nat. Acad. Sci. USA 68 (1971) 160.
11. Oesch, F., Jerina, D.M., Daly, J.W. and Rice, J.M. Chem.-Biol. Interactions 6, (1973) 189.
12. Mitchell, J.R., Jollow, D.J., Gillette, J.R. and Brodie, B.B. Drug Metab. Dispos. 1 (1973) 418.
13. Benedict, W.F., Gielen, J.E. and Nebert, D.W. Int. J. Cancer 9 (1972) 435.
14. Grover, P.L., Sims, P., Huberman, E., Marquardt, H., Kuroki, T. and Heidelberger, C. Proc. Nat. Acad. Sci. USA 68 (1971) 1098.
15. Cookson, M.J., Sims, P. and Grover, P.L. Nature New Biol. 234 (1971) 186.
16. Somogyi, A., Banerjee, S., Jacobson, M.M., Spranger, J., Achor, L., Kuntzman, R. and Conney, A.H. Proc. Amer. Ass. Cancer Res. 14 (1973) 111.
17. Ames, B.N., Durston, W.E., Yamasaki, E. and Lee, F.D. Proc. Nat. Acad. Sci. USA 70 (1973) 2281.

18. Oesch, F. and Daly, J. Biochim. Biophys. Acta 227 (1971) 692.
19. Jerina, D.M., Daly, J.W., Witkop, B., Zaltzman-Nirenberg, P. and Udenfriend, S. Arch. Biochem. Biophys. 128 (1968) 176.
20. Nebert, D.W. and Gielen, J.E. Fed. Proc. 31 (1972) 1315.
21. Gielen, J.E. and Nebert, D.W. Science 172 (1971) 167.
22. Gielen, J.E. and Nebert, D.W. J. Biol. Chem. 246 (1971) 5189.
23. Nebert, D.W. and Gielen, J.E. J. Biol. Chem. 246 (1971) 5199.
24. Benedict, W.F., Gielen, J.E., Owens, I.S. Niwa, A. and Nebert, D.W. Biochem. Parmacol. 22 (1973) 2766.
25. Pitot, H.C., Peraino, C., Morse, P.A., Jr. and Potter, V.R. Natn. Cancer Inst. Monogr. 13 (1964) 2299.
26. Richardson, U.I., Tashjian, A.H. and Levine, L. J. Cell Biol. 40 (1969) 236.
27. Bernhard, H.P., Darlington, G.J. and Ruddle, F.H. Devl. Biol. 35 (1974) 83.
28. Ikawa, Y., Niwa, A., Tomatis, L., Baldwin, R.W., Gazdar, A.F. and Chopra, H.C. Proc. Am. Assoc. Cancer Res. 14 (1973) 109.
29. Owens, I.S. and Nebert, D.W. J. Biol. Chem. (1974) in press
30. Owens, I.S. and Nebert, D.W., manuscript submitted for publication.
31. Sladek, N.E. and Mannering, G.J. Biochem. Biophys. Res. Commun. 24 (1966) 668.
32. Alvares, A.P., Schilling, G., Levin, W. and Kuntzman, R. Biochem. Biophys. Res. Commun. 29 (1967) 521.
33. Nebert, D.W. J. Biol. Chem. 245 (1970) 519.
34. Gielen, J.E., Goujon, F.M. and Nebert, D.W. J. Biol. Chem. 247 (1972) 1125.
35. Nebert, D.W., Gielen, J.E. and Goujon, F.M. Mol. Pharmacol. 8 (1972) 651.
36. Nebert, D.W., Robinson, J.R. and Kon, H. J. Biol. Chem. 248 (1973) 7637.
37. Nebert, D.W. and Kon, H. J. Biol. Chem. 248 (1973) 169.
38. Goujon, F.M., Nebert, D.W. and Gielen, J.E. Mol. Pharmacol. 8 (1972) 667.
39. Poland, A.P., Glover, E., Robinson, J.R. and Nebert, D.W. J. Biol. Chem. 249 (in press).
40. Nebert, D.W., Heidema, J.K., Strobel, H.W. and Coon, M.J. J. Biol. Chem. 248 (1973) 7631.

41. Gunsalus, I.C., Tyson, C.A., Tsai, R. and Lipscomb, J.D. Chem.-Biol. Interactions 4 (1971) 75.
42. Hildebrant, A.G., Leibman, K.C. and Estabrook, R.W. Biochem. Biophys. Res. Commun. 37 (1969) 477.
43. Creaven, P.J., Parke, D.V. and Williams, R.T. Biochem. J. 96 (1965) 879.
44. Conney, A.H., Levin, W., Jacobson, M. and Kuntzman, R. in Microsomes and Drug Oxidations, (eds. Gillette, J.R., Conney, A.H., Cosmides, G.J., Estabrook, R.W., Fouts, J.R. and Mannering, G.J., Academic Press, New York 1969) pp. 279-302.
45. Zampaglione, N., Jollow, D.J., Mitchell, J.R., Stripp, B., Hamrick, M. and Gillette, J.R. J. Pharmacol. Exper. Therap. 187 (1973) 218.
46. Frommer, U., Ullrich, V. and Orrenius, S. FEBS Lett. (in press).
47. Nebert, D.W., Benedict, W.F. and Kouri, R.E. in Model Studies in Chemical Carcinogenesis (eds. Ts'o, P.O. and DiPaolo, J.A.) Marcel-Dekker, Inc., New York, in press (1973).
48. Felton, J.S. and Nebert, D.W. Fed. Proc. 33 (1974) 596.
49. Zampaglione, N.G. and Mannering, G.J. J. Pharmacol. Exper. Ther. 185 (1973) 676.

Note: We would like to thank the publishers of Mol. Pharmacology for the permission to reproduce Figures 2, 3, 4, 5A, 5B and 6.

SUSCEPTIBILITY OF MAMMALIAN CELLS IN VITRO TO NEOPLASTIC TRANSFORMATION BY CHEMICAL CARCINOGENS

J.A. DiPaolo

The concept of utilizing in vitro models for the study of carcinogenesis is not as new as many may think. As early as 1925, Carrel (1) suggested that the best way to study the malignant state would be to transform cells in vitro, prove that they were able to produce a sarcoma or carcinoma and then to study the differences between the normal strain and the transformed strain. Subsequently, a number of different laboratories reported the transformation of chick embryo cells by both physical and chemical agents (2-6). Later it was realized that the results of these experiments had probably been due to accidental contamination of the cultures with chicken tumor viruses (7). Today a number of tissue culture systems exist in which there is confidence that the transformation by chemical carcinogens is not spurious (8-13). Specific chemical markers and/or statistical analysis provide evidence of direct alteration by the external agent.

If one side of the problem of carcinogenesis is concerned with the nature of the carcinogenic insult, it is quite logical to think that the other side must include the origin of the cells being cultured as well as the methodology used to establish and maintain the cells in culture. The standard procedures used in tissue culture have been aimed at obtaining logarithmic growing cells. In retrospect, most individuals have been interested in establishing uncontaminated cell lines, those that lack slow growing bacteria and PPLO. As a result of the techniques in general use, normal specialized cells die out after a relatively short period in vitro and are replaced with undifferentiated cells called fibroblasts. In some situations, cell lines from tumors may retain specific differentiated characters both functionally and morphologically (14-15). Some tumor derived cell lines, however, are undifferentiated and may have arisen from tissue stroma.

The simplest explanation for the difference between the cells in the explant and those finally established in culture may be that although the cultures were derived from specific parenchymal cells, they adopt similar form in morphology and exhibit wide biochemical differences not related to tissue or species of origin but because of the tissue culture conditions. Without homeostasis all unnecessary genes are turned off. Differentiated tumor cells represent the exceptional class in which special functional genes cannot be turned off, thus providing one of the reasons for defining these specific cells as tumor cells. It should be pointed out that another explanation for the similarity of normal cultured cells had been advanced by L.M. Franks (16) who suggests that many normal cultures are derived from endothelial cells or pericytes that are present in the original explant.

No ideal system exists for biological research and this seems particularly true of problems dealing with carcinogenesis. The techniques of Puck (17) have made possible the development of quantitative methods which in turn have permitted the development of quantitative systems for studying the phenomenon of carcinogenesis. Such systems facilitated the separation of primary events leading to neoplastic transformation caused by a carcinogen from those that are incidental by-products. It has thus become possible to state that the process of transformation which leads to neoplasia may be inductive rather than selective and, in the case of chemical carcinogenesis, may be direct rather than indirect, indicating that one need not suggest that chemical carcinogens act by viral activation.

Thus far, our laboratory has reported successful transformation of embryo or fetal derived cells of rat (18), guinea pig (19), and Syrian hamster (12,13). In addition, chemicals have been utilized to demonstrate transformation with a mouse cell line, Balb/3T3 (20). We have unpublished evidence that Chinese hamster embryo cells may also be transformed by chemical carcinogens. Results with all systems except for the guinea pig are considered to be quantitative. A dose response relationship within certain limits exists between the chemical carcinogen concentration and the number of cells that are transformed (21). The transformation of guinea pig cells in culture by the chemical carcinogens in our laboratory differs from those in other mammalian cell models that we have used in studies of carcinogenesis. The morphological transformation of guinea pig cells does not occur for 4 or more months following exposure of the cells to a chemical carcinogen. Furthermore, morphological transformation may preceed demonstration of the ability of the cells to grow as tumors (22).

The results obtained with the quantitative in vitro systems may be contrasted with those thus far obtained with guinea pig cells. With the former case, a few of the cells exposed to a chemical carcinogen undergo rapid transformation and soon after can be demonstrated to be neoplastic since they produce tumors in appropriate hosts. In the case of the guinea pig cells, treatment with a chemical carcinogen results in an alteration of cells that makes them different from the control cells but not equivalent to what is referred to as transformation since these cells do not exhibit the random criss-cross or haystack pattern associated with transformation. Subsequently, these cells will undergo the typical transformation but at this point still fail to produce tumors when injected into guinea pigs that have been conditioned by pretreatment with radiation. It is only after guinea pig cells have acquired additional properties that they can produce progressively growing tumors in guinea pigs.

In order to obtain a rapid quantitative method, cells must be seeded so that they may form rapidly growing discrete colonies. When cell strains such as primary or secondary hamster embryo cells are used, colony formation is facilitated by the use of a feeder layer. Ordinarily an irradiated feeder layer made by X-irradiating hamster or rat embryo cells is used. In a 60 mm petri dish the feeder layer consists of approximately 60,000 cells, a relatively large number of cells compared to the number of hamster cells to be seeded. These cells may grow in volume but should not multiply. The seeded cells may be placed in the petri dish as many as three days prior to the hamster cells. If a cell line such as Balb/3T3 is used, no feeder layer is required and a respectible cloning efficiency of even 50% may be obtained by using a low number of loging cells. The carcinogen may be added as soon as the cells are stretched out and, if less than 24 hr have passed, one is assured that very few cells will have had the opportunity to divide. Depending upon the specific carcinogen as well as upon the concentration used, the chemical may be removed after 30 min or permitted to stay in the cultured medium for the entire remaining period of culturing which may extend for an additional six to ten days depending upon the system. Subsequently, the dishes are fixed and stained with Giemsa for morphological study and scoring. The total number of colonies obtained from the number of cells seeded constitutes the cloning efficiency. The percent of transformation may be reported on a per dish basis, relative to the total number of colonies scored, or on the basis of the number of cells originally inoculated. As stated previously, the transformed colonies exhibit piling up or criss-crossing as a result of loss of contact inhibition. Such alterations are not seen in controls.

These piled up colonies can be isolated and shown to produce tumors under conditions where spontaneous transformation is not seen (23,24). The results obtained with mouse cells derived from Balb mice and the cells from Syrian hamsters have proven that transformation is an inductive rather than a selective phenomenon and that dose-response relation occurs with a number of different carcinogens. Thus cells in culture may be used to answer a number of biological questions related to the phenomena of chemical carcinogenesis.

Studies with classical carcinogens lead to a number of conclusions. For example, with Syrian hamster cells and benzo(a)pyrene that required activation, transformation is independent of toxicity and transformation increases with concentration of carcinogen; in terms of transformation there appears to be no safe dose (21). Using statistical analysis of the transformation frequency, it has been possible to conclude that the distribution of the transformants is Poisson in distribution and thus consistent with the one hit hypothesis. In analogous studies with Balb/3T3 and DMBA treatment for 48 hrs. a linear relationship was found to exist between 0.025 and 0.5 ug DMBA/ml of medium. This also was interpreted as being a one hit response since the slope was 0.96. The number of transformed colonies per dish showed Poisson distribution and again implied that transformation occurred by induction. The question of whether viral activation by chemical carcinogens is necessary for transformation has been debated for many years. In the case of Balb/3T3, it is known that there may be induction of mouse C type viruses from cloned lines of virus free cells, occurring as a result of treatment with 5-bromodeoxyuridine. The cloned lines of normal and transformed Balb/3T3 cells used by us have been found to be negative for murine leukemia virus complement-fixing antigen, but no conclusion can now be reached as to the possibility that activation of a latent oncogenic virus is responsible for the neoplastic transformation in the Balb/3T3 system.

The use of fresh, diploid, Syrian hamster cells appears to present a different situation. These cells have been tested for a variety of viruses ranging from Reo to Sandai to known oncogenic viruses such as polyoma, SV40, and the adenos. The oncogenic viruses are all known to transform hamster cells. These hamster cells are also negative for hamster leukemia virus and it has been impossible to demonstrate any reverse transcriptase activity (25).

The hamster cells transformed by chemical carcinogens as well as the Balb/3T3 cells are capable of being further transformed by other agents and under these circumstances acquire additional properties associated with the agent. These experiments raise questions as to the significance of the released type C viruses of Balb/3T3 as well as any other virus that may be stirred up when the state of the cells is abruptly changed from control to transformed and then to neoplastic.

The results obtained with in vitro transformation unquestionably correlate with in vivo studies (13). Quantitative systems can be used to show that the different numbers of transformations observed were related to the known carcinogenic potency of the compounds tested. Strong or potent in vivo carcinogens produced a large number of transformations and weak carcinogens caused only a few transformations while non-carcinogens failed to produce transformation. Furthermore, there is definitive evidence that transformed cell lines do produce tumors. In the Syrian hamster system that has been described, over 95% of the transformed cell lines produce tumors; with the mouse Balb/3T3 system, 100% of the transformed cell lines produce tumors. The type of tumor produced varies from a frank fibrosarcoma to an anaplastic tumor with an occasional indication of secretory material. In our laboratory, we have felt it was best to be conservative and to consider all of them as being sarcomas.

The main point has been that these systems are reproducible and lend themselves to the desired endpoint of tumor production. Whether or not the mammalian cell model has a cell type counterpart of significance in the human population is unimportant since it facilitates the study of possible factors that may influence the incidence of transformation.

Transformation with the aromatic amines represents an interesting situation. The parent compound, N-2-fluorenyl-acetamide (FAA), may or may not be carcinogenic depending upon the species of animals tested. This compound failed to transform guinea pig cells as anticipated (22). With Syrian hamster cells, it was an extremely weak carcinogen (26). The hydroxylated compound N-hydroxy derivative was more active than the parent compound and is known to induce cancer in species not affected by the parental compound, FAA. As a transforming chemical, it was more effective than the parental compound when hamster cells were used. The hydroxy derivative converts to an ester to form the N-acetoxy-FAA, which is considered to be the metabolite closer to the proximate carcinogenic agent. This compound produced significant number of transformations of hamster and Balb/3T3 cells, 15% and 17% respectively. When applied to guinea pig cells, N-acetoxy-FAA eventually causes transformation which in turn produces tumor formation when the cells are injected into conditioned guinea pigs. Thus neoplastic transformation of non-specific cells can be produced with compounds that are known to be associated predominantly with a specific target organ.

In the course of studying diverse carcinogens, it was found that some, such as diethylnitrosamine and urethan, were incapable of causing transformation when applied directly to cells in culture. Because a reliable bioassay system for determining the potential carcinogenicity of chemicals should give no false negative results, a host mediated in vivo - in vitro combination bioassay with mammalian cells was developed (27). In this system the pregnant animal is injected with a chemical 2-4 days prior to removal of the fetuses. Fetuses exposed in utero are subsequently excised and the cells processed and the transformation studied as in the standard in vitro system by examination of colony. Neoplastic transformation can be again verified by tumor production. Under these circumstances, a number of compounds, including urethan and diethylnitrosamine, have been shown to be carcinogenic. This system is currently being investigated to determine whether differentiated cells in addition to the fibroblasts have been transformed as a result of the transplacental insult. Cells of various types that mimic transformation are being isolated and will be tested to see whether they first of all produce carcinomas.

As previously stated, the results with fetal strain 2 guinea pig cells have differed from those obtained with the quantitative systems. Cells obtained from fetuses 32-49 days old are prepared in the manner similar to that described above except that the cells are obtained from an individual fetus and cultured at a 38.5° C as opposed to 37° C for most mammalian cells. The cells are treated with the chemical at the second subpassage for 5 days. It is also possible to use the host mediated assay in which the cells are exposed in utero by single maternal interperitoneal injection of the chemical (mg/100 gm maternal body weight). This is usually done 48 hr prior to the excision of the fetus for culture of the cells. With these two procedures, 14 cell strains have been developed that have been exposed to a number of carcinogens including polycyclic hydrocarbons, nitrosoamine, and aminofluorene derivatives as well as aflatoxin-B_1.

Some of the cell strains have been subcultured and observed for 24 months for alteration in morphology and for ability to produce tumors when reinjected into guinea pigs. An additional six cell strains were obtained following exposure to noncarcinogens such as pyrene, anthracene, phenanthrene and acetylaminofluorene which for the guinea pig is not carcinogenic. These control cultures retained normal, control type cell morphology during ten months of continous culturing. In addition, an acetone treated strain has remained nontumorigenic and contact inhibited for over 125 passages in the last 20 months.

The 14 cell strains that were exposed to a carcinogen exhibited varying types of morphological alterations referred to here as alteration (A) or transformation (T). The A type deviates from control oriented cells but the deviation is not considered sufficient to be classified as transformed. True transformation, consistent with what has been reported for other in vitro systems is indicated by the letter t. Only 6 of the carcinogen exposed cell strains produced tumors when inoculated into recipient syngeneic guinea pigs. All 6 tumorigenic cultures demonstrated transformation in vitro; 4-15 months of post-carcinogen treatment were required before the morphological alteration became apparent. With guinea pig cells, the altered cell morphology referred to as transformation is not sufficient to indicate that the cells are capable of producing tumors. When 10^8 cells of the other lines that exhibited transformed morphology were inoculated into newborn guinea pigs irradiated with a midline dose of 450r total body cobalt irradiation, no tumors were produced. No cell strains treated with a noncarcinogen or solvent have exhibited transformation or the ability to grow as tumors.

This has raised the question as to what in vitro index or indices can be used to indicate that the cells have neoplastic potentiality. For this purpose, four tumorigenic and four nontumorigenic cells strains have been compared in terms of cloning efficiency, saturation density, doubling time and ability to grow in agar. Whereas transformed lines are not serum dependent in terms of colony formation, three of the four nontumorigenic strains are serum dependent. Saturation density does not distinguish the tumorigenic from the nontumorigenic cultures since one line transformed by diethylnitroamine is more like the nontumorigenic cell line than the other tumorigenic lines. Furthermore this same line transformed by diethylnitrosamine has a doubling time similar to those of the nontumorigenic cultures. Thus, neither serum dependent growth saturation density nor doubling time are reliable indices of the tumorigenicity of the cells. The only parameter that appears to correlate well with the ability of the cells to grow as a tumor is the ability to form colonies progressively in 0.35% agar. Thus, in this system transformation is considerably delayed for at least four or more months following exposure of the chemical to a carcinogen; the subsequent ability of the transformed cells to grow as a tumor may not occur until months later.

 The experimental models described illustrate a number of factors concerning morphological transformation that eventually lead to neoplasia. The phenotypic alteration from control to the transformed state may be telescoped into a relatively short time period of days or may be extensive and require several months to occur. The transformed cells are profoundly different from control cells in a number of characteristics including culture patterns *in vitro*, adhesiveness to the substrate, saturation density, cloning efficiency and growth in agar. Transformed cells need not possess all of these properties and although the correlation between these changes and transformation is high, it is not necessarily universal. For example, it is known that not all virus transformed cells are immortal, that certain cell lines may be aneuploid and contact-inhibited and yet produce tumors while others are contact-inhibited and do not produce tumors. An example of the former is the BHK/13 cell line and an example of the latter is a Balb/3T3 line. The only indubitable proof that transformed cells are neoplastic is their ability to transform the cells to form progressively growing tumors. Thus far, it is our opinion that the best correlation between *in vitro* carcinogenesis and confirmation *in vivo* by tumor production is the production of colony growth in soft agar. It is also possible that change in nuclear/cytoplasmic ratio may be an indication of epithelial type cells that will produce tumors as well as grown in agar.

 The events leading to the formation of transplantable cells after treatment with chemicals of fresh cells may appear to differ to a certain extent from one species to another; but the events occurring in a transformation are probably the same except that they are delayed in time. Part of the evidence for such a concept is the study with the Balb/3T3 line. This permanent line has already undergone a certain number of permanent hereditary changes and yet there is no indication that the number of primary events is different from that observed using fresh Syrian hamster cells. It has yet to be established whether any of the specific phenotypic differences between euploid and aneuploid transformation determine the biological character of malignancy. Information is lacking concerning the factors which limit the growth of normal differentiated cells in vivo or in vitro.

References

1. Carrell, A. J. Amer. Med. Assoc. 84 (1925) 157.
2. Fischer, A. Compt rend Soc. biol. 94 (1926) 1217.
3. Laser, H. Arch. exper. Zellforsch 6 (1928) 142.
4. Bisceglie, V. Arch. exper. Zellforsch 6 (1928) 161.
5. Des Ligneris, M.J.A. Compt. rend. Soc. biol. 120 (1935) 777.
6. Des Ligneris, M.J.A. Compt. rend. Soc. biol. 121 (1936) 1579.
7. Begg, A.M. Lancet 2 (1929) 697.
8. Berwald, Y. and Sachs, L. J. Natl. Cancer Inst. 35 (1965) 641.
9. Huberman, E. and Sachs, L. Proc. Natl. Acad. Sci. U.S. 56 (1966) 1123.
10. Chen, T.T. and Heidelberger, C. J. Natl. Cancer Inst. 42 (1969) 915.
11. Chen, T.T. and Heidelberger, C. Intern. J. Cancer 4 (1969) 166.
12. DiPaolo, J.A. and Donovan, P.J. Exptl. Cell Res. 48 (1967) 361.
13. DiPaolo, J.A., Donovan, P.J. and Nelson, R.L. J. Natl. Cancer Inst. 42 (1969) 867.
14. Davidson, E.H. Advanc. Genet. 12 (1964) 143.
15. Sato, G., Augusti-Tocco, G., Posner, M. and Kelly, P. Recent Progr. in Hormone Res. 26 (1970) 539.
16. Franks, L.M. and Cooper, T.W. Int. J. Cancer 9 (1972) 19.
17. Puck, T.T., Marcus, P.I. and Cieciura, S.J. J. Exp. Med. 103 (1956) 273.
18. Olinici, C.D. and DiPaolo, J.A. J. Natl. Cancer Inst. (In Press).
19. Evans, C.H., Nelson, R.L. and DiPaolo, J.A. Proc. Amer. Assoc. Cancer Res. 14 (1973) 76.
20. DiPaolo, J.A., Takano, K. and Popescu, N.C. Cancer Res. 32 (1972) 2686.
21. DiPaolo, J.A., Donovan, P.J. and Nelson, R.L. Nature New Biol. 230 (1971) 240.
22. Evans, C.H. Amer. Assoc. Cancer Res. 15 (1974) 40.

23. DiPaolo, J.A., Nelson, R.L. and Donovan, P.J. Science 165 (1969) 917.
24. DiPaolo, J.A., Nelson, R.L. and Donovan, P.J. Cancer Res. 31 (1971) 1118.
25. DiPaolo, J.A. In Ts'o, P.O. and DiPaolo, J.A. (Eds.) Chemical Carcinogenesis. New York, Marcel Dekker, Inc. (1974) 443.
26. DiPaolo, J.A., Nelson, R.L. and Donovan, P.J. Nature New Biol. 235 (1972) 278.
27. DiPaolo, J.A., Nelson, R.L., Donovan, P.J. and Evans, C.H. Arch. Path. 95 (1973) 380.

THE ENZYMOLOGY OF DNA REPAIR AND ITS RELATION TO
CARCINOGENESIS

J.L. Van Lancker and T. Tomura

Introduction

From its inception at the beginning of life replicating DNA, for better or for worse, became the target of various agents, UV light, ionizing radiation and possibly small chemical agents present in the atmosphere. When DNA became in some way associated with other macromolecules to form the replicative and transcribing unit of the cell, the integrity of genotype and phenotype depended upon the immutability of the DNA.

Since, at least on the surface of the earth, exposure to UV and ionizing radiation was inescapable, persistance of life required that the living unit accomodate itself either with extensive mutability or create defense mechanisms against DNA damage. In the course of evolution, bacteria have developed various mechanisms for repairing DNA; they include photoreactivation, base excision and post-replicative repair (for review (1)). There is evidence that some of these mechanisms remain at least in some mammalian cells (2,3,4).

This paper will address itself to DNA repair in mammalian cells. Although little is known of either mode of repair in mammalian cells, the repair mechanisms assume particular significance because their total or partial absence may cause disease. In the mammalian cells as in bacteria the "unit of determination of specificity", that is to say the molecular machinery required for the elaboration of a specific protein, is composed of three groups of macromolecules: DNA, RNA and proteins and therefore, the integrity of DNA replication and transcription depends upon the immutability of the DNA.

DNA Repair and Disease

Let's consider the options available to a cell with damaged DNA (Figure I). The damage can either be repaired or not. If repaired, the repair may be integral and the life of the cell goes on unaltered, but the repair could also be faulty and bring about such changes as double strand breaks and base replacement which may or may not be compatible with replication of the DNA. If the DNA is not repaired

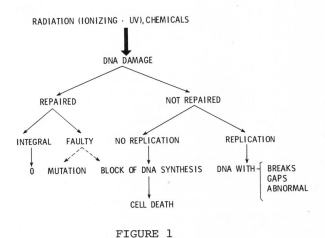

FIGURE 1

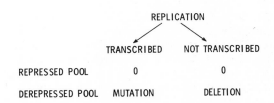

FIGURE 2

it may or may not replicate. In absence of replication, one may expect blocking of DNA synthesis and ultimately cellular death. If the damaged DNA is replicated, the new DNA may contain breaks, gaps or point mutations. Such lesions may or may not be compatible with transcription. The consequences for the cell phenotype depend upon whether the damaged DNA belongs to the repressed or derepressed pool (Figure 2).

Damage to DNA which belongs to the repressed pool will not affect the phenotype. Damage to DNA which belongs to derepressed pool will result in transcription defects such as deletion or point mutation and possibly reprogramming of transcription such as is seen in some benign or malignant tumors. The ultimate consequences for the organism will vary depending upon whether it is the DNA of a germ or a somatic cell that is afflicted.

It is not difficult to imagine how the alteration (Table I) of DNA described above could in the germ cell lead to sterility, inborn defects and chromosomal anomalies and how in the somatic cell they would lead to necrosis or to alterations of gene expression that might be benign in that they provide little survival advantage as in the case of benign tumors or some forms of aging or malignant in that they provide a considerable survival advantage over the cells of the host (5,6). Moreover, if the cell dies, is it not conceivable that at least portions of the damaged DNA will resist hydrolytic attacks and as such could become immunogenic and lead to the formation of the antigen antibody complexes? Are some forms of lupus, diseases in which DNA is not repaired?

DNA Repair and Carcinogenesis

Time will tell how much of the speculation is real, but what is certain is that absence of repair may lead to cell death or somatic mutation and the possibility that at least some cancers result from somatic mutation cannot be excluded and therefore, the study of DNA repair is particularly relevant to mechanisms of carcinogenesis.

Clearly we cannot even begin to suspect what specific macromolecular alterations convert a normal cell into a cell which has acquired such survival advantage over all the cells of the organism that it ultimately kills the host. Moreover, it cannot be excluded that this transformation occurs in a sequence of steps, some of which might even be reversible and the pattern of which might not always be identical (for review 7). But even if the potential

A. Germ Cell
   Sterility
   Inborn Errors
   Chromosomal Anomalies

B. Somatic Cell
   Cell Death
   Alteration of Gene Expression
      Benign - Benign tumors
               Some Forms of Aging
   Malignant

TABLE I. CONSEQUENCE OF DAMAGE TO DNA

1. Strand Breaks
     Single
     Double

2. Base Alterations

3. Crosslinks
     Intrastrand
     Interstrand
     Between DNA and other Macromolecules

TABLE II. TYPE OF DAMAGE OBSERVED IN DNA

pathogenic mechanisms for cancer may be extremely complex and varied, some unifying observations have emerged. For example, most carcinogens, physical (such as UV, light or ionizing radiation), chemical (with the possible exception of plastic polymers) react with DNA molecules and often cause it to be altered (for review 8,9).

In addition, the survival advantages acquired by the cancer cell in the course of transformation are transferred from one generation of cells to another, and therefore, they must in some way have been imprinted in the replicative and transcribing units; namely, the DNA. Whether the DNA alterations result from inappropriate repair remains to be shown, but what is certain is that an accurate understanding of the pathogenesis of cancer will require that the role of DNA repair in radiation, chemical and even viral carcinogenesis be clarified.

Various Approaches to the Study of DNA Repair in Mammalian Cells

To understand repair we must first find out what type of injuries can be expected to occur in DNA. The injuries caused by chemical and physical agents include single strand breaks, double strand breaks, formation of cross links and base alterations (Table II).

Repair of strand breaks are usually studied by sedimentation on neutral or alkaline sucrose gradients and I am sure that we will hear more of that approach in this symposium in the papers from the laboratories of Dr. Potter and Dr. Farber, who were pioneers in the study of DNA repair after the administration of carcinogen. Repair of cross-linked DNA has been investigated in bacteria and studies in mammalian cells have barely begun.

There is an extensive literature on repair of base alterations in bacteria and mammalian cells in which the degree of repair is evaluated by measuring the incorporation into DNA of tritium labeled thymidine in absence of DNA replication. Such an approach has been most helpful in identifying circumstances in which the repair mechanism was absent and in searching for the areas of DNA that are preferentially repaired, either because they are more susceptible to damage or because they are more accessible to the repair enzymes (10).

Enzymology of DNA Repair in Mammalian Cell

A detailed molecular understanding of the events of

DNA repair requires in vitro dissection of the macromolecular events that take place. In bacteria two schemes have been proposed for the excision repair of base alterations. In the first, nicking of the DNA at a point close to the damaged base is followed by an exonucleolytic excision of the damaged base, patching of the denuded area with complementary bases through the action of the DNA polymerase and final insertion of the new complementary segment within the DNA strand by polynucleotide ligase. In a simplified scheme by Kornberg and his associates, the exonucleolytic attack and the polymerization are both accomplished by a single enzyme DNA polymerase I (1).

Although it is claimed that in bacteria an endonuclease highly specific for thymine dimers can be isolated, it is not inconceivable that some of the bacterial repair enzymes simply recognize distortion of the DNA strand caused by the base alteration. In fact, this possibility has not been excluded for the bacterial enzyme purified by Kaplan et al. from micrococcus luteus (11). But what about the mammalian enzymes?

The mammalian cell is subject to so many forms of insults that can cause base alteration (UV light, x-radiation and a multitude of chemicals), that it seems unlikely that each cell stores endonucleases specific for each type of base damage. Thus, one can assume that the specificity of the enzyme will be mainly directed toward the strand distortion caused by the base alteration. Moreover, because no organ escapes exposure to deleterious chemicals the enzyme must be ubiquitous. For these reasons one could expect the enzyme to be present in rat liver, an organ which has the distinction of having taught us most of what we know of mammalian biochemistry.

Although the enzyme is extremely unstable, it was successfully purified to electrophoretic homogeneity (12). After approximately 25 different purifications the following facts were established, most of which have been published (12). (1) The enzyme causes single strand breaks in UV irradiated DNA or in DNA containing acetylaminofluorene or bromobenzanthracene bound to the bases.
(2) The enzyme also causes an increment in single strand breaks to appear in x-irradiated DNA when such DNA is sedimented on alkaline sucrose gradients (Figure 3).
(3) The enzyme is ineffective on single stranded or apurinic DNA. (4) The enzyme is not inhibited by caffeine, but is inhibited by the two cocarcinogens and anthraline kindly given to us by Dr. Van Duuren (Figure 4,5). This effect was first believed to result from chelation of magnesium.

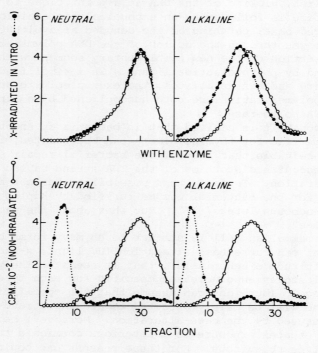

FIGURE 3

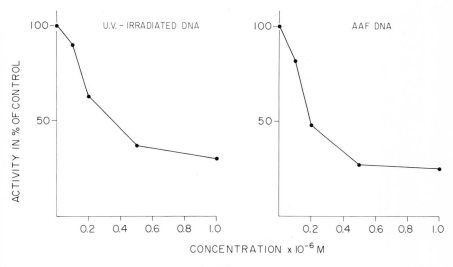

FIGURE 4

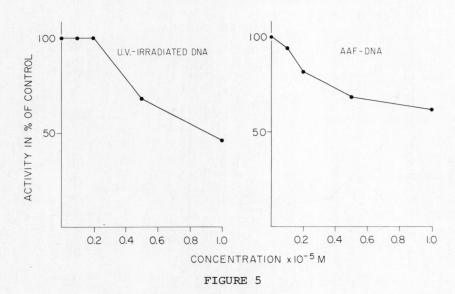

FIGURE 5

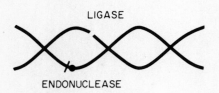

FIGURE 6

However, although magnesium stimulates activity, it is not absolutely indispensable for activity and secondly, when Dr. Van Duuren kindly provided us with a phorbol derivative which is not a cocarcinogen, inhibition of enzyme activity was not observed. (5) When DNA containing tritium labeled thymine dimers or C^{14} labeled acetylaminofluorene is sequentially treated with the endonuclease and then subjected to the exonucleolytic activity of DNA polymerase, the labeling is recovered after appropriate chromatographic separation with thymine dimers or with acetylaminofluorene bound to guanosine (12). (6) Inasmuch as alkaline phosphatase treatment is necessary to secure the binding of DNA polymerase, it seems logical to assume that the nick caused by the endonuclease is of the $3'PO_4$ type.

These observations raise a number of important questions: for example, what role does the enzyme play in the repair scheme? Are there other enzymes nicking DNA in similar fashion? Are the effects of the endonuclease always favorable? Is the damaged DNA repaired integrally? Let's consider these questions one by one.

Except for the fact that the enzyme purified from rat liver is likely to be the one missing in xeroderma pigmentosum (10), we know nothing of its role in vivo. Such studies require experiments in which the activity of the enzyme in vivo is closely correlated with biological processes. Experiments done in cooperation with Dr. Maher in which transforming DNA to which carcinogens are bound have been used are in progress. The transforming DNA which has lost its transforming properties is nicked with the repair endonuclease and introduced into a mutant bacteria devoid of the enzyme. If the nicking is indispensible for repair, the transforming properties should reappear. To date Dr. Maher's data is encouraging but not conclusive. The unusual instability of the enzyme and the presence of a multitude of other endonucleases, render crude assay very difficult. Preliminary evidence does, however, suggest that the repair endonuclease activity is increased after administration of acetylaminofluorene.

Although this is the first repair endonuclease that has extensively been purified from mammalian cell, it is most likely that other repair endonucleases with different specificities exist as well. Verly (13) has purified from rat liver an endonuclease which attacks the apurinic sites of DNA.

It may well be that the endonucleolytic attack is not always advantageous to the DNA integrity. That the endonuclease might add to the damage caused by radiation is suggested by experiments in which the effects of in vivo and in vitro irradiation of DNA were compared. Although it is of interest that x-irradiated DNA acts as a substrate for the mammalian repair endonuclease, this was not surprising to us because we were the first to report that endonuclease purified from micrococcus luteus nicked x-irradiated DNA (14), a finding confirmed in Setlow's laboratory (15,16). It is, however, not known how the enzyme effects the x-irradiated DNA. It is likely that the nicking is aimed at base distortion, but the nature of the base alterations have not been identified.

However, the use of x-irradiated DNA as a substrate suggests that under some circumstances the endonuclease can lead to the formation of double strand breaks. When DNA is extracted from the liver of irradiated animals and centrifuged in alkaline and neutral sucrose gradients according to a scheme described in detail, the most critical observation is the appearance of a dose dependent peak at the low molecular weights in neutral sucrose gradients. Thus suggesting the formation of double strand breaks. For a number of reasons that have been discussed in a separate publication, it is unlikely that the appearance of double strand breaks is an artifact (17). In contrast as already mentioned when DNA is irradiated in vitro and placed on sucrose gradients, only single strand breaks are observed. However, if DNA x-irradiated in vitro is subjected to the action of the endonuclease and then placed on neutral sucrose gradient, a new peak suggesting development of double strand breaks appears. These findings can be interpreted in the following way: (Figure 6) x-irradiation causes the appearance of single strand breaks and base alterations. Both are potentially repairable. The single strand breaks by the action of the ligase, the base alteration by excision repair. If the base alteration is located on one strand, opposite to a single strand break on the other strand and the endonuclease reaches the altered base before the ligase has repaired the break on the opposite strand, then double strand breaks develop. Whether such double strand breaks can be repaired in mammalian cells remains to be seen. The hypothesis, recently proposed by Farber and Sarma (18) which postulate that the induction process in cancer, results from the formation of double strand breaks, with a high probability of error in repair, emphasizes the significance of the presence of double strand breaks.

Finally is the repair integral? This problem can be looked at in two different ways. The first concerns the restrictions imposed upon repair either by the nature of the substrate, by the presence of inhibitors, by the absence of enzyme or by the inaccessibility of the substrate. There are no different answers to any of these questions except for the fact that the total absence of the enzyme results in absence of repair, at least in most cases. But there is indirect evidence that repair is not uniform. In vivo experiments with regenerating liver have brought us to propose that after x-radiation all DNA is damaged, but only derepressed DNA is rapidly repaired while repressed DNA is not (19). Direct confirmation in in vitro systems of this hypothesis is not yet available, but there have been, however, repeated reports that damage of repair is not uniform throughout chromatin whether it be after the administration of carcinogens or ionizing radiation (20).

Conclusion

Clearly, damage to DNA has considerable consequences for the mammalian cell and although repair of DNA damage may not always be beneficial to the cell's economy, it occurs in mammalian cells as in bacteria. One potential enzymic mechanism for repair of mammalian DNA damage by UV light, x-radiation and some carcinogens has been described. It is likely that many others exist; their clarification will help to understand not only the pathogenesis of cancer, but possibly of other disease mechanisms.

References

1. Howard-Flanders, P., Br. med. Bull. 29 (1973) #3 226.
2. Cleaver, J.E., Nucleic Acid-Protein Interactions-Nucleic Acid Synthesis in Viral Infection (1971) p 87, Amsterdam, North-Holland Publ. Co.
3. Steward, B.W. and Farber, E. Cancer Res. 33 (1973) 3209.
4. Damjanov, I., Cox, R., Sarma, D.S.R. and Farber, E. Cancer Res. 33 (1973) 2122.
5. Van Lancker, J.L. Proceedings of the 9th International Congress de Gastro-Enterologie, Regeneration Hepatique, Paris, France (1972) p 99.
6. Van Lancker, J.L. World Symposium on Chemical Carcinogens (1974) p. 427, Dekker, Baltimore, Md. in press.
7. Farber, E. Cancer Res. 33 (1973) 2537.
8. Miller, J.A. Cancer Res. 30 (1970) 559.
9. Heidelberger, C. Fed. Proc. 32 (1973) 2154.
10. Cleaver, J.E. Proc. Natl. Acad. Sci. 63 (1969) 428.
11. Kaplan, J.C., Kushner, S.R. and Grossman, L. Proc. Nat. Acad. Sci. USA 63 (1969) 144.

12. Van Lancker, J.L. and Tomura, T. Biochim. et Biophys. Acta, in press.
13. Verly, W.S. and Paquette, Y. Can. J. Biochem. 51 (1973) 1003.
14. Van Lancker, J.L. Proc. 11th Annual Meeting Amer. Soc. Cell Biol. (1971) 311.
15. Paterson, M.C. and Setlow, R.B. Proc. Natl. Acad. Sci. 69 (1972) 2927.
16. Setlow, R.B. and Carrier, W.L. Nature New Biol. 241 (1973) 170.
17. Van Lancker, J.L. and Tomura, T. Cancer Res. 34 (1974) 699.
18. Farber, E. and Sarma, D.S.R. Control Processes in Neoplasia (1974) 173, Academic Press, Inc.
19. Collins, T., David, F. and Van Lancker, J.L. Fed. Proc. 31 (1972) 641.
20. Natarajan, A.T. and Schmid, W. Chromosoma 33 (1971) 48.

CERTAIN ASPECTS OF CHEMICAL CARCINOGENESIS IN VITRO USING ADULT RAT LIVER CELLS

P.T. Iype, T.D. Allen and D.J. Pillinger

While animal tumours induced in vivo are predominantly carcinomas and are tissue-specific, the tumours produced by injection of chemically transformed cells in vitro are mainly sacromas. This is due to the widespread use of fibroblasts or mixed cultures as the material for in vitro transformation studies. A number of investigators have developed epithelial cell systems, especially rat liver cells [1-7] to study malignant transformation in vitro. In some of these studies the untreated control cells underwent spontaneous transformation and produced tumours in vivo. In others, tumours induced by the chemically transformed cells were of mixed type i.e. fibrosarcoma as well as carcinoma.

Our studies have been directed to find out whether malignant transformation can be induced in well-characterised parenchymal liver cells by chemical carcinogens. We have selected rat liver cells as the material for investigation because hepatic cells are involved in the metabolism of various distal chemical carcinogens and since considerable information is already available on the changes in rat liver during chemical carcinogenesis in vivo. The cells from adult animals are recognised to be less susceptible to spontaneous transformation, hence the selection of adult animals.

There are at present a number of methods available for the isolation of cells from intact liver. We have used the collagenase and hyaluronidase perfusion method[8] with a few modifications for the initial isolation of cells. Rather than attempting to prepare a large quantity of cells as is generally the case where immediate biochemical studies are to be made, we have isolated relatively few viable cells with emphasis on minimal damage since these cells were to be grown in culture. Cell suspensions were prepared from 6-8 weeks old Wistar rats and plated in Ham's F10 medium supplemented with 10% foetal bovine serum. The plated cells were polygonal in shape, highly granular and many were binucleate. Cells with very few granules were also observed after a few days in the primary culture. Cells from areas in the dishes which were rich in granular cells, were preferentially isolated, pooled and cultured. Binucleate cells were not seen in any significant level after a few days in culture. These cells were epithelial in their morphology and exhibited con-

tact inhibition of growth and maintained these characteristics even after a number of months in continuous culture.
Fig. 1 shows some of the light microscopic pictures of the cultured cells. Figs. 1a and 1b are stained preparations of the liver cells. Figs. 1c and 1d are phase contrast micrographs of normal hepatic and hepatoma cells respectively. The epithelial nature of the non-malignant cultured liver cell is clearly seen. The nucleocytoplasmic ratic in the tumour cell is high compared to that of the non-malignant cells. Scanning electron micrographs (Iype and Murphy, unpublished) revealed a smooth cell surface in normal cells, whereas the membrane of the hepatoma cells is characterized by a much rougher surface presumably due to a higher membrane activity. These observations have been confirmed by transmission electron microscopy of cells prepared in situ and sectioned either vertical to, or parallel with, the plastic surface. The normal cell is extremely flattened and well-spread being a maximum of 5μ in thickness and a diameter of up to 50μ. Conversely the tumour cell is rounded (diameter 15-20μ) and has only a small surface in contact with the substratum. In parallel sections of normal cells (Fig. 2) the main features are closely packed well-spread cells with an overall polygonal outline and slight irregularities on the plasma membrane which is also characterized by desmosomal connections to adjacent cells. These connections can be visualised with both cell surface replicas (Fig. 3a) and in sections to consist of closely apposed membranes with electron dense thickening and intracellular fibrils (Fig. 3b). The nucleus has a smooth circular outline with little peripheral condensed chromatin and 2 or 3 characteristic nucleoli.
Fig. 4a is a vertical section through the nuclear region and shows a smooth flattened profile with a prominent nucleolus. The cytoplasm in normal cells is characterized by numerous mitochondria and considerable areas of endoplasmic reticulum which bears many ribosomal rosettes in surface profile (Fig. 4b). The occurrence of large amounts of rough endoplasmic reticulum is consistant with the protein synthesizing capacity of these cells. Microfibrils are also prominent in the cytoplasm of normal cells distributed with a regular orientation in the cytoplasmic periphery close to the plasma membrane (Figs. 3a, b). Vertical sections of the hepatoma cells (Fig. 5a) show a rounded profile as opposed to the flattened nature of the normal cell. There is also a very irregular nuclear profile and increased nucleo-cytoplasmic ratio in the hepatoma cells. A parallel section (Fig. 5b) shows, in surface view, the irregular nature of the nuclear profile and increased nucleo-cytoplasmic ratio in the hepatoma

CHEMICAL CARCINOGENESIS IN VITRO

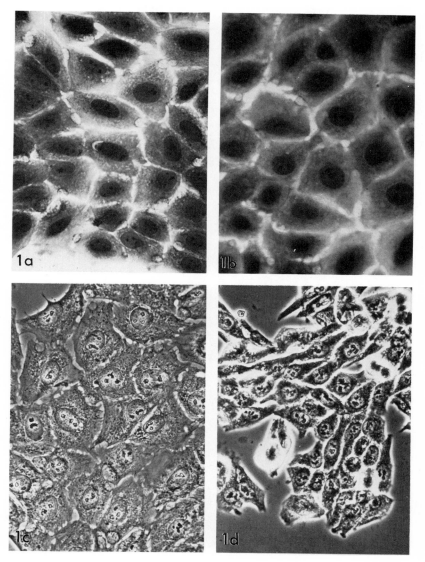

FIG. 1a & b: <u>Normal rat hepatic cells grown in vitro, fixed, and and stained with haematoxylin and eosin (bright field illumination, x 2,000)</u>. FIG. 1c: <u>Normal rat hepatic cells grown in vitro (phase contrast, x 2,000)</u>. FIG. 1d: <u>A rat derived hepatoma cell line grown in vitro</u>. These cells are considerably smaller than their normal counterparts in FIG. 1a,b, and c. (phase contrast, x 2,000).

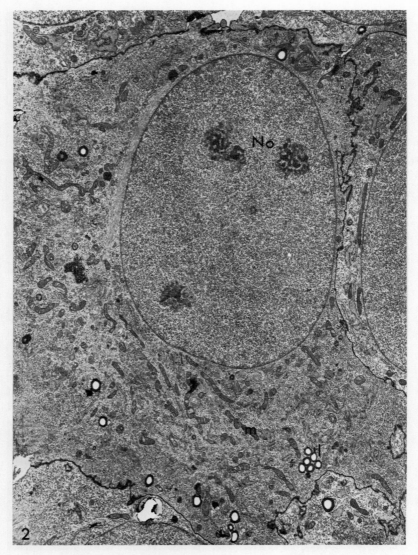

FIG. 2: Low power micrograph of a section through normal liver cell cut parallel to the substratum. The cells are tightly packed, with slight irregularities in the plasma membrane. The nucleus has a smooth circular outline with 3 prominent nucleoli (No). The cytoplasm contains numerous mitochondria (m) and a few lipid (l) droplets (x 5,500).

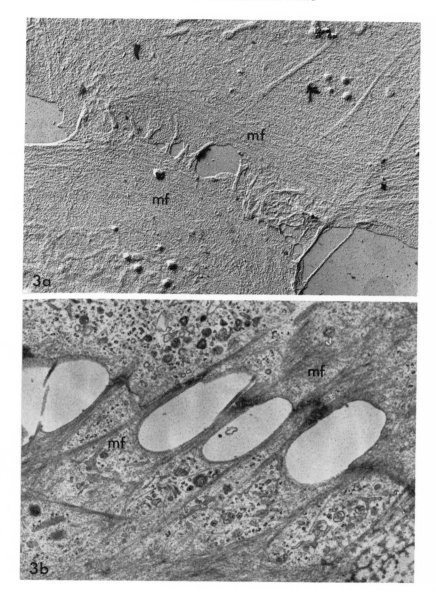

FIG. 3a: <u>Cell surface replica of adjoining region of liver cells in culture</u>. The membrane from the adjacent cells are in close association. Orientated sub membrane microfilaments (mf) are seen at the periphery of the cells. (x 6,500). FIG. 3b: <u>Parallel section through a similar region as in Fig. 3a</u>. Well defined desmosomal contact, membrane thickening and microfilaments (mf) can be seen. (x 17,200).

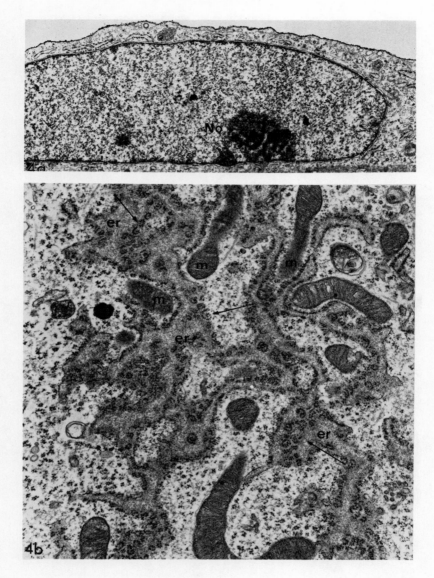

FIG. 4a: <u>Vertical section through a cell cut perpendicular to the substratum.</u> The flattened nuclear profile is visible with a nucleolus (No) in contact with the nuclear membrane (x 9,450). FIG. 4b: <u>Parallel section through the cytoplasmic region of normal liver cell,</u> showing several areas of endoplasmic reticulum (er) bearing numerous ribosomes and closely surrounding many mitochondria (m). (x 21,000).

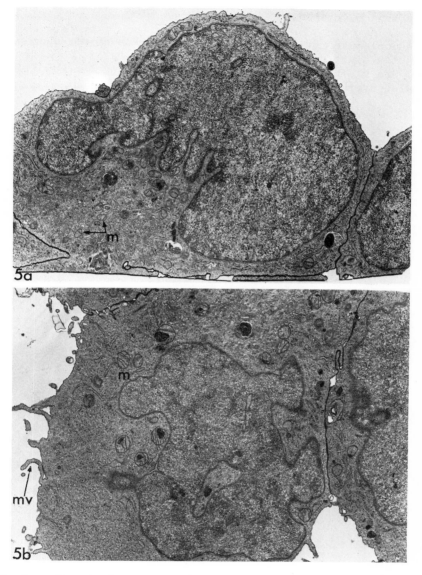

FIG. 5a: <u>Vertical section through cultured hepatoma cells shows the rounded profile of these cells with minimal attachment to the substratum.</u> The nuclear profile is highly irregular and there is an increased nucleo-cytoplasmic ratio compared to the normal cells. Mitochondria (m). (x 12,750).
FIG. 5b: <u>Parallel section through an area of cultured hepatoma cells.</u> The cytoplasm of these cells show a reduced amount

FIG. 5b: (Continued)

of endoplasmic reticulum and reduced numbers of mitochondria (m) with electron - dense inclusions. Several microvilli (mV) are visible at the cell membrane. The nuclear profile is also extremely irregular. (x 11,500).

cells. A parallel section (Fig. 5b) shows, in surface view, the irregular nature of the nuclear profile and also many microvilli on the cell surface. The cytoplasm contains sparse endoplasmic reticulum compared to normal cells and fewer mitochondria which contain dense inclusions. Microfilaments are also visible in hepatoma cells but in fewer numbers, lacking orientation and distributed irregularly through the entire cytoplasm. The ultrastructural characteristics, and especially the occurrence of desmosomes, emphasises the epithelial nature of our cell-lines. Although these cultured epithelial cells qualify as ideal material for carcinogenesis studies in vitro, we have characterised them antigenically and biochemically so as to validate their use as liver cells. Using membrane immunofluorescence techniques, we have studied the antigenic profile of freshly isolated as well as cultured rat liver cells. The occurrence of organ-specific antigens on the isolated cells and the retention of these antigens on the cells cultured for 6 months have been demonstrated[9]. These results indicated that the cultured cells are of hepatic origin. No embryonic antigens were detected on the adult liver cell lines[5]. The aminoazo dye-binding h protein has been detected on the cell-surface[9] at a level similar to that found in freshly isolated liver cells. Synthesis of serum albumin is one of the characteristic biochemical properties of intact liver[10]. Production of albumin has previously been reported in cell-cultures derived from liver[11,12]. In contrast, cell-lines from non-liver sources including cultured fibroblasts do not have the capacity to synthesise albumin [11,13]. Single cell-suspensions from two of the normal cell-lines were incubated in HEPES buffered Hanks balanced salt solution containing C^{14} -labelled Chlorella protein hydrolysate (Radiochemical Centre, Amersham; Specific activity 57mC/mA; final concentration 0.33μC/ml) for 5 hours at 37°C. When the total soluble proteins were chromatographed on Sephadex G-200 column 3 main optical density peaks were observed (Fig. 6). From a control run using rat serum, it was found that peak 2 represented the position where albumin was eluted. When the pooled proteins from each group were dialysed and lyophilised, the fraction obtained from peak 2 showed the highest specific activity. Moreover, the electrophoretic (polyacrylamide gel) mobility of the fraction from peak 2 was found to be similar to that of rat serum albumin (Fig. 6).

The inducibility of tyrosine aminotransferase (TAT) is a specific differentiated function[14] and has been shown in adult and foetal rat liver in vivo using various inducing agents[15]. Induction of TAT has also been shown in cell cultures[16] and studied in detail in hepatoma cell cultures (inter alia, 14,

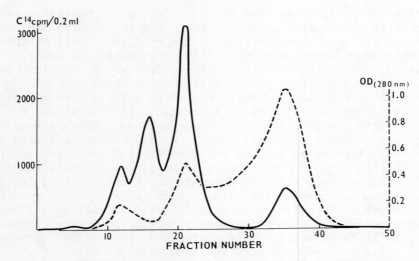

FIG. 6: Elution profile of soluble proteins from rat liver cells fractionated on Sephadex G-200 column showing optical density at 280 nm (----) and distribution of radioactivity (——) with respect to fraction number.

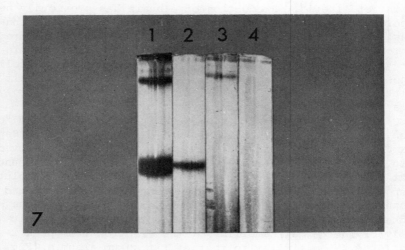

FIG. 7: Polyacrylamide gel separation of rat serum and also proteins corresponding to optical density peaks in FIG.1b. Gel.1. Rat serum; Gel.2. Peak 2; Gel.3. Peak 1; Gel.4. Peak 3.

15,17). Confluent monoalyers of the cell-lines were treated with dexamethasone phosphate or hydrocortisone 21-sodium succinate at a final concentration of 10^{-5}M for 24 hours. In some cases, insulin or L-leucine was added with or without the steroid inducers. The method used for TAT assay was a modification of the method of Diamondstone[18] and the results are given in Table 1. There was a 3-fold increase in specific activity when freshly isolated parenchymal cells were used. However, when these cells were incubated for 24 hours, there was a considerable decrease in the enzyme activity although treatment with dexamethasone phosphate or hydrocortisone hemisuccinate increased the enzyme activity 8 to 9-fold. The TAT activity of our established cell-lines was found to be 2.3 - 4.3 units in the various media used. These values were comparable to that of a hepatoma cell line (HTC).

The 3-fold increase in the specific activity of TAT we have observed in the isolated cells compared to liver tissue may have been caused by the removal of all the connective tissues and non-parenchymal cells which form a considerable part of the liver tissue, thereby concentrating the TAT-containing parenchymal cells. Induction of TAT has been shown by Haung and Ebner[19] in freshly isolated rat liver cells. We have confirmed this property of freshly isolated rat liver cells using dexamethasone phosphate or hydrocortisone hemisuccinate. There was no induction of TAT by steroid inducers in our cell-lines whereas in HTC there was a 7-fold induction. This could be attributed to the availability of specific glucocorticoid receptors which have been detected in rat hepatoma cell cultures[17]. Although our cell-lines have been shown to contain liver specific cell surface antigen[9], ultrastructurally the cell surface differs from that of hepatoma cells (Figs. 2-5). Since our cell-lines attach firmly to the culture dishes and exhibit a higher degree of contact inhibition, and have different surface properties, the steroids may not reach the receptor-molecules. The induction of TAT in freshly isolated normal liver cells would suggest that until the cells are attached to the substratum, the binding sites for steroids may be accessible. This reasoning is supported by the fact that hepatoma cells do not attach firmly to the substratum. It has been shown that for cultured hepatoma cells which have the necessary receptors for steroids, the TAT activity can be modified by the amount of leucine or the presence of insulin in the culture medium. Addition of extra leucine or insulin to Ham's F10 (which has a low concentration of leucine and a higher concentration of glutamic acid which could act as a product inhibitor for TAT reaction) did not induce TAT activity, further suggesting

Table I. The Effect of Inducers On the Tyrosine Aminotransferase Activity* of Hepatic and Hepatoma Cells

Sample	Control	Hydrocortisone 21-sodium Succinate 10^{-5}M	Dexamethasone phosphate 10^{-5}M
Liver	5.13	–	–
Primary cells (freshly prepared)	17.15	–	–
Primary cells (incubated for 24 hrs.)	5.91	46.0	40.0
Liver cell-lines	3.99	2.99	–
	2.30	–	1.7
	4.04	6.32	–
Liver cell-lines+leucine	3.15	1.52	–
Liver cell-lines+insulin	4.31	1.64	–
HTC	3.0	21.2	–

* Specific activity is defined as mu moles of product formed per minute per mg of protein.

that steroid receptors may not be accessible in our cell-lines from normal rats. It should be noted that apart from freshly isolated liver cells (which were unattached), and in one case where the cultured liver cells were aneuploid[16] and probably malignant[20] it was not possible to induce TAT in "normal" liver cells in vitro. The inducibility of TAT could be regulated by changes in the cell surface and is therefore manifested in vitro systems only after such changes as in hepatoma cell cultures. We are at present testing this hypothesis by experimentally modifying the cell-surface of the cultured liver cells (by trypsin or colcimide) and studying the inducibility of TAT by steroids.

These rat liver parenchymal cell-lines do not show any of the known properties of transformed cells and do not produce tumours when injected into syngeneic rats. The effects of the chemical carcinogens N-methyl-N-nitrosourea (MNU) or aflatoxin B1 were examined. We have used MNU, because of its direct action on cells and since it is known to produce fibroblast transformation in vitro[21]. No significant morphological changes were seen at the light microscope level, after treatment with a non-toxic dose (50-100µg/ml) but usually a large number of cells in mitosis was present after 24 hours. Piled up colonies were not formed. Forty to fifty days after the brief exposure of MNU (2 hrs.) some of the dishes showed foci of highly refractile cells, which were later found to have the capacity to grow in soft agar. Control cells of corresponding age did not produce soft agar colonies, however, after 62 weeks of continuous culture, one of the cell-lines did grow in soft agar. Early sub-cultures of cells stored in liquid nitrogen were used in subsequent experiments.

Aflatoxin at a final concentration of 0.5µg/ml produced toxic effects in the cells. Three weeks after this treatment areas in the culture dishes contained multinucleated cells. Cells preferentially isolated from the refractile foci, formed soft agar colonies. However, when very early sub-cultures were used the carcinogen-treatment did not induce soft agar colony forming capacity within 6 weeks.

We have also tested the antigenic changes induced in the carcinogen-treated cell-lines[5]. MNU-treated cell-lines elicited antibody production (as determined by membrane immunofluorescence tests in vitro) when injected into a highly inbred strain of rats from which the liver cells were originally isolated. These antisera were used to test for the presence of new surface antigens on the carcinogen-treated cells; they not only reacted with the immunising cells but also with some of the cells treated with other chemical carcinogens. The treated cells were injected into rats after various periods in culture. The intervals between the carcinogen-treatment and intravenous injection varied from 6-11 weeks. Lung nodules were found 1-2 months after the injection and the tumours were identified histologically as undifferentiated carcinomas. Carcinoma induction has been shown in the neonatal animal by other investigators[6,7] and a study of this kind is being made with our cell-lines.

Although we have demonstrated chemical carcinogenesis in vitro, a number of points have to be studied. These include the effect of carcinogens in vitro on dividing and non-dividing liver cells (certain chemical carcinogens differ in their tumour-producing capacity in normal and regenerating liver[22]; also transformation rates in a cell-system can differ during the cell-cycle[23]) and synergistic effects due to the environmental pH. An acid pH of the culture medium has been shown to produce "transformation"[12] and an alkaline pH results in an increase in cell-fusion[24] which is also implicated in carcinogenesis[25]. We are experimenting on these problems using aflatoxin B1 on rat liver cells.

Some of the characteristics of tumour cells can be traced to their modified cell surface. Most of the criteria used in the assessment of transformation of fibroblasts in vitro are manifestations of cell-membrane modifications induced by the carcinogens. These criteria of fibroblastic transformation may not occur during epithelial cell transformation because of the difference in the morphology and functions of these cells. We have therefore studied some of the properties of normal liver cells and in vivo induced rat hepatomas. Since it is known that Cytochalasin B exerts a differential effect on normal and virally transformed fibroblasts[26,27], we have checked it in our system. In the normal hepatic cells, as in the normal fibroblasts, only bi-nucleated cells were produced in response to Cytochalasin B whereas in hepatoma cells a number of nuclei were seen. Using this procedure, many of the carcinogen-treated cell-lines are now being studied at intervals after treatment. Other changes during chemically induced transformation in vitro being studied include ultrastructural changes of the plasma membrame and biochemical changes mediated through the cell-surface.

This work was supported by grants from the Cancer Research Campaign and the Medical Research Council.

References

1. Katsuta, H. and Takaoka, T. Japan J. Exp. Med. 35 (1965) 231.
2. Sato, J. and Yabe, T. Japan J. Exp. Med. 35 (1965) 445.
3. Namba, M., Masuki, H. and Sato, J. Japan J. Exp. Med. 39 (1969) 253.
4. Toyoshima, K., Niasa, Y., Ito, N. and Tsubura, Y. Gann 61 (1970) 557.
5. Iype, P.T., Baldwin, R.W. and Glaves, D. Br. J. Cancer 27 (1973) 128.
6. Williams, G.M., Elliot, J.M. and Weisburger, J.H. Cancer Res. 33 (1973) 6060.
7. Montesano, R., Saint-Vincent, L. and Tomatis, L. Br. J. Cancer 28 (1973) 215.
8. Howard, R.B., Christenson, A.K., Gibbs, F.A. and Pesh, L.A. J. Cell Biol. 35 (1967) 675.
9. Iype, P.T., Baldwin, R.W. and Glaves, D. Br. J. Cancer 26 (1972) 6.
10. Peters, T. J. Biol. Chem. 237 (1962) 1181.
11. Kaighn, M.E. and Prince, A.M. Proc. Natl. Acad. Sci. U.S. 68 (1971) 2396.
12. Borek, C. Proc. Natl. Acad. Sci. U.S. 69 (1972) 956.
13. Halpern, M. and Rubin, H. Exptl. Cell Res. 60 (1970) 86.
14. Lee, K.L. and Kenney, F.T. J. Biol. Chem. 246 (1971) 7595.
15. Butcher, F.R., Becker, J.E. and Potter, V.R. Exptl. Cell Res. 66 (1971) 321.
16. Gerschenson, L.E., Andersson, M., Molson, T. and Okigaki, T. Science 170 (1970) 859.
17. Baxter, J.D. and Tomkins, G.M. Proc. Natl. Acad. Sci. U.S. 65 (1970) 709.
18. Diamondstone, T.I. Analyt. Biochem. 16 (1966) 395.
19. Haung, Y.L. and Ebner, K.E. Biochem. Biophys. Acta. 191 (1969) 161.
20. Oshiro, Y., Gerschenson, L.E. and DiPaolo, J.A. Cancer Res. 32 (1972) 877.
21. Frei, J.V. and Oliver, J. J. Natl. Cancer Inst. 47 (1971) 857.
22. Craddock, V.M. J. Natl. Cancer Inst. 47 (1971) 889.
23. Bertram, J.S. and Heidelberger, C. Cancer Res. 34 (1974) 526.
24. Crose, C.M., Koprowski, H. and Eagle, H. Proc. Natl. Acad. Sci. U.S. 69 (1972) 1953.

25. Elkort, R.J., Handler, A.H., Kibrick, S. and Kleinman, L. Eur. J. Cancer 8 (1972) 259.
26. Wright, W.E. and Hayflick, L. Exptl. Cell Res. 74 (1972) 187.
27. Kelly, F. and Sambrook, J. Nature New Biol. 242 (1973) 217.

MECHANISMS OF CHEMICAL CARCINOGENESIS ANALYZED IN RAT LIVER
AND HEPATOMA CELL CULTURES

*I. Bernard Weinstein, Nobuo Yamaguchi, Jan Marc Orenstein,
Ronald Gebert and M. Edward Kaighn*

Introduction

    Because of the importance of chemical carcinogens in the causation of human cancer (1), it is desirable to develop a variety of tissue culture systems which can be employed either for the assay of potential chemical carcinogens in the human environment or to facilitate studies on the mechanism of cell transformation by chemicals. Since the report by Berward and Sachs (2) on transformation of hamster embryo cultures by polycyclic hydrocarbons, a number of investigators have reported transformation of cells in culture by various chemical carcinogens [for a review of this subject see DiPaolo et al. (3)]. Most of these studies have employed either embryo cultures of mixed morphology or fibroblastic cell lines. Since the majority of human tumors are of epithelial origin, it is important to also develop epithelial cell culture systems.

    For these reasons we have established and extensively characterized several epithelial cultures which we believe will be useful for analyzing the mechanism of action of chemical carcinogens. In these studies we have concentrated on cultures established from normal rat liver, and from hepatomas induced in the rat with aromatic amine carcinogens, for the following reasons: 1) the metabolism and cellular interactions of the hepatic carcinogen N-2-acetylaminofluorene have been elucidated in greater detail than with any other carcinogen (1,4,5); 2) normal fetal and adult rat liver and a spectrum of rat hepatomas have been extensively characterized biochemically (6,7); 3) it is possible to readily obtain cloned epithelial cultures from both normal adult rat liver (8-12) and from chemically induced hepatomas (13-16). (Also see related papers in this Symposium).

The abbreviations used are: BudR, 5-bromo-2-deoxyuridine; DMSO, dimethyl sulfoxide; TS, temperature sensitive.

MORPHOLOGY AND GROWTH PROPERTIES OF LIVER AND HEPATOMA EPITHELIAL CELLS

Cultures from normal rat liver (Table 1) had extremely flat cells with a cuboidal or polygonal shape. They formed mosaic-like colonies with sharply circumscribed borders and grew into a monolayer of tightly packed but non-overlapping cells (17). Phase microscopy revealed that most cells contained 2-5 prominent nucleoli per cell. There were occasional binucleate cells and multinucleate giant cells. The cytoplasm frequently contained numerous granules which gave a positive stain for lipid with oil-red-O. These are probably related to the prominent autophagic vacuoles seen in the cytoplasm of these cells by electron microscopy.

In the confluent monolayer the cells were frequently arranged in a cord-like pattern. When scraped from the growth surface the cells tended to come off the plate in cord-like clusters which were maintained even after centrifugation. These cords are reminiscent of the arrangement of liver parenchymal cells in vivo.

Time lapse microcinematography studies of the normal liver cell cultures (done in collaboration with Dr. Peter Riddle of the Imperial Cancer Research Fund, London) indicated that these cells, in contrast with fibroblasts, do not have locomotion. They are extremely adhesive to each other and, after cell division, daughter cells remain immediately adjacent to each other and the colony edge. The latter properties explain the tendency of clones in sparse cultures to form islands with extremely smooth borders (17). This is in contrast to fibroblast clones which have ragged borders because daughter cells tend to migrate away from the clone. In subconfluent cultures, although cell division was most active at the peripheral edge of colonies, dividing cells were observed at the center of colonies and in cells that were in close contact on all sides with other cells. In the latter case dividing cells would rise up from the growth surface, transiently leaving an open space. After division the two daughter cells would squeeze into the mosaic, slightly displacing the adjacent cells, enlarging the size of the colony, and preserving the monolayer. These morphologic and growth characteristics are typical of epithelial cells (17,18) and clearly distinguish our cells from fibroblasts.

As measured by cloning efficiency, the optimum concentration of fetal calf serum was 3-5%. In contrast to most fibroblast cultures, higher concentrations of serum actually inhibited growth. A low serum dependence has been described for other epithelial cultures (19,20) and it is possible that

MECHANISMS OF CHEMICAL CARCINOGENESIS

	Cells	Origin	Rat Strain	Morphology	Growth in Agar	EM Evidence for Virus
Normal	K-9	Normal liver	Sprague-Dawley	Epithelial	−	−
	K-16	"	Sprague-Dawley	"	−	−
	K-22	"	Sprague-Dawley	"	−	−
	KF-7	"	Fischer 344	"	−	−
	KF-9	"	Fischer 344	"	−	−
Transformed In Vivo	B-1	Hepatoma 5123	Buffalo	Epithelial	+	+
	H-4	Reuber hepatoma	AXC	"	+	+
	W-14	Primary hepatoma	Fischer 344	"	+	+
	W-15	Primary hepatoma	Fischer 344	"	+	+
	HTC	Hepatoma 7288C	Buffalo	"	+	+
Spontaneously Transformed	K-63	Normal liver	Buffalo	Epithelial	+	+

TABLE 1: RAT LIVER AND HEPATOMA CELL CULTURES

this may be a general characteristic of rodent epithelial cells in cultures. Growth curves indicated minimum doubling times, during the exponential phase, of 18-24 hours and a saturation density of approximately $1-2 \times 10^5$ cells/cm^2. None of the normal cultures grew in 0.34% soft agar in contrast to the hepatoma cells (see below). Chromosome studies of the K-9 and K-16 cells, performed by Dr. O.J. Miller, indicated a grossly normal rat karyotype with a large number of tetraploid cells, but a careful quantitative analysis and chromosome banding studies have not been done.

Criteria for Transformation of Epithelial Cells. The morphologic criteria often used for assessing transformation of cells in culture by either viruses or chemicals include increased refractility, piling-up of cells (i.e., loss of a monolayer), a criss-cross arrangement of cells, and decreased serum dependence (20,21,37). In general, these criteria were established from studies on fibroblast cultures. As described above, the normal rat liver epithelial cells have growth properties which are quite different from those of fibroblasts. Loss of contact inhibition (contact inhibition of cell division) (20) is frequently emphasized as a criterion for transformation of fibroblast cultures. Clearly this criterion cannot be reliably applied to epithelial cultures since normal epithelial cells have little or no locomotion and continue to divide even when in close contact with other cells. It was therefore necessary to establish reliable criteria for assessing transformation of epithelial cultures.

Our approach was to study the growth properties of cell cultures derived from slowly or moderately rapidly growing rat hepatomas induced in vivo by aromatic amine carcinogens (Table 1) insisting that the cultures retain sufficient features to establish that the cells were of epithelial origin. We found that these cultures retained many of the properties described above for normal rat liver epithelial cells, particularly in confluent cultures. Sparse cultures of the hepatoma cells did reveal a tendency towards greater pleomorphism in size and shape, a greater number of elongated cells, and more ragged colony edges than the normal liver cell cultures. Therefore, these cultures have been designated "epithelioid" (Table 1). Although some cultures displayed a tendency to form patches of multiple layers of cells when confluent plates were refed, "piling up" was not a constant and reliable feature of the hepatoma lines. Morphologic criteria for distinguishing hepatoma cultures from normal rat liver cultures were, therefore, not completely objective or reliable. We found, however, that growth in soft agar suspension (22) does provide a simple, reproducible and objective criterion for transformation of our cells. When $10^4 - 10^5$ cells were placed in agar suspension none of the normal cultures derived from rat liver grew, whereas numerous colonies (greater than 10^2) were obtained with all of the hepatoma cultures (Table 1). As many as 10^6 normal cells plated in soft agar failed to give a single colony. Nor did growth of the normal cell line K-16 occur when the agar medium contained either fetal calf, chicken, dog, or monkey serum at concentrations ranging from 5-20%. It is of interest that although transformed fibroblasts usually give spherical colonies in soft agar, the clones obtained from certain cultures (W-14, W-15 and K-63) grew as a flat two-dimensional sheet, particularly when the clone was small.

Although derived from the liver of a normal rat, cell line K-63 also grew in soft agar. The culture had a more fibroblastic appearance than our other liver cultures, tended to pile-up in multiple layers, and was tumorigenic. We concluded, therefore, that this represents a case of spontaneous transformation during cell culture (see also 10 and 23). K-63, like the cell lines derived from hepatomas, also produces types A and C viral particles (see below).

In general, the tumor cell lines had somewhat shorter doubling times and higher saturation densities than did the normal cell cultures but these proved to be less reliable criteria than growth in soft agar.

Ultrastructure of the Normal and Hepatoma Cultures. Extensive cell-cell contact with interdigitation of plasma membrames and junctional complexes were common features in the ultrastructure of the normal cells. The cytoplasm contained varying amounts of smooth and rough endoplasmic reticulum with only an occasional Golgi apparatus. Pinocytosis was also a common feature of these cells. Autophagic vacuoles containing myelin figures and an impressively dilated rough endoplasmic reticulum were characteristics of these cells, particularly in confluent cultures. Glycogen granules were seen in the cytoplasm of KF-7 cells but were not found in the other normal cell cultures. In contrast to the hepatoma cultures (see below), extensive studies of the normal liver cultures failed to reveal evidence of viral particles (Table 1).

The hepatoma cells frequently had a greater diameter, more microvilli, and larger nuclei and nucleoli than the normal cells. The most prominent feature in the ultrastructure of all of the hepatoma cultures (Table 1) was the clear-cut evidence for production of types A and C RNA viruses. These findings are described in greater detail below.

Are the Normal Epithelial Cultures Derived From Liver Parenchyma?

Several groups have sought methods for culturing normal adult liver cells which retain the in vivo properties of liver parenchyma (8-13, 16,23-26, and related papers in this Symposium). Published procedures have utilized a variety of cell dissociation techniques including: mechanical dissociation alone (10) or in association with the chelating agent sodium tetraphenyloboron (12); enzymatic disociation with either trypsin (11) alone or hyaluronidase plus collagenase (26), or trypsin plus collagenase (8,9,16,25).
In our experience epithelial cultures can be reproducibly obtained from rat liver by: 1) utilizing procedures which employ collagenase and avoid excessive trypsinization of the tissue, 2) plating the primary culture in a low concentration of fetal calf serum (3-5%), since this favors the growth of epithelial cells over fibroblasts, and 3) plating the cells at a low density so that epithelial clones which emerge can be picked up before they are overgrown by adjacent fibroblasts. The details of this method have been published by Kaighn (9).
Although the cultures we have obtained from normal rat liver clearly have an epithelial morphology, it would be desirable to establish whether or not they are derived from liver parenchymal cells. Since the hydrocortisone inducible enzyme tyrosine aminotransferase (TAT) is characteristic of adult liver, we have assayed this enzyme in extracts obtained from our various cell cultures (Table 2). An extract of normal rat liver and of the HTC line (kindly provided by Dr. Brad Thompson) were used as positive controls and two non-liver cell lines, NRK and 3T3, were used as negative controls. It is apparent that even after exposing the culture to dexamethasone (1×10^{-5} to 1×10^{-6} M for 18 hours either in the presence or absence of fetal calf serum) the normal cell cultures were negative for this enzyme. Two hepatoma cell lines, B-1 and W-15, were also negative. On the other hand, when grown under identical conditions, the hepatoma cell lines H-4 and HTC had high levels of inducible TAT. An in vitro chemically transformed rat liver cell line, W8, and the cell lines 3A and 5E transformed by the murine sarcoma virus (17) also had negligible enzyme levels (Table 2).
A number of investigators have established epithelial cell

	Enzyme Level (Δ A331/15 min/mg protein)	
Cells	Dexamethasone	
	(−)	(+)
Liver	10.23	−
K-16	0.17	0.46
K-22	−	0.49
KF-7	0.24	0.16
KF-9	0.18	0.21
B-1	−	0.44
W-15	−	0.44
H-4	−	29.60
HTC	3.38	18.10
W-8	0.32	0.18
3A	0.33	0.23
5E	0.31	0.28
NRK	−	0.33
3T3	−	0.50

Cultures were grown in the absence or in the presence of $10^{-6}M$ dexamethasone for 18 hours and were assayed in duplicate for tyrosine aminotransferase by a modification of the method of Diamondstone (27).

TABLE 2: ASSAYS OF TYROSINE AMINOTRANSFERASE

cultures from rat hepatomas which synthesize albumin or other serum proteins or have hydrocortisone inducible TAT (13,14, 28-30). On the other hand, there is no well-documented publication that we are aware of which gives evidence that non-tumorigenic epithelial cell lines from normal adult rat liver perform similar liver specific functions. Morphology itself is not reliable since our normal epithelial cells that lack TAT, the hepatoma cells which also lack TAT, and the hepatoma cells which have high levels of TAT are morphologically similar. In addition, epithelial cells can undergo marked changes in morphology following changes in the growth medium, cell density, serial passage or virus transformation (17). Kaighn and Prince (25) have described the paradox presented by certain human liver cell lines which have a fibroblastic appearance yet synthesize human serum albumin.

In view of these difficulties we cannot say with certainty whether the epithelial cultures of normal rat liver we and other groups have been studying were truly derived from liver parenchymal cells. Their morphology is consistent with this. It is our working hypothesis that this is the case and that the cells do not synthesize certain liver specific proteins due to deficiencies in the growth medium, the selection of parenchymal cell variants capable of rapid growth in vitro, and/or partial reversion to a fetal pattern of gene expression. Consistent with the latter possibility are studies indicating that liver cell cultures similar to ours have certain fetal rather than adult liver isozymes (31).

Possible Mechanisms of Chemical Carcinogenesis

Whether or not the normal cell lines described in this paper are of parenchymal origin, they do provide highly useful epithelial cultures for analyzing the mechanisms by which chemical carcinogens bring about cellular transformation. Four current major theories of chemical carcinogenesis are listed in Table 3. These include: 1) Somatic Mutation; 2) Cell Selection; 3) Viral Activation, and 4) Aberrant Differentiation. Thus far, studies in the intact animal have not provided conclusive evidence for or against any one of these mechanisms. I would now like to illustrate how cell culture systems may provide more direct answers to these questions.

The somatic mutation theory is the most popular explanation for chemical carcinogenesis. I must emphasize, however, that this theory, although it has the advantage of simplicity, remains unproven. Indeed, the relatively high efficiency of

1. Somatic Mutation: change in cell DNA

2. Cell Selection: including suppression of immune surveillance

3. Virus Activation or Enhancement

4. Aberrant Differentiation: epigenetic

TABLE 3: POSSIBLE MECHANISMS OF CHEMICAL (AND RADIATION) CARCINOGENESIS

1. Causative. The carcinogen activates latent RNA viruses and the virus is then responsible for transforming cells.

2. Symptomatic. A major effect of the carcinogen is to cause aberrations in the control of gene expression in cells. As part of this aberration there is a de-repression of latent viral genes. The synthesis of these viruses, however, does not contribute to the malignancy of the tumor cell. In this sense the synthesis of viruses by tumors may have no more significance than the frequent expression of fetal genes in tumors.

3. Cofactor. Latent viruses are de-repressed for reasons mentioned in 2, and the virus then contributes some but not all of the information required for cell transformation.

TABLE 4: BIOLOGIC ROLE OF RNA VIRUSES IN CHEMICAL CARCINOGENESIS

transformation (10%) obtained with chemical carcinogens in certain cell culture systems (2,3,32), when compared to the very low frequency of specific induced mutations about 10^{-4}) in mammalian cell culture systems (33,34) speaks strongly against conventional random mutation as the basis of chemical carcinogenesis.

Data obtained with cell culture systems also indicate that carcinogens can directly transform normal cells into tumor cells, thus providing evidence against the theory that carcinogens act only via cell selection or enhancement of the growth of pre-existing or dormant tumor cells (2,3,32).

The third possibility listed in Table 3 has to do with viruses, and I would now like to discuss this aspect in some detail.

Discovery of RNA Viruses in Hepatoma Cultures

There is currently considerable interest in the possibility that chemicals and radiation induce cancer via the activation of latent oncogenic viruses or enhancement of the transforming activity of exogenous viruses (35). We previously reported the detection of A and C type RNA viruses in cell cultures established from rat hepatomas induced with aromatic amine carcinogen, and the effects of BudR and DMSO on the synthesis and maturation of these viral particles (16,36).

We have now studied five independently isolated rat hepatoma cell lines, and all are positive for RNA virus production both by the ^{3}H-uridine labeling technique (16) and by electron microscopy Table 1. It is of interest that a spontaneous transformant (K-63) obtained from a normal adult rat liver is also positive for virus production. Previous investigators have also described spontaneous in vitro transformation of rat liver epithelial cells in culture (10, 38) but, to our knowledge, these were not scored for virus production. Five independently isolated normal epithelial cell lines established from adult rat liver have thus far been negative for evidence of production of similar viruses. By morphology, the majority of the particles seen in the hepatoma cultures are enveloped A and intracisternal A particles, and true C particles are less common. Intracytoplasmic A particles and B particles, typical of the mouse mammary tumor system, were never observed in our normal or hepatoma cell cultures.

It was of interest to look for similar particles in hepatomas grown in the rat. Thus far we have been unable to detect by electron microscopy viral particles in tissue preparations obtained from Morris hepatomas 5123C and 7288C grown in the rat. On the other hand, viral particles were readily detected by ^{3}H-uridine labeling and electron microscopy in the HTC cell line which was originally established from hepatoma 7288C. Electron micrographs of HTC cells revealed numerous C particles, budding particles, enveloped A particles, and a few intracisternal A particles. It is of interest that tumor tissue taken from the hepatoma cell line 7288 CTC, which was established by injecting the HTC cell line back into Buffalo rats (H. Morris, personal communication), also revealed similar viral particles by electron microscopy. It appears, therefore, that growth of chemically induced hepatoma cells in cell culture frequently induces the production of latent RNA viral particles, and that the synthesis of these particles is not necessarily suppressed when the cells are passaged back into the rat.

What is the significance of the presence of viral particles in our hepatoma cultures? Thus far we have no evidence that these particles can replicate in normal cells, transform cells in culture, or induce tumors in normal animals. Intracisternal A particles and enveloped A particles have been previously described in a variety of spontaneous and chemically induced tumors but, to our knowledge, a biologic function for these particles has never been described.

We are quite certain that the viral particles we have detected are not due to laboratory contamination for the following reasons: 1) particles have been detected not only in cultures established from transplantable hepatomas but also from primary hepatomas (16); 2) parallel normal cultures established from adult rat liver and maintained in the same laboratory have been negative for similar particles; 3) attempts to intentionally infect normal cultures with particles obtained from hepatoma cells have been repeatedly negative (unpublished studies), and 4) other investigators have detected similar particles in hepatomas which were grown in vivo and never cultured in vitro (39-41). It is likely, therefore, that these particles and/or information related to their synthesis reside(s) in the chemically induced hepatomas and that their expression is favored by growth of the hepatoma cells in vitro. It is well known that cell culture frequently induces the synthesis of latent DNA and RNA tumor viruses (42,43).

Obviously, much more work must be done to determine the possible role of viral particles in hepatocarcinogenesis and chemical carcinogenesis in general. Several possibilities are listed in Table 4.

Although it is attractive to believe that RNA viruses may play a "causative" role in chemical carcinogenesis, we cannot at the present time rule out the possibility that they play a more passive role, perhaps due to de-repression of latent viral genomes. The latter process may simply be symptomatic of a more general aberration in the control of gene expression which appears to characterize neoplasia (6,7,44,45).

In nucleic acid hybridization studies utilizing a DNA probe made from the rat hepatoma associated virus, we plan in future studies to determine at what stage of hepatocarcinogenesis the expression of viral-related nucleic acids can be detected, and this may help to decide whether the virus expression is cause or effect.

Temperature Sensitive (TS) Mutants of Chemically Transformed Rat Liver Epithelial Cells

Mutants of chemically transformed cells which are temperature sensitive (TS) in the expression of transformation could afford a powerful tool for analyzing the mechanism of transformation, since one could at will turn off or on the transformed phenotype. A careful analysis of such mutants would also help to distinguish primary from secondary phenomena associated with the tumor phenotype. A few TS mutants of oncogenic viruses or of cells transformed by oncogenic viruses have been previously described in which maintenance of the transformed state is temperature sensitive (46-51). We have recently isolated five TS mutants of chemically transformed epithelial cells (52), and I would now like to briefly describe some of the properties of these mutants.

A chemically transformed cell line, W8, was obtained by treating K-16, a normal rat liver epithelial cell line, with a single exposure to N-acetoxy AAF. W8 cells, in contrast to K-16, readily grow in soft agar and are tumorigenic. A clonal isolate of W8 cells obtained by successive colony isolation in soft agar and in methocel at 41° was mutagenized with N-methyl-N-nitrosoguanidine, and cells which had a decreased capacity for growth in confluent cultures maintained at 41° were selected by FudR. Following this selection, 147 clones were picked and each one was scored for growth in agar at 36° and 41°. Five of these clones proved to be temperature sensitive in the sense that their cloning efficiency in agar at 41° was markedly reduced when compared to that at 36°.

	Growth in Agar[a]			Growth in Liquid Medium[b] Saturation Density
Cell Lines	Plating Efficiency (%)		Ratio (40°/36°)	Ratio (40°/36°)
	40°	36°		
K-16 (Normal)	0	0	--	0.96
W-8 (Transformed)	5.8	6.0	0.97	0.72
TS 223	0.03	8.6	0.003	0.10

[a] 10^3-10^5 cells were plated in agar at 36° and 40°C, and the number of colonies (8 or more cells per colony) counted at 14 days.

[b] Cells were grown in liquid medium at 36° or 40°C until confluent, and the total number of cells attached to the plate were counted.

TABLE 5. TEMPERATURE SENSITIVE (TS) MUTANT OF CHEMICALLY TRANSFORMED EPITHELIAL CELLS

The properties of one of these mutants, TS223, and that of the parental cells are described in Table 5. The normal K-16 cell line failed to grow in agar at either 40° or 36°, whereas the transformed wild type W8 cells grew in agar about equally well at both temperatures. On the other hand, TS223 had about a 300-fold reduction in plating efficiency in agar at 40° when compared to that obtained at 36°.

A temperature variation of 36° to 40°C had little effect on the saturation densities of K-16 or W8. On the other hand, the saturation density of TS223 at 40°C was only one-tenth that obtained at 36° (Table 5). Phase microscopy indicated that at 40° the plateau in saturation density occurred as soon as the culture had grown to a confluent monolayer. At 36° however, the number of cells attached to the plate continued to increase even after the culture was confluent. This phase of growth was associated with the appearance of patches of cells which were piled up into multiple layers. Scanning electron microscopy of these cells revealed very striking changes in cell shape and cell surface structure. At 40°C almost all of the TS223 cells were extremely flat and polygonal and had relatively few microvilli. At 36°, however, there were numerous patches of spherical piled-up cells and the surfaces of these cells were completely covered with microvilli. The normal K-16 cells grown at 36° or 40°C had the above described flat appearance, whereas the wild type transformed cells displayed the spherical type of cells at both 36° and 40°C.

Growth curves of TS223 at 36° and 40°C are given in Figure 1. Although there was little difference in the exponential growth rate at the two temperatures, there was almost a 10-fold difference in the final saturation density. The saturation densities remained at these plateaus even though fresh medium was added. If, however, the temperature of the 40° culture was then shifted to 36° after a lag of about one day, the number of cells began to increase until it reached the level of the culture originally grown at 36°. What is even more remarkable is that when the culture originally grown at 36° was shifted to 40°, the opposite effect was observed (Fig. 1.).

These results demonstrate that the temperature sensitive defect in this mutant is readily reversible by upward or downward shifts in temperature, as would be expected for a mutation that affects expression of the transformed state rather than cell replication per se or cell viability. Although the lesion in TS223 is most likely due to a single step mutation, the mutation appears to be highly pleiotropic in its effects since the temperature shift alters several

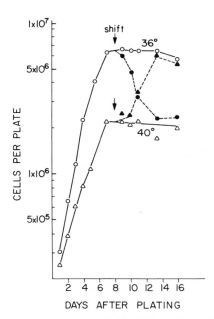

FIG. 1:

Growth curves of TS 223 at 36° (o―o) and 40 (△―△). At the time indicated by the arrows, one of a duplicate set of 36° cultures was shifted to 40° (●―●) and a 40° culture was shifted to 36° (▲―▲).

properties including ability to grow in agar, saturation density in liquid medium, cell shape and cell surface structure. We have found that like W8 cells TS223 is tumorigenic when injected into newborn rats, but we cannot determine whether it loses tumorigenicity at 40° since we cannot maintain rat body temperature at that level for the several weeks required for a tumorigenicity assay. Studies are in progress with this mutant in which we hope to elucidate the physiologic and biochemical events required for maintenance of the transformed state in chemically transformed cells.

The Theory of Aberrant Differentiation

The fourth possibility listed in Table 3, namely, that carcinogens act at the epigenetic level by producing aberrations in differentiation that do not involve conventional mutagenesis, although attractive conceptually, is a theory which is difficult to approach experimentally. In large measure this is due to the fact that at the present time, as biologists, we are fairly ignorant of the molecular basis of differentiation and its stability. In addition there is a paucity of experimental approaches and criteria which can be applied in cell culture to accurately distinguish between somatic mutation and an altered yet stable state of differentiation. This area is one of the most intriguing and challenging in contemporary cell biology, and it may well be that research on the fundamental mechanism by which chemical agents transform normal cells into tumor cells will provide a window into the entire area of molecular mechanisms of differentiation, both normal and aberrant.

Summary

Cell cultures have been establsihed from normal adult rat liver and rat hepatomas induced <u>in vivo</u> with aromatic amine carcinogens. The morphology and growth properties of these cultures indicate that they have characteristics of epithelial rather than fibroblast cells. Several criteria used to score for transformation of fibroblasts were not satisfactory for distinguishing normal epithelial cells from hepatoma cells in culture. Growth in agar, however, provides a simple and objective method of scoring for transformed epithelial cells, since only the tumorigenic cells grow in agar. Since none of the normal cultures had hydrocortisone inducible tyrosine aminotransferase, we lack definitive evidence that they are derived from liver parenchymal cells.

The outstanding feature in the ultrastructure of the hepatoma cells in culture was the presence of A and C type viral particles. Whereas five hepatoma cultures and a spontaneously transformed normal liver cell line were positive for these particles, both by electron microscopy and ^{3}H-uridine labeling, five independently isolated cell cultures from normal adult rat liver were negative for evidence of RNA virus production even when grown in the presence of BudR. The significance of these viral particles is not clear at the present time.

We have succeeded in isolating the first temperature sensitive mutants of chemically transformed epithelial cells. These have properties of tumor cells when grown in culture at 36° but adopt several properties of normal cells when grown at 40°C. These changes are reversible when the temperature is shifted either up or down. These mutants provide a valuable tool for analyzing biochemical properties specific to the normal or tumor phenotype in chemically tranformed cells.

These findings and data from the literature are discussed in terms of several current theories on the mechanism of chemical carcinogenesis.

The authors thank Miss Ulrike Stadler and James Chi for valuable technical assistance, and Moshe Rosen for his expert assistance with the electron microscope. We are indebted to Dr. Harold Morris for providing several transplantable rat hepatomas to us and for sharing with us his extensive information related to the properties of these tumors. We thank Dr. Brad Thompson for providing the HTC cell line.

References

1. Miller, J.A., Cancer Res. 30, (1970) 559.
2. Berwald, Y., and Sachs, L., J.Nat.Cancer Inst. 35, (1965) 641.
3. DiPaolo, J.A., Takano, K., and Popescu, N., Cancer Res. 32, (1972), 2686.
4. Nelson, J.H., Grunberger, C., Cantor, C.R and Weinstein, I.B. J.Mol.Biol. 62, (1701), 331.

5. Weinstein, I.B. and Grunberger, D. World Symposium on Model Studies in Chemical Carcinogenesis, Johns Hopkins Univ. School of Hygiene and Public Health, Baltimore, Maryland, 1972, in press.
6. Potter, V.R. Cancer Res. 21 (1971) 1331.
7. Weinhouse, S. Cancer Res. 32 (1972) 2007.
8. Coon, H.G. J. Cell Biol. 39 (1968) 29A.
9. Kaighn, M.E. Tissue Culture Methods and Applications, New York and London, Academic Press (1973) p. 54-58.
10. Sato, J., Namba, M., Usui, K. and Nagano, D. J. Exptl. Med. 38 (1968) 105.
11. Williams, G.M., Weisburger, E.K. and Weisburger, J.H. Exptl. Cell Res. 69 (1971) 106.
12. Gerschenson, L.E., Okigaki, T., Andersson, M., Molson, J. and Davidson, M.B. Exptl. Cell Res. 71 (1972) 49.
13. Pitot, H.C. and Jost, J.P. Cell, Tissue and Organ Culture, NCI Monograph 26 (1967) 145.
14. Ohanian, S.H., Taubman, S.B. and Thorbecke, G.J. J. Nat. Cancer Inst. 43 (1969) 397.
15. Rommel, F.A., Goldlust, M.B., Bancroft, F.C., Mayer, M.M. and Tashjian, A.H., Jr. J. Immunol. 105 (1970) 396.
16. Weinstein, I.B., Gebert, R., Stadler, U.C., Orenstein, J.M. and Axel, R. Science 178 (1972) 1098.
17. Bomford, R. and Weinstein, I.B. J. Nat. Cancer Inst. 49 (1972) 379.
18. Federoff, S. J. Nat. Cancer Inst. 38 (1967) 607.
19. Castor, L.N. J. Cell Physiol. 72 (1968) 161.
20. Dulbecco, R. Nature 227 (1970) 802.
21. Eagle, H. and Levine, E.M. Nature 213 (1967) 1102.
22. MacPherson, I. and Montagnier, L. Virology 23 (1964) 291.
23. Borek, C. Proc. Nat. Acad. Sci. USA 69 (1972) 956.
24. Iype, P.T., Baldwin, R.W. and Gloves, D. Brit. J. Cancer 26 (1972) 6.
25. Kaighn, M.E. and Prince, A.M. Proc. Nat. Acad. Sci. USA 68 (1971) 2396.
26. Muller, M., Schreiber, M., Kartenback, J. and Schreiber, G. Cancer Res. 32 (1972) 2568.
27. Diamondstone, T.I. Analyt. Biochem. 16 (1966) 395.
28. Potter, V.R. Cancer Res. 32 (1972) 1998.
29. Levisohn, S.R. and Thompson, E.B. J. Cellular Physiol. 81 (1972) 225.
30. Richardson, V.I., Tashjian, A.H., Jr. and Levine, L. J. Cell Biol. 40 (1969) 236.
31. Walker, P.R., Bonney, J., Becker, J.E. and Potter, V.R. In Vitro 8 (1972) 107.

32. Heidelberger, C. Advances Cancer Res. 18 (1973) 317.
33. Kao, F.T. and Puck, T.T., J. Cellular Physiol. 80 (1972) 41.
34. Chasin, L.A. Cell 2 (1974) 37.
35. Todaro, G.J. and Huebner, R.J. Proc. Nat. Acad. Sci. USA 69 (1972) 1009.
36. Orenstein, J.M. and Weinstein, I.B. Cancer Res. 33 (1973) 1998.
37. Stoker, M.G.P. and Rubin, H. Nature 215 (1967) 171.
38. Oshiro, Y., Gerschenson, L.E. and DiPaolo, J.A. Cancer Res. 32 (1972) 877.
39. Dalton, A.Y. Cellular Control Mechanisms and Cancer Amsterdam: Elsevier (1964) pp 211-225.
40. Karasaki, S. Cancer Res. 29 (1969) 1313.
41. Locker, J., Goldblatt, P.J. and Leighton, J. Cancer Res. 28 (1968) 2039.
42. Klein, G. Proc. Nat. Acad. Sci. USA 69 (1972) 1056.
43. Lieber, M.M., Benveniste, R.E., Livingston, D.M. and Todaro, G.J. Science 182 (1973) 56.
44. Markert, C.L. Cancer Res. 28 (1968) 1908.
45. Weinstein, I.B. Genetic Concepts and Neoplasia. 23rd Annual Symposium on Fundamental Cancer Research at the University of Texas, Houston, The Williams and Wilkins Company, Baltimore, Maryland (1970) pp 380-408.
46. Martin, G.S. Nature 227 (1970) 1021.
47. Bader, J.P. J. Virol. 10 (1972) 267.
48. Biquard, J.M. and Vigier, P. Virology 47 (1972) 444.
49. Kawai, S. and Hanafusa, H. Virology 46 (1971) 470.
50. Toyoshima, K. and Vogt, P.K. Virology 39 (1969) 930.
51. Renger, H.C. and Basilico, C. J. Virol. 11 (1973) 702.
52. Yamaguchi, N. and Weinstein, I.B. Proc. Amer. Assoc. Cancer Res. (1974).

THE PROTECTIVE EFFECT OF 7,8-BENZOFLAVONE AND STEROID HORMONES AGAINST AFLATOXIN B_1 AND 7,12-DIMETHYLBENZ(A)ANTHRACENE-INDUCED CYTOTOXICITY IN CULTURED RAT CELLS

Arthur G. Schwartz

Summary

   Within a series of methyl and ethyl-substituted benze(a)-anthracenes of varying carcinogenicities, there was a good correlation between carcinogenicity and capacity to inhibit acid-insoluble H^3-thymidine incorporation in both cultures of rat liver epithelial cells and lung fibroblasts. 7,8-Benzoflavone, a competitive inhibitor of the mixed function oxidase which is believed to activate the polycyclic hydrocarbons, protects cultured rat liver cells against the DMBA-induced depression in acid-insoluble H^3-thymidine incorporation, as well as against the depression in H^3-thymidine incorporation produced by two mycotoxin carcinogens, aflatoxin B_1 and sterigmatocystin. Estrogenic steroids, which others have shown to competitively inhibit the mixed function oxidase, also protect against DMBA and aflatoxin-induced inhibition in H^3-thymidine incorporation. Dehydroepiandrosterone, an adrenal androgenic steroid which has been reported to be produced in subnormal amounts in women with benign and malignant breast tumors and in men with a variety of types of cancer, is also protective, while related adrogenic steroids showed significantly less protective effect. The possible relationship of the protective effects of both the estrogenic steroids and dehydroepiandrosterone to clinical data in humans is discussed.

   Tissue culture systems offer certain advantages for studying the mechanism of action of chemical carcinogens, in that one can work with a controlled cellular environment not possible in vivo. We have been interested primarily in two classes of carcinogens, the polycyclic hydrocarbons, which were initially discovered as carcinogenic constituents of coal tar (1), and aflatoxin B_1 and sterigmatocystin, two mycotoxins which have been implicated in the very high incidence of hepatoma which occurs in natives of sub-Sahara Africa (2). Both of these groups of carcinogens have been shown by several investigators to induce malignant transformation and to be markedly cytotoxic to cultured rodent fibroblasts (3,4) and rat liver epithelial cells (5).

We have used two culture systems in our studies with these carcinogens. One is a line of liver epithelial cells (E-3) established from an adult Sprague-Dawley rat and that has been propagated in culture for over two years. This line is apparently similar to cultures of rat liver epithelial cells that several investigators have succeeded in growing, and which display few liver-specific functions (6). Its usefulness, however, is in that unlike other established rodent cell lines, it has retained a high degree of sensitivity to the cytotoxic effect of the polycyclic hydrocarbons. We have also used cultured strains of adult rat lung fibroblasts, which are also highly sensitive to the cytotoxic effect of hydrocarbon carcinogens. All cultured cells are propagated as monolayers in iminmal essential medium (7) supplemented with 10% calf or fetal calf serum, with added organic buffers (8), adjusted to pH 7.6.

Structural Requirements for Carcinogenicity in the Benz(a) anthracene Hydrocarbons.

Benz(a)anthracene, the parent molecule for a group of carcinogenic polycyclic hydrocarbons, is itself non-carcinogenic (9). Methyl substitution at positions 6,7,8 or 12 converts benz(a)anthracene into an active carcinogen, while methyl substitution at other positions is ineffective (9). Ethyl, or higher alkyl substitution, is also ineffective. 7,12-Dimethylbenz(a)anthracene is probably the most potent of the polycyclic hydrocarbon carcinogens, while 7,12-DEBA is non-carcinogenic (9).

DMBA treatment (10^{-5} - 10^{-7} M) of either diploid lung fibroblasts or liver epithelial cells is cytotoxic, resulting in an inhibition of growth and marked depression in acid-insoluble H^3-thymidine incorporation (10). The depression in H^3-thymidine incorporation may be a result of overall cytotoxicity, but it may also represent a somewhat more specific effect of DMBA. Slaga et al have found that application of initiating doses of DMBA to mouse skin epidermis depressed DNA synthesis for 24 hrs, but was without effect on RNA and protein synthesis. We have also observed that treatment of cultured liver cells with DMBA for 24 hrs depresses acid-insoluble H^3-thymidine to a much greater extent than C^{14}-uridine incorporation (unpublished observations).

Figure 1 shows the inhibition in acid-insoluble H^3-thymidine incorporation produced by treatment of rat lung fibroblasts (A) or liver epithelial cells (B) with methyl and ethyl-substituted benz(a)anthracenes of varying carcinogenicities (in vivo carcinogenicity data taken from Huggins et al

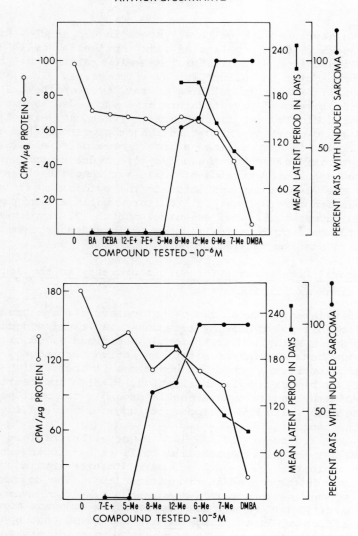

FIG. 1. <u>Correlation between cytotoxicity of polycyclic hydrocarbons & reported carcinogenicity (from data of Huggins, et al, 9).</u> Rat liver epithelial cells (A) or lung fibroblasts (B) were inoculated into T-15 flasks, & after several days' growth (medium changed every 2 days), cultures received media containing indicated polycyclic hydrocarbon at 10^{-5} (A) or 10^{-6} M (B). After 24 hr, cultures were pulsed with H^3-thymidine for 90 min, & the acid-insoluble radioactivity/µg protein was determined (O——O) (24). Carcinogenicity is indicated as % rats with induced sarcoma (●——●) & mean latent period in days (■——■). Me, methyl; et, ethyl.

(9)). There is good correlation between the reported in vivo carcinogenicity of a particular substituted benz(a)anthracene and its capacity to inhibit H^3-thymidine incorporation in both cell lines.

Protective Effect of 7,8-Benzoflavone

There is much evidence that the relatively chemically inert polycyclic hydrocarbons must be metabolized to active carcinogenic and cytotoxic metabolites by a NADPH-requiring mixed-function oxidase found in the endoplasmic reticulum of cells (12). 7,8-Benzoflavone, a competitive inhibitor of the mixed-function oxidase, has been found by others to protect cultured hamster fibroblasts against the cytotoxic effect of DMBA (13), and to protect mice against DMBA-induced skin tumorigenesis (14). We have confirmed that 7,8-benzoflavone protects agains the DMBA-induced depression in acid-insoluble H^3-thymidine incorporation in cultured rat liver cells (Table 1).

7,8-Benzoflavone was also found to protect against the depression in acid-insoluble H^3-thymidine incorporation produced by treatment of cultured liver cells with aflatoxin B_1 or sterigmatocystin (Tables 2 and 3). This result suggests that these two mycotoxin carcinogens are activated by the same, or similar enzyme, that activates DMBA. The 7,8-benzoflavone protected cultures also give no morphological evidence of cytotoxicity, which is very apparent in the carcinogen treated cultures after 24 hours (unpublished observations).

Protective Effect of Estrogen

A structural similarity between polycyclic hydrocarbon carcinogens and steroid hormones was noted many years ago (16). Yang et al (16) constructed Stuart-Briegleb molecular models for a number of polycyclic hydrocarbon carcinogens and found them sterically similar to steroids. Other evidence, such as the induction of selective adrenal necrosis in the rat by DMBA (17), also suggests a relationship between these carcinogens and steroid hormones.

Several steroids were tested for a possible protective effect against the DMBA-induced depression in H^3-thymidine incorporation in cultured liver cells, and estradiol-17β was found to be protective (Table 4). Other steroids (progesterone, testosterone and hydrocortisone showed little, if

Benzoflavone Concentration (M)	cpm/100 μg protein with DMBA concentration			
	10^{-6} M	2×10^{-7} M	4×10^{-8} M	0
0	220 ± 12	3,190 ± 50	11,100 ± 350	18,850 ± 350
10^{-6}	359 ± 60	2,710 ± 180	8,750 ± 1,210	17,300 ± 570
10^{-5}	5,920 ± 119	15,900 ± 235	20,200 ± 650	21,200 ± 330
2×10^{-5}	8,170 ± 365	18,850 ± 150	22,400 ± 95	22,500 ± 225

E-3 Cells were plated in T-25 (Falcon flasks). After 6 days of growth, cells were treated with various test media and pulsed for 90 min with H^3-thymidine (24). Values are expressed as cpm/100 μg of protein, and are the mean ± S.E. for 3 separately treated cultures.

TABLE 1. PROTECTIVE EFFECT OF 7,8-BENZOFLAVONE ON DMBA-INDUCED DEPRESSION IN H^3-THYMIDINE INCORPORATION.

Benzoflavone Concentration (M)	cpm/100 μg protein with aflatoxin concentration			
	10^{-6} M	2×10^{-7} M	4×10^{-8} M	0
0	412 ± 64	2,260 ± 210	10,300 ± 870	13,300 ± 570[a]
10^{-6}	14,900 ± 895	16,900 ± 1,250	13,800 ± 1,410	18,800 ± 875[b]
10^{-5}	14,600 ± 200	15,900 ± 850	15,700 ± 1,060	15,600 ± 390
2×10^{-5}	14,100 ± 930	15,900 ± 410	16,100 ± 380	15,200 ± 930

E-3 Cells were plated in T-25 flasks. After 6 days of growth, cells were treated with various test media and pulsed with H^3-thymidine (24). Values are expressed as cmp/100 μg of protein, and are the mean ± S.E. for 3 separately treated cultures (except where otherwise indicated).

[a] Mean of 2 separate cultures.
[b] Mean of 4 separate cultures.

TABLE 2. PROTECTIVE EFFECT OF 7,8-BENZOFLAVONE ON AFLATOXIN-INDUCED DEPRESSION IN H^3-THYMIDINE INCORPORATION.

	cmp/100 μg protein with sterigmatocystin concentration			
Benzoflavone Concentration (M)	10^{-6} M	2×10^{-7} M	4×10^{-8} M	0
0	350 ± 15	915 ± 60	1,410 ± 95	16,800 ± 220
10^{-6}	2,350 ± 210	11,800 ± 780	18,300 ± 450	19,800 ± 490
10^{-5}	6,890 ± 700	17,400 ± 430	20,400 ± 340	20,700 ± 275
2×10^{-5}	8,320 ± 700	18,300 ± 160	20,100 ± 650	20,000 ± 220

E-3 cells were plated in T-25 flasks. After 2 days growth, cells received various test media and were pulsed with H^3-thymidine (24). Values are cpm/100 μg of protein and are the mean ± S.E. for 3 separately treated cultures.

TABLE 3. PROTECTIVE EFFECT OF 7,8-BENZOFLAVONE ON STERIGMATOCYSTIN-INDUCED DEPRESSION IN H^3-THYMIDINE INCORPORATION.

	cpm/100 µg protein with DMBA concentration	
Steroids	4×10^{-7}	0
None	690 ± 17	10,800 ± 205[a]
Estradiol-17β (10^{-6} M)	887 ± 37	10,600 ± 133
Estradiol-17β (10^{-5} M)	1,350 ± 103[b]	10,200 ± 570
Estradiol-17β (2×10^{-5} M)	1,670 ± 67[b]	6,980 ± 398
Testosterone (10^{-6} M)	751 ± 28	10,700 ± 265
Testosterone (10^{-5} M)	790 ± 28	10,300 ± 80
Testosterone (2×10^{-5} M)	788 ± 43	9,880 ± 90
Progesterone (10^{-6} M)	703 ± 17	11,200 ± 422
Progesterone (10^{-5} M)	805 ± 18	11,700 ± 79
Progesterone (2×10^{-5} M)	1,000 ± 67[c]	12,000 ± 868
Hydrocortisone (10^{-6} M)	620 ± 47	9,700 ± 300
Hydrocortisone (10^{-5} M)	600 ± 12	10,500 ± 430
Hydrocortisone (2×10^{-5} M)	605 ± 50	10,600 ± 198

E-3 cells were plated on T-15 flasks. After 6 days of growth, cultures were treated with test media and pulsed with H^3-thymidine (24). Values are the mean ± S.E. of 3 separate cultures and are expressed as cpm/100 µg of protein.

[a] Mean of 4 separate cultures.
[b] Significantly higher than DMBA-treated cultures without steroid ($p > 0.001$).
[c] Significantly higher than DMBA-treated cultures without steroid ($p > 0.005$).

TABLE 4. EFFECT OF STEROIDS ON DMBA-INDUCED DEPRESSION IN H^3-THYMIDINE INCORPORATION.

any, protective activity. This protective effect of estrogen may be related to the observation of Bates (18) that male mice are more susceptible to DMBA-induced skin cancer than are females, while removal of the ovaries of the females significantly enhanced their susceptibility. Several investigators have reported that estradiol-17β competitively inhibits the mixed function oxidase which activates the polyclic hydrocarbons (19,20) and estradiol-17B was found to reduce the rate of conversion of H^3-DMBA to its epoxide by rat liver microsomes (21). Competitive inhibition of the activating enzyme by estradiol-17B may explain the protective effect of estrogen against the DMBA-induced depression in H^3-thymidine incorporation in cultured liver cells as well as the increased susceptibility of male mice to DMBA-induced skin cancer, as suggested by Nebert et al (19).

Estradiol-17B also protects against the aflatoxin B_1-induced inhibition in H^3-thymidine incorporation in cultured liver cells (Table 5). Progesterone is also somewhat protective, while testosterone and hydrocortisone have no apparent protective effect. This observation is also consistent with animal data, which indicate a greater susceptibility of male than female rats to aflatoxin B_1-induced liver cancer (22).

Modified Estrogens

Introduction of two additional double bonds into the B ring of estrone (equilenin) markedly reduces estrogenic activity (23). However, equilenin, as well as the similarly modified estradiol-17B (dihydroequelinin), was more protective against aflatoxin B_1-induced depression in H^3-thymidine incorporation than were their parent estrogens (Table 6). There is apparently no correlation between estrogenic potency and protective activity against aflatoxin B_1.

Possible Relationship to Clinical Observations

I have previously suggested that the decline in plasma estrogen levels in women with age may be a contributing factor in the rising incidence of cancer with age (24). Burch, Byrd and Vaughn have undertaken a retrospective study on the incidence of cancer in a group of women who have been on long-term estrogen therapy following hysterectomy. Their latest reports, in which they have followed 737 women for approximately 10,000 patient years of estrogen therapy,

ANTHRACENE — INDUCED CYTOTOXICITY

Steroid	cpm/100 µg protein with aflatoxin concentration	
	2×10^{-7} M	0
None	620 ± 27	9,640 ± 164
Estradiol-17β (10^{-6} M)	792 ± 17	11,400 ± 205
Estradiol-17β (10^{-5} M)	2,170 ± 49[a]	11,000 ± 178
Estradiol-17β (2×10^{-5} M)	3,520 ± 39[a]	11,300 ± 106
Testosterone (10^{-6} M)	590 ± 30	10,200 ± 382
Testosterone (10^{-5} M)	689 ± 36	10,400 ± 300
Testosterone (2×10^{-5} M)	838 ± 40	11,400 ± 166
Progesterone (10^{-6} M)	865 ± 44	11,200 ± 46
Progesterone (10^{-5} M)	1,330 ± 38[a]	11,100 ± 96
Progesterone (2×10^{-5} M)	2,220 ± 16[a]	11,500 ± 374
Hydrocortisone (10^{-6} M)	865 ± 348	9,300 ± 66
Hydrocortisone (10^{-5} M)	508 ± 30	9,300 ± 122
Hydrocortisone (2×10^{-5} M)	615 ± 16	9,250 ± 286

E-3 cells were plated in T-15 flasks and, after 6 days of growth, were treated with test media and pulsed with H^3-thymidine (24). Values (cpm/100 µg of protein) are the mean ± S.E. of 3 separate cultures.

[a] Significantly higher than aflatoxin treated cultures without steroid ($p < 0.001$).

TABLE 5. EFFECT OF STEROIDS ON AFLATOXIN-INDUCED DEPRESSION IN H^3-THYMIDINE INCORPORATION.

Steroid	cpm/100 µg protein with aflatoxin concentration	
	2×10^{-7} M	0
None	620 ± 27	9,650 ± 164
Estrone (10^{-5} M)	1,760 ± 55[a]	10,350 ± 390
Estrone (2×10^{-5} M)	3,860 ± 170[a]	10,700 ± 60
Equilenin (10^{-5} M)	3,620 ± 520[b]	6,850 ± 197
Equilenin (2×10^{-5} M)	5,900 ± 200[a]	7,750 ± 330
Estradiol-17β (10^{-6} M)	782 ± 17	11,400 ± 210
Estradiol-17β (10^{-5} M)	2,070 ± 49[a]	11,000 ± 178
Estradiol-17β (2×10^{-5} M)	3,520 ± 39[a]	11,300 ± 106
Dihydroequilenin (10^{-6} M)	905 ± 54	7,650 ± 308
Dihydroequilenin (10^{-5} M)	2,840 ± 224[a]	7,000 ± 96
Dihydroequilenin (2×10^{-5} M)	4,840 ± 188[a]	6,020 ± 78

E-3 cells were plated in T-15 flasks, and after 6 days of growth, test media were added and cultures were pulsed with H^3-thymidine (24). Values (cpm/100 µg of protein) are mean ± S.E. of 3 separate cultures.

[a] Significantly higher than aflatoxin-treated cultures without steroid ($p < 0.001$).
[b] Significantly higher than aflatoxin-treated cultures without steroid ($p < 0.005$).

TABLE 6. EFFECT OF MODIFIED ESTROGENS AND NATURAL ESTROGENS ON AFLATOXIN-INDUCED INHIBITION IN H^3-THYMIDINE INCORPORATION.

	cpm/100 µg protein with aflatoxin molar concentration		
Androstenedione Concentration (M)	2×10^{-7}	10^{-7}	0
0	200 ± 8	495 ± 80	4530 ± 10
2×10^{-5}	295 ± 5	635 ± 40	3830 ± 390
4×10^{-5}	360 ± 40	760 ± 20[b]	3670 ± 250
6×10^{-5}	540 ± 60	840 ± 120[a]	3350 ± 410
8×10^{-5}	590 ± 55[c]	925 ± 20[d]	4410 ± 480
Etiocholanolone concentration (M)			
0	200 ± 8	495 ± 80	4530 ± 10
2×10^{-5}	305 ± 4	600 ± 20	4780 ± 320
4×10^{-5}	360 ± 35[a]	730 ± 50[b]	5020 ± 35
6×10^{-5}	455 ± 9	700 ± 135[a]	5940 ± 140
8×10^{-5}	535 ± 35[c]	790 ± 12[d]	4190 ± 290
Testosterone concentration (M)			
0	200 ± 8	495 ± 80	4530 ± 10
2×10^{-5}	220 ± 8	530 ± 40	3260 ± 120
4×10^{-5}	315 ± 20[b]	600 ± 15[c]	2440 ± 270
6×10^{-5}	370 ± 20[b]	685 ± 2[c]	3670 ± 515
8×10^{-5}	405 ± 35[c]	730 ± 3[d]	3140 ± 230
Androsterone concentration (M)			
0	200 ± 8	495 ± 80	4530 ± 10
2×10^{-5}	310 ± 40	700 ± 130	4890 ± 10
4×10^{-5}	470 ± 15	1270 ± 130	5000 ± 180
6×10^{-5}	715 ± 5	1310 ± 20	5400 ± 200
8×10^{-5}	1080 ± 50[b]	1790 ± 35[c]	6590 ± 60

TABLE 7. Continued

Dehydroepiandrosterone concentration (M)			
0	200 ± 8	495 ± 80	4530 ± 10
2×10^{-5}	380 ± 20	730 ± 5	3150 ± 85
4×10^{-5}	610 ± 55	1070 ± 20	3700 ± 285
6×10^{-5}	990 ± 130	1450 ± 30	4650 ± 130
8×10^{-5}	1660 ± 30	2530 ± 10	4950 ± 630

E-3 cells were plated in T-15 flasks. After 5 days of growth, cells were treated with various test media and pulsed with H^3-thymidine (24). Values are expressed as cpm/100 μg of protein, and are the mean ± S.E. for 2 separately treated cultures.

a,b,c,d Significantly less than comparably treated culture with dehydroepiandrosterone; $p < 0.05$ (footnote a); $p < 0.025$ (footnote b); $p < 0.005$ (footnote c); $p < 0.001$ (footnote d).

TABLE 7. PROTECTIVE EFFECT OF ANDROGENIC STEROIDS ON AFLATOXIN-INDUCED DEPRESSION IN H^3-THYMIDINE INCORPORATION IN RAT LIVER CELLS.

	cpm/100 µg protein with aflatoxin molar concentration	
Androstenedione concentration (M)	5×10^{-8}	0
0	230 ± 10	4110 ± 350[a]
6×10^{-5}	375 ± 30[c]	5590 ± 260
8×10^{-5}	265 ± 75	5500 ± 470
10^{-4}	406 ± 8[c]	6070 ± 400
Etiocholanolone concentration (M)		
0	230 ± 10	4110 ± 350[a]
6×10^{-5}	250 ± 11[d]	3360 ± 200
8×10^{-5}	290 ± 20[d]	3140 ± 300
10^{-4}	350 ± 2[c]	3410 ± 190
Testosterone concentration (M)		
0	230 ± 10	4110 ± 350[a]
6×10^{-5}	415 ± 35[b]	3940 ± 300
8×10^{-5}	510 ± 40	3830 ± 200
10^{-4}	590 ± 35[b]	3910 ± 80
Andosterone concentration (M)		
0	230 ± 10	4110 ± 350[a]
6×10^{-5}	415 ± 23[c]	4280 ± 345
8×10^{-5}	470 ± 15[b]	4470 ± 315
10^{-4}	773 ± 100	4680 ± 260
Dehydroepiandrosterone concentration (M)		
0	230 ± 10	4110 ± 350[a]
6×10^{-5}	600 ± 7	5460 ± 95
8×10^{-5}	682 ± 30	5290 ± 200
10^{-4}	1050 ± 85	4590 ± 130

TABLE 8. Continued

Rat fibroblasts were plated in T-15 flasks. After 2 days, cells were treated with various test media and pulsed with H^3-thymidine (24). Values are expressed as cpm/100 µg of protein, and are the mean ± S.E. for 3 separately treated cultures.

[a] Mean of 2 cultures.
[b,c,d] Significantly less than comparably treated culture with dehydroepiandrosterone; $p < 0.01$ (footnote b); $p < 0.005$ (footnote c); $p < 0.001$ (footnote d).

TABLE 8. PROTECTIVE EFFECT OF ANDROGENIC STEROIDS ON AFLATOXIN-INDUCED DEPRESSION IN H^3-THYMIDINE INCORPORATION IN RAT LUNG FIBROBLASTS.

indicate a normal incidence of breast cancer but a very significant reduction in the incidence of other forms of cancer (25,26).

Dehydroepiandrosterone

Much indirect evidence suggests that steroid hormones may play a role in the genesis of breast cancer in humans (27). Bulbrook et al (28) and Kumaoka et al (29) have reported that women with this disease secrete significantly less of the androgen metabolites, androsterone and etiocholanolone, than do healthy control subjects. A prospective study on a large population of women clinically free of breast cancer has shown that the urinary excretion rates of androsterone and etiocholanolone are subnormal in women who subsequently develop breast cancer, with a linear increase in the probability of developing the disease as the excretion rate of etiocholanolone fell from 1 mg per 24 hrs to 500 µg per 24 hrs (30).

Urinary androsterone and etiocholanolone are derived principally from four plasma precursors, dehydroepiandrosterone, dehydroepiandrosterone-sulfate, androstenendione and testosterone (31). Poortman et al (32) found that the urinary production rates of dehydroepiandrosterone and dehydroepiandrosterone-sulfate were significantly lower in breast cancer patients than in control women, while the production rates of androstenedione and testosterone were not significantly different. Likewise, Brennan et al (33) have reported that premenopausal women with benign breast tumors, who are reported to have a greater probability of subsequently developing breast cancer than normal women, have significantly lower plasma concentrations of dehydroepiandrosterone-sulfate than matched controls. Similarly, Sonka et al (34) have found that men and women with a variety of types of non-advanced cancer have significantly lower plasma concentrations of dehydroepiandrosterone-sulfate than normal controls.

We have found that dehydroepiandrosterone also protects against the aflatoxin B_1-induced depression in H^3-thymidine incorporation in cultured rat liver cells and lung fibroblasts, while related androgenic steroids (androsterone, etiocholanolone, testosterone and androstenedione) showed significantly less protective effects (Tables 7 and 8). A similar protective effect of dehydroepiandrosterone was seen against sterigmatocystin and DMBA-induced depression in H^3-thymidine incorporation (unpublished observations).

It appears well established that dehydroepiandrosterone is a potent non-competitive inhibitor of glucose-6-phosphate

Etiocholanolone concentration (M)	cpm/100 μg protein with aflatoxin molar concentration	
	2×10^{-7}	0
0	485 ± 28	5190 ± 135
4×10^{-5}	515 ± 38	5100 ± 505
6×10^{-5}	630 ± 53	5050 ± 110
8×10^{-5}	660 ± 35	4150 ± 190
Dehydroepiandrosterone concentration (M)		
0	485 ± 28	5190 ± 135
4×10^{-5}	770 ± 53	5330 ± 170
6×10^{-5}	1020 ± 55	4845 ± 345
8×10^{-5}	1450 ± 96	5200 ± 415
Epiandrosterone concentration (M)		
0	485 ± 28	5190 ± 135
4×10^{-5}	970 ± 39[b]	5230 ± 445
6×10^{-5}	1910 ± 18[c]	6700 ± 110
8×10^{-5}	2620 ± 69[c]	5180 ± 630[a]

E-3 cells were plated in T-15 flasks. After 2 days of growth, cells were treated with various test media and pulsed with H^3-thymidine (24). Values are expressed as cpm/100 μg of protein, and are the mean ± S.E. for 3 separately treated cultures.

[a] Mean of 2 cultures.
[b,c] Significantly greater than comparably treated culture with dehydroepiandrosterone; $p < 0.05$ (footnote b); $p < 0.001$ (footnote c).

TABLE 9. COMPARATIVE PROTECTIVE EFFECTS OF EPIANDROSTERONE AND DEHYDROEPIANDROSTERONE ON AFLATOXIN-INDUCED DEPRESSION IN H^3-THYMIDINE INCORPORATION IN RAT LIVER CELLS.

dehydrogenase from a variety of mammalian tissues (35) and it has been suggested that this steroid may regulate the activity of this enzyme in vivo (36). Glucose-6-phosphate dehydrogenase controls the activity of the hexose monophosphate shunt, which is the primary source of extra-mitochondrial NADPH (37). Since NADPH is a necessary co-factor for the mixed-function oxidase which activates DMBA, aflatoxin B_1 and sterigmatocystin, dehydroepiandrosterone may protect by reducing the intracellular levels of this co-factor. Support for this notion comes from an observed greater protective effect of epiandrosterone than dehydroepiandrosterone against the aflatoxin B_1-induced inhibition in the H^3-thymidine incorporation (Table 9). The former steroid, which lacks the Δ-5,6 double bond of dehydroepiandrosterone, is also a more effective inhibitor of glucose-6-phosphate dehydrogenase than its parent steroid (38).

In conclusion, we have found that estrogenic steroids and dehydroepiandrosterone protect cultured rat cells against the depression in H^3-thymidine incorporation produced by treatment with either aflatoxin B_1, sterigmatocystin, or DMBA. These same steroid hormones have also been implicated, by a variety of clinical studies, in human cancer.

I thank Joanne Lorenz for excellent technical assistance. Supported by grant CA 14661 from NIH and General Research Support, Temple University Medical School.

References

1. Kennaway, E. L. Biochem. J. 24 (1930) 497.
2. Enomoto, M. and M. Saito Ann. Rev. Microbiol. 26 (1972) 279.
3. Berwald, Y. and L. Sachs Nature 200 (1963) 1182.
4. Starikova, V.B. and J.M. Vasiliev Nature 195 (1962) 42.
5. Williams, G.M., J. M. Elliott and J.H. Weisburger Cancer Res. 33 (1973) 606.

6. Coon, H.G. J. Cell Biol. 39 (1968) 29a.
7. Eagle, H. Science 130 (1959) 432.
8. Eagle, H. Science 174 (1971) 500.
9. Huggins, C.B., J. Pataki and R.C. Harvey Proc. Nat. Acad. Sci. U.S.A. 58 (1967) 2253.
10. Schwartz, A.G. Cancer Res. 33 (1973) 2431.
11. Slaga, T.J., G.T. Bowden, B.G. Shapas and R.K. Boutwell Cancer Res. 34 (1974) 771
12. Heidelberger, C. Fed. Proc. 32 (1973) 2154.
13. Diamond, L. and H.V. Gelboin Science 166 (1969) 1023.
14. Kinoshita, N. and H.V. Gelboin Proc. Nat. Acad. Sci. U.S.A. 69 (1972) 824.
15. Inhoffen, H.H. Progress in Organic Chemistry (1953) 131.
16. Yang, N.C., A.J. Castro, M. Lewis and T.W. Wong Science 134 (1961) 386.
17. Huggins, C.B. and G. Morii J. Exp. Med. 114 (1961) 741.
18. Bates, R.R. J. Nat. Cancer Inst. 41 (1968) 559.
19. Nebert, D.W., L.L. Bausserman and R.R. Bates Intern. J. Cancer 6 (1970) 470.
20. Spencer, T. Enzymologia 43 (1972) 301.
21. Booth, J., G.R. Keysell and P. Sims Biochem. Pharmacol 23 (1974) 735.
22. Wogan, G.N. and P.M. Newberne Cancer Res. 27 (1967) 2370.
23. Korenman, S.G. Steroids 13 (1969) 163.
24. Schwartz, A.G. Cancer Res. 34 (1974) 10.
25. Byrd,Jr., B.F., J.C. Burch and W.K. Vaughn Ann. Surg. 177 (1973) 626.
26. Burch, J.C., B.F. Byrd, Jr. and W.K. Vaughn Am. J. Obstet. Gynecol. 118 (1974) 778.
27. McMahon, B. J. Nat. Cancer Inst. 50 (1973) 21.
28. Bulbrook, R.D., J.L. Hayward and C.C. Spicer Lancet ii (1971) 395.
29. Kumaoka, S., N. Sakauchi and O. Abe J. Clin. Endocr. 28 (1968) 667.
30. Bulbrook, R.D., J.L. Hayward and C.C. Spicer Lancet ii (1971) 395.
31. MacDonald, P.C., A. Chapdelaine, O. Gonzalez, E. Gurpide, R.L. VandeWiele and S. Lieberman J. Clin. Endocr. 25 (1965) 1557.
32. Poortman, J., J.H.H. Thijssen and F. Schwarz J. Clin. Endocr. Metab 37 (1973) 101.
33. Brennan, M.J., R.D. Bulbrook, N. Desphande, D.Y. Wang and J.L. Hayward Lancet i (1973) 1076.
34. Sonka, J., M. Vitkova, I. Gregorova, Z. Tomosova, J. Hilgertova and J. Stas Endokrinologie 62 (1973) 61.
35. Lopez, S.A. and A. Rene Proc. Soc. Exp. Biol. Med. 142 (1973) 258.

36. Kurtenbach, P., P. Benes and G.W. Oertel Eur. J. Biochem. 39 (1973) 541.
37. Flatt, J.P. J. Lipid Res. 11 (1970) 131.
38. Oertel, G.W. and P. Benes J. Steroid Biochem 3 (1972) 493.

Note: We would like to thank the publishers of Cancer Research for the permission to reproduce all the Figures and Tables.

THE STUDY OF CHEMICAL CARCINOGENESIS USING CULTURED RAT LIVER CELLS

Gary M. Williams

There are two major considerations that recommend liver as a source from which to derive cell cultures for the study of chemical carcinogenesis. The first is that liver possesses all of the known enzyme systems for metabolizing chemical carcinogens (1). The second is that the action of carcinogens on the liver has been so extensively studied that the potential for comparison of cell culture findings with those in vivo is very great.

The first consideration lead to the hope that cultures form liver could be used for screening carcinogens. For this reason, I began in the laboratory of Dr. John Weisburger the study of cultures derived from rat liver. The Fisher strain of rat was chosen because of its broad sensitivity to chemical hepatocarcinogens. A method was developed for culturing from livers of ten day old animals epithelial-like cells free of fibroblasts (2). These cultures were screened for their sensitivity to carcinogens and it was found that they were converted to malignant cells following treatment with a variety of carcinogens, most of which required metabolic activation (3). Montesano et al. (4) have used this system and confirmed our results. The converted cells gave rise to adenocarcinomas upon inoculation into hosts (3), indicating that they were indeed derived from an epithelial cell type.

These cultures possessed some properties of hepatocytes which were accentuated at confluency (5), but lacked the unique features of differentiated parenchymal cells. Since the livers of ten day old rats are not functionally mature, we (6) have begun using epithelial cultures started from adult rat liver dissociated after perfusion with collagenase and hyaluronidase by the method of Seglen (7). This method afforded a large yield of parenchymal-type cells. Furthermore, most of these cells were viable as evidenced by almost universal adaptation to cell culture by 24 hours. Interestingly, however, only a small proportion were capable of sustained replication. It is not known whether this is due to an inherent property of the non-proliferating cells or to a deficiency in the culture conditions. The initiated lines, which we refer to as ARL (for adult rat liver) were composed of polygonal cells morphologically similar to those obtained previously (2). We find that an enriched medium (6) is

helpful in maintaining the polygonal morphology. These
lines have not been extensively studied for liver functions
but have displayed some properties of hepatocytes. For
example, in one line of three examined, Dr. Kiwamu Okita
found both albumin and alpha-fetoglobulin production by
immunofluorescence and immunoprecipitation in agar. Only one
line has been examined for specific cytosol tyrosine aminotransferase by Dr. G. Litwack, who found 0.0012 units (corrected by about 40% for the activity of soluble aspartate
aminotransferase acting on tyrosine and oxaloacetate), which
is about 1% of the activity of adult liver. These results
are hopeful indicators that appropriate culture techniques
may be devised to preserve hepatocytic function in culture.

The two oldest lines ARL 6 and ARL 7 were found by Dr.
Pen-Ming Ming after 13 months and 4 months in culture respectively to have near diploid karyotypes. The chromosome
numbers were in a range of 41 to 46 with a mode of 43 for
ARL 6 and a mode of 44 for ARL 7. Both lines had approximately 25% structural abnormalities. The lines have not
yet been tested for tumorigenicity and as previously described (3) the usual morphologic criteria for transformation
of fibroblasts do not apply to epithelial-like cells. I
presently use growth in soft agar to quantitate transformation in the cultures (8). This method has the advantage
that seeded cells are immobilized and thus there is no possibility as in cell culture that one transformed colony might
be a metastasis from another. Approximately 0.01-0.03% of
the cells of untreated lines were capable of growing as colonies of greater than 8 cells after 7 days in soft agar. In
this assay 4.0-6.0% of malignant cells of cultured hepatomas
formed colonies (8).

These cultures are presently being used to study various
aspects of chemical carcinogenesis. It is widely accepted
that interaction of chemical carcinogens with DNA may be the
basis for malignant conversion (9). One way of studying this
interaction is by the determination of single-strand breaks
in DNA induced by carcinogens (10,11).

DNA breaks can be observed in ^{3}H-thymidine-labeled DNA
which is alkali denatured and subjected to alkaline sucrose
gradient centrifugation. The cellular DNA was labeled by
adding 0.15 µCi/ml ^{3}H-thymidine to sparse cultures 24 hr after
seeding. The cultures were grown to confluency in the presence of the ^{3}H-thymidine and then were switched to control
medium containing only 2% fetal calf serum overnight before
use. This manipulation allowed incorporation of all the thymidine into full length DNA, but did not permit enough replication to give rise to a signficant proportion of cells

without labeled DNA. Autoradiography (5) revealed that
under these conditions over 95% of the cells contained radioactive DNA. Therefore, the results reflect the situation in
the whole culture. Incidentally, there was no cytoplasmic
labeling as might occur with mycoplasma infection. The cultures were gently washed with a balanced salt solution to
remove detached cells before harvesting. As a consequence,
the analysis of breaks did not include those in detached
dead or severely damaged cells. The attached cells were then
gently detached, pelleted through complete medium and washed
twice. The cells were suspended and subjected to alkaline
sucrose gradient centrifugation as described by Cox et al.
(12). The centrifugation was sufficient to sediment the
bulk DNA of control cells onto a dense sucrose cushion while
smaller molecular weight broken DNA of treated cells was
distributed in the gradient.

The direct acting carcinogen nitrosomethylurea produced
a dose-related fragmentation of DNA which was maximal at
15 min after treatment (fig. 1). It is not known how much
of the DNA was actually broken in the cells, but regardless
the method allows an appreciation of the effect of the carcinogen on DNA. Further incubation after maximal fragmentation revealed a shift of the sedimentation profile toward
that of the bulk DNA (fig. 2) indicating fragments of higher
molecular weight. This result is due to repair of some of
the DNA damage giving rise to breaks. After 24 hours incubation, there was almost complete restoration of the normal
sedimentation pattern. With concentrations of carcinogen
which did not produce significant cell killing (and thereby
loss of DNA) the return of the original number of cpm to the
sedimentation position of bulk DNA was observed. Thus, the
repair of the DNA damage which gave rise to the breaks could
be followed over time.

In collaboration with Dr. Richard O. Michael the ability
of a variety of carcinogens to produce breakage has been studies. Direct-acting carcinogens such as N-acetoxyfluoreny-
lacetamide and methyl methanesulfonate readily produced
breaks, but procarcinogens requiring metabolism gave equiv-
ocal or negative results. We hypothesized that in view of
the observed rapid repair that a low level of activation
might result in so little damage to DNA that the cell could
repair it as fast as it occurred; thus no breaks would be
observed. We had documented that chloroquine could inhibit
the repair of methyl methanesulfonate induced breaks (13)
and therefore examined whether chloroquine would inhibit
repair of procarcinogen-induced breaks sufficiently for them
to accumulate to detectable levels.

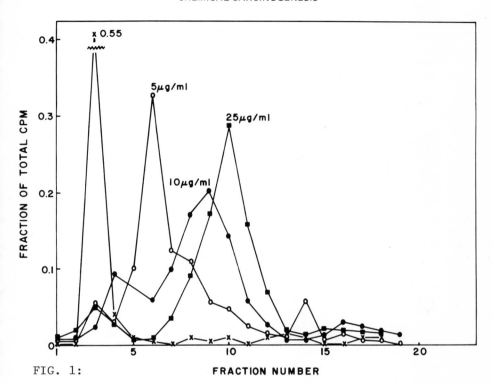

FIG. 1:

Single-strand DNA breaks induced by various doses of nitrosomethylurea after 15 min exposure.

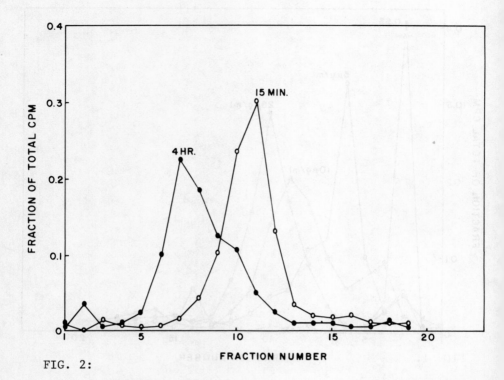

FIG. 2:

Repair of nitrosomethylurea-induced single-strand DNA breaks.

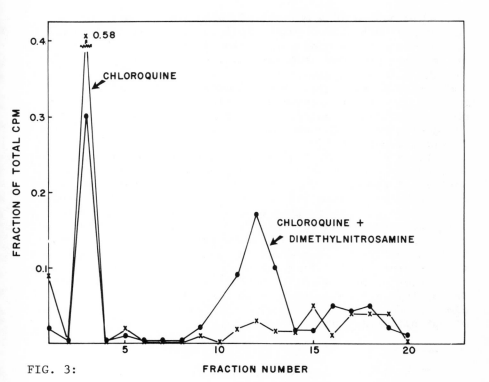

FIG. 3:

Single-strand DNA breaks induced by dimethylnitrosamine in the presence of chloroquine.

Chloroquine (0.4 mM) was added to the cultures for 1 hour prior to treatment with carcinogens and allowed to remain present for 4 hours of carcinogen treatment. Chloroquine alone produced no breakage. Treatment with 2×10^{-5}M dimethylnitrosamine, N-2 fluorenyl acetamide, aflatoxin B_1 and dimethylbenzanthracene by themselves also yielded no breaks. But in the presence of chloroquine breakage was obtained with all four procarcinogens (fig. 3). These carcinogens were previously found to produce malignant conversion in similar cultures (3).

It is noteworthy that whereas nitrosomethylurea produced fragmentation of all the DNA the procarcinogens broke only a portion of the DNA. This probably indicates that some of the cells were activating the carcinogens while others were not.

We are presently attempting to determine if these results were truly due to inhibition of repair or to some other effect of chloroquine. But these preliminary data suggest that the cultures can generate a "genotoxic" product (9) from the procarcinogens. Thus, this manipulation may make possible the detection of low levels of activation in these cultures. If so, this system would provide, as originally hoped for, a simple, inexpensive, extremely rapid assay for the screening of potential carcinogens.

This work was supported by Research Grant #BC-133 from the American Cancer Society, Inc.

References

1. Weisburger, J.H. and Williams, G.M. In Cancer; A Comprehensive Treatise. Becker, F. (Ed.) Plenum Press, N.Y. in preparation.
2. Williams, G.M., Weisburger, E.K. and Weisburger, J.H. Exptl. Cell Res. 69 (1971) 106.
3. Williams, G.M., Elliot, J.M. and Weisburger, J.H. Cancer Res. 33 (1973) 606.
4. Montesano, R. Saint Vincent, L. and Tomatis, L. Br. J. Cancer 28 (1973) 215.
5. Williams, G.M., Stromberg, K. and Kores, R. Lab. Invest. 29 (1973) 293.
6. Williams, G.M. and Gunn, J.M. Exptl. Cell Res. In Press.
7. Seglen, P.O. Exptl. Cell Res. 76 (1973) 25.
8. Williams, G.M. Second, U.S.-Japan Cooperative Medical Science Program on Methods for Evaluating Mutagenesis and Carcinogenesis, Charleston, S.C., Nov. 1973.

9. Brookes, P., Druckrey, H., Lagerlöf, B., Litwin, J. and Williams, G. Ambio Special Report, No. 3, 15, (1973).
10. Andoh, T. and Ide, T. Cancer Res. $\underline{32}$ (1972) 1230.
11. Horikawa, M., Fukuhara, M., Suzuki, F., Nikaido, O. and Suguhara, T. Exptl. Cell Res. $\underline{70}$ (1972) 349.
12. Cox, R., Damjanov, I., Abanobi, S.E. and Sarma, D.S.R. Cancer Res. $\underline{33}$ (1973) 2114.
13. Michael, R.O. and Williams, G.M. Mutation Res. (In Press).

INTERNATIONAL SYMPOSIUM ON GENE EXPRESSION AND CARCINOGENESIS IN CULTURED LIVER: SUMMARY AND SOME PERSPECTIVES

Emmanuel Farber

No summary of such a conference as this can be truly meaningful or possible, since so many different basic aspects of biology were covered. Obviously, we all attended the conference, not primarily because of our keen interest in any one single phase of cell biology or cell pathology, be it hormone action, cell differentiation, control of protein synthesis, carcinogenesis or neoplasia, but because of an overriding interest in one property of one cell system - the differentiation of liver parenchymal cells and its maintenance under in vitro conditions. Since this is the focal point for the whole conference, I shall restrict my comments to this target and to this alone.

It is clear that a good beginning has been made but only a beginning. Certain patterns seem to stand out clearly, as do certain questions and problems. Some of these are:

(a) Isolated non-proliferating parenchymal liver cells can be prepared from normal animals and these can be induced to retain some important differentiated functions for several days. The further refinements in the modulation of the environment used for such preparations should no doubt lead to highly reproducible systems for liver cell maintenance in the near future. The imminent development of such systems which retain important properties of liver, such as ability to respond to variations in dietary and hormonal influences, will open up another important approach in our understanding of the mechanisms of the physiologic and pathologic response patterns.

(b) Any attempt to grow parenchymal cells from liver so far seems to result in the breakdown of the differentiated state. Although several investigators have obtained and even reported cultures of fetal, neonatal or adult liver cells that demonstrate one or more properties distinctive of liver, such as production of albumin or alpha-fetoprotein, ability to respond to some hormones, etc., these appear to occur at random. Although they are obviously useful for some purposes, the lack of reproducibility makes this system today anything but attractive for a major commitment in a research effort.

What is the basis for this loss of reproducibility?

Some investigators are of the opinion that a major factor is the nature of the cell being selected for growth. Are we initiating growth of a truly differentiated hepatocyte that subsequently loses its selective programming, are we selecting for growth an as yet unidentified stem cell which by chance occasionally undergoes differentiation or are we growing biliary epithelial cells? Other investigators believe that the type of cell growing in these "liver cell cultures" is an hepatocyte. If this be true, then presumably the hepatocyte selected for growth lacked many of the differentiated characteristics or more likely lost these with the onset of proliferation. In vivo, proliferating normal or neoplastic hepatocytes retain many of the specialized biochemical activities seen in mature liver. If one assumes the growing cell came from an hepatocyte, why does the proliferating hepatocyte in vitro often lose its differentiated characteristics? Is it mainly a reflection of our failure to discover the appropriate environment or is there some more basic reason concerned with unrestricted growth and metabolic programming? Some clues appear to be coming from cell fusion experiments. Not only hepatoma cells but also some normal liver cells can maintain at least some differentiated functions in close association with some other cells. In my view, this lead should be pursued vigorously, since it offers the possibility of obtaining answers to some of the fundamental aspects of liver cell differentiation. Also, the need for specific markers again becomes apparent. In liver, there is evidence that antigens may exist that are specific for ductular cells as compared to hepatocytes. Also, there is evidence that biliary epithelial cells may contain enzymes which are absent from hepatocytes. What about selective isozymes for the different cell populations in the liver? It seems to me that the ignorance in this phase of liver biology may become one of the rate-limiting steps in the development of differentiated liver cell cultures.

(c) Why are hepatoma cells, as a group, in culture able to retain more easily at least some of the differentiation markers they share in common with normal hepatocytes? In general, hepatoma cells have a simple organization with fewer coordinated specialized functions. Yet, some retain many of the enzymes of specialized function, e.g. urea cycle enzymes. It would be very interesting and important to study whether highly coordinate systems, like urea synthesis, or if possible, gluconeogenesis, retain their coordinate function when fused with appropriate cells, since the evidence so far seems to favor noncoordinate regulation in the few cell hybrids studied.

(d) Is the loss of differentiated functions in normal cells *in vitro* due to an effect at the level of transcription, post transcription, translation or post translation? In my opinion, this type of problem should receive high priority, since its solution, if a single solution exists, may point immediately to new ways to encourage the growing liver cells from losing their differentiated properties. Again, the early data on extinction of properties in hybrids indicate that nuclear fusion is not needed. Perhaps the major site of control is translational or posttranslational.

As a corollary, should we be thinking of using "chemical freeze storage"? For example, by temporarily inhibiting certain metabolic reactions, especially those at key points in the control systems, perhaps we might be able to prevent the "downhill" course and thereby preserve some intact functions, especially the more complex coordinated ones. Conceivably, low concentrations of inhibitors of transcription or of translation, especially the short acting reversible ones, could be used while the cells are being isolated and manipulated prior to initiating conditions for maintenance or growth.

(e) The success of the development of reliable differentiated liver cell cultures seems to be intimately bound up with the progress in the isolation and identification of discrete growth factors from serum, etc. The finding of a serum tripeptide that has growth promoting factors seems potentially very exciting. Conceivably, this is only one of many, either naturally-occurring or man-made. One cannot help but think of how active pharmacologically and physiologically many small peptides are - bradykinin, angiotension, oxytocin, etc., etc. Along the same lines of thinking, are there special mechanisms for turning off growth as well as turning it on or is this due simply to the progressive loss of substrates or growth promotors? Of course, this question still remains as pivotal in the analysis of the regeneration of liver after injury or surgical removal.

(f) One of the major objectives of *in vitro* biology is the elucidation of mechanisms of biological phenomena. This naturally presupposes that the phenomenom one is studying in an *in vitro* situation is the same or at least close to that occurring *in vitro*. However, it is becoming evident that *in vivo* biology is only one segment of the potential spectrum of any biological system. It may be that many phenomena, or at least major modulations of phenomena, may occur only *in vitro* and not be seen at all under the relatively strict homeostasis characteristics of the intact organism. The apparent discrepancies between responses to hormones, etc. by cells in culture and by the intact organism are illustrative. I do not think we have to worry about this now, until the systems are better developed. However, since many hopes are placed on *in vitro* systems in the analysis of complex biological phenomena, such as carcinogenesis, hormonal regulations, development and differentiation, we must always keep this potential handicap in the back of our minds.

Along these lines, my own strong prejudice is to "go mechanistic" as early as possible. I believe that the experience from the *in vivo* teaches us that often contradictory or irreconcilable effects of some agent can be clarified if we understand mechanism. Hopefully, this will remain one of the major objectives of the work on liver cell culture.

In conclusion, it is fair to say that a significant beginning has been made in the development of useable *in vitro* liver cell systems, resting or growing. However, it is also evident that major conceptual and perhaps technical advances have yet to be made before the systems can be exploited for the study of mechanisms of many important biological problems such as the control of differentiation, endocrine control of biological behavior and carcinogenesis and neoplasia, to name but three. Hopefully, the next conference will bring us closer to our major objective.

SUBJECT INDEX

A

3 A cells, 447
A particles, 445, 450, 451
Aflatoxin, 465-477
Alanine aminotransferase, 346, 355
Albumin, 59, 126, 127, 294, 296-299, 342-345, 433, 481
Alcohol dehydrogenase, 313, 355
Alcohol, liver cell protein synthesis effects of, 55, 58
Aldehyde dehydrogenase, 314, 321, 322
Aldolase, 105-108, 242, 355
Amino acid transport, 26, 32, 36-39, 53, 190-204
Androgens, 111-113, 176, 178, 467, 471-474
Arginase, 168-171
Argininosuccinate lyase, 168-171
Argininosuccinate synthetase, 168-171
ARL cells, 480
Aryl hydrocarbon hydroxylase 378-401

B

B-1 cells, 447
Balb/3T3 cells, 403-411
Benz(a)anthracene, 378-400, 402-411, 461-477
Benzo(a)pyrene, 130, 379-380
Benzoflavone, 460-479
Branched chain amino acid transaminase, 232-248
BRL cells, 355
BRLC-GAI cells, 220-226

C

C particles, 445, 450, 451

Carbamoyl phosphate synthetase, 168-171
Cellular communication in culture, 62, 74-80
Chemical carcinogenesis, 402-411, 425-459, 480-487
Chloroquine, 482-487
Cholesterol, synthesis of, 127, 297, 300
Chromosomes, 73, 105, 240, 241, 243, 313, 327, 328, 334, 340, 355, 444
Complement, 171-177
Corticosterone, metabolism of, 110, 111
Cyclic AMP, 205-219, 363
Cytochrome P450, 126-129, 378-400

D

Degradation, protein, 264-281
Dexamethasone, 85, 153, 193-195, 200, 221, 325-327, 329, 330, 332, 435-437, 446, 447
Dimethylbenzanthracene, 460-479
Dimethylnitrosamine, 485
DNA repair, 412-424
    single strand breaks, 481-487
    synthesis, 87, 88, 123, 161-165, 185, 304-306, 365, 412-420, 463-477

E

5 E cells, 447
Epinephrine, 220, 221, 385
ERL-2-Cl-3 cells, 385
Estrogens, 463-477

F

Fibroblasts
    guinea pig, 402-411

hamster, 402-411
human, 326, 328, 329
mouse, 311-323, 358-377, 403-411, 447
rat lung, 461, 474
FLC cells, 312-323
Folic acid, 40-43
Fusion, somatic cell, see Hybrids

G

Gangliosides, 81
Gucagon, 39, 127, 132, 221
Glucokinase, 161
Glucose-ATP phosphotransferase, 242
Glucose-6-phosphatase, 129
Glycogen, 127, 132, 314
Glycolytic enzymes 109, 110
Growth factors, 282-310
Guinea pig, cultured cells of, 403-411

H

H 4 II EC 3 cells, 347
Hepa 1a cells, 333-338, 385
Hepatomas, cultured lines of, 46-61, 69, 74, 168-180, 190-204, 206-219, 243, 300, 304, 325, 326, 333, 358-377, 385, 441-457
Histones, phosphorylation of, 249-263
in cell cycle, 253-256
HTC cells, 190-204, 249-263, 312-323, 358-377, 436, 446, 447
Hybrids, somatic cell, 311-377
activation of differentiated functions in, 341, 343-345
extinction of differentiated functions in, 318, 325-332, 347, 348, 358-377
recovery of extinguished functions in, 327, 330, 348-350
Hydrocortisone, 87, 88, 177, 243, 467-470

I

Insulin, 37, 127, 196-198, 221, 293, 307, 436

K

K 16 cells, 446, 447, 452-456
K 22, KF-7, KF-9 cells, 446, 447

L

L cells, 311-323, 358-377
Lectins, 81-84
Leucocytes, human 338
Lipids, synthesis of, 296, 297
Liver cells, in culture, 1-46, 62-93, 95-167, 181-189, 220-226, 238-241, 282-291, 378, 383-385, 390, 441-459, 461, 480-487
culture methods, 2, 3, 8, 11, 24-30, 47, 48, 64, 68, 95-98, 120-122, 138-141, 190, 191, 287, 425, 446, 480, 481
microsomal functions in culture, 119, 123, 126-129
mitochondria, 182-184
morphology of cultured cells, 5-7, 9, 10, 12-14, 48, 63, 65-67, 75, 78, 100-104, 124, 153-159, 181-189, 426-433, 442, 443, 445
neoplastic transformation in culture, 68, 70, 71, 402-411, 425, 444-445
nonparenchymal cells, 5, 7, 99
primary culture, 8, 11, 24-61, 98, 99, 119-167
regenerating, 161-165, 251, 264-285
Methylcholanthrene, 393-395
Metyrapone, 393, 396
MH_1C_1 cells, 385

N

N-acetoxyfluorenylacetamide, 406, 407, 441, 482
Neoplastic transformation, 68, 70, 71, 90, 402-411, 425-459
NRK cells, 447

O

Ornithine carbamoyl transferase, 168-171

SUBJECT INDEX

Ornithine decarboxylase, 152, 153
Ovalbumin, 298

P

PC cell lines, 239
Phenobarbital, 379, 384
Phosphodiesterase, 211
Phosphoenolpyruvate carboxykinase, 205, 217
Phospholipids, synthesis of, 300, 301
P-nitroanisole O-demethylase, 130
Polyglutamates, 27
Progesterone, effects on DNA synthesis, 467-470
  metabolism of 113-115
Protein
  degradation, 264-281
  synthesis, 34, 46-61, 123, 142-151, 198, 205-219, 267-276, 304, 390
  synthesis, inhibitors of, 56, 57, 198, 390
Protein kinase, 215, 216
Pteroylglutamate, 27
Pyruvate kinase, 161

R

Reuber H35 cells, 206, 325-332, 347-357, 385, 447
Revertants, temperature sensitive, from transformed cells, 452-456
Ribosomes, 268-272, 397
RLC-GA1 cells, 181-189, 220-226
RLN cell lines, 240-243
RNA, synthesis, 35, 60, 152, 153, 159, 160, 198, 304-306
  viruses, 443-459

S

Sterigmatocystin, 466

Steroid hormones, amino acid transport and, 39, 190-204,
  enzyme induction by, 26, 32, 33, 221, 227-231, 243, 325-332, 358-377, 435-437, 446, 447, 481
  inhibitory effects of, 84-88, 181-184, 363-365
  metabolism of by cultured cells, 110-115, 176, 178
  morphological effects of, 181-189
  protective effects against cytotoxicity, 460-479
  receptors for 329, 360-376, 435
Syrian hamster, cells, 403, 409

T

3T3 cells, 447
Temperature sensitive transformed cells, 452-456
Testosterone, metabolism of, 111-113, 176, 178
  protective effect of, 467, 471-474
TRL-2-Cl-2 cells, 385
Tryptophan pyrrolase, 242, 318
Tyrosine, 227-231
Tyrosine aminotransferase, 26, 32, 33, 207-215, 220-231, 238, 242, 243, 325-332, 355, 363-365, 435-437, 446, 447, 481

U

Urea cycle, 168-171

V

Viruses, RNA type, 443-459

W

W-8 cells, 447, 452-456
W-15 cells, 447